POWER ORTHOGONAL POLYNOMIALS

POWER ORTHOGONAL POLYNOMIALS

YING GUANG SHI

Nova Science Publishers, Inc.
New York

For permission to use material from this book please contact us:
Telephone 631-231-7269; Fax 631-231-8175
Web Site: http://www.novapublishers.com

NOTICE TO THE READER

Library of Congress Cataloging-in-Publication Data
Shi, Ying Guang.
Power orthogonal polynomials / Ying Guang Shi.
p. cm.
Includes bibliographical references and index.
ISBN 1-59454-855-2
1. Orthogonal polynomials. I. Title.
QA404.5.S4 2006
515'.55–dc22 2005036589

Published by Nova Science Publishers, Inc. ✤ New York

Contents

Preface

The topic of orthogonal polynomials is a vast one, and is the subject of the excellent monograph by Szegő and of the treatises by Freud, Nevai, Dunkl and Xu, and others. Closely related to orthogonal polynomials is the Gaussian quadrature formula with simple nodes. In the works of Jacobi and Christoffel there appeared the idea of quadrature formulas involving multiple nodes. In 1930 Bernstein found that the nth Chebyshev polynomial of the first kind is the solution of L^m extremal problem with respect to the Chebyshev weight for every $m > 0$. Later in 1948 Tchakaloff and in 1954 Turán investigated quadrature formulas with multiple nodes. They leads to power orthogonal polynomials and Gaussian quadrature formulas with multiple nodes. They have been studied by many mathematicians since, such as Popoviciu, Kis, Stancu, Stroud, Hobby, Rice, Ossicini, Ghizzetti, Micchelli, Rivlin, Rosati, Karlin, Pinkus, Barrow, Bojanov, Gautschi, Sharma, Braess, Dyn, Lubinsky, Saff, Gori, Lo Cascio, Milovanović, Spalević, Yang, Wang, Xu, Shi, and others.

It turns out that the theory of power orthogonal polynomials is very difficult. Orthogonal polynomials and power orthogonal polynomials are as different as, let us say, the linear and nonlinear problems, although the former is a special case of the latter.

This book has been influenced by the important and excellent book of Szegő in style and taste. The book is largely self-contained.

In Chapter 1 we begin with listing the basic results of orthogonal polynomials, Jacobi, Laguerre, and Hermite polynomials, and collecting some frequently used theorems and formulas.

As a base and useful tool, the representation and quantitative theory of Hermite interpolation is putted in Chapter 2.

The theory of power orthogonal polynomials begins in Chapter 3: existence, uniqueness, Characterizations, properties of zeros, and continuity with respect to the measure and the indices.

Chapter 4 deals with Gaussian quadrature formulas and their convergence.

Chapter 5 is devoted to the theory of Christoffel type functions, which is related to Gaussian quadrature formulas and is one of the important contents of power orthogonal polynomials.

The explicit representation of power orthogonal polynomials is an interesting problem and is discussed in Chapter 6.

Chapter 7 is a detailed treatment of zeros of power orthogonal polynomials.

Chapter 8 is devoted to bounds and inequalities of power orthogonal polynomials.

In Chapters 9 and 10 we study asymptotics of general polynomials and power orthogonal polynomials, respectively. The former is a base of the latter.

In Chapter 11 we discuss convergence of power orthogonal series, Lagrange and Hermite interpolation, and two positive operators constructed by power orthogonal polynomials.

In Chapter 12 we investigate Gaussian quadrature formulas for extended Chebyshev spaces.

In Chapter 13 we give construction methods for power orthogonal polynomials and Gaussian quadrature formulas; we also provide numerical results and numerical tables.

The author is grateful to President and Editor-in-Chief Frank Columbus, and Vice-President Nadya S. Columbus for providing this possibility of working.

The work is supported in part by the National Natural Science Foundation of China (No.10571049) and by Hunan Provincial Natural Science Foundation of China (No. 05JJ30011).

Ying Guang Shi
Department of Mathematics
Hunan Normal University
Chansha, Hunan, China

Chapter 1

Preliminaries

1.1 Notations

Throughout of this book, we will use the standard notations.

We denote by $\mathbf{N}$, $\mathbf{N}_0$, $\mathbf{N}_1$, or $\mathbf{N}_2$ the set of positive, nonnegative, odd, or even integers, respectively. For convenience we accept the symbol $\mathbf{N}_{02} = \{0\} \cup \mathbf{N}_2$. We also denote by $\mathbf{R}$, $\mathbf{R}^+$, or $\mathbf{C}$ the set of real, nonnegative, or complex numbers, respectively.

The symbol $\mathbf{P}_N$ or $\mathbf{T}_N$ stands for the set of algebraic or trigonometric polynomials of degree at most N, respectively, here $N \in \mathbf{N}_0$. Let $\widehat{\mathbf{P}}_N = \{P(x) = x^N + c_{N-1}x^{N-1} + \cdots + c_1 x + c_0,\ c_{N-1}, \ldots, c_1, c_0 \in \mathbf{R}\}$. The symbol ∂P denotes the exact degree of the polynomial $P \neq 0$, i.e., $P \in \mathbf{P}_{\partial P} \setminus \mathbf{P}_{\partial P-1}$. The symbol Δ denotes an (open or closed) interval and Δ^0 its interior part. For a Lebesgue measurable set S its Lebesgue measure is denoted by $|S|$. For $P \in \mathbf{P}_N$ and $S \subset \mathbf{R}$ the symbol $Z(P,S)$ or $Z'(P,S)$ is defined by the number of real zeros of P in S not counting or counting multiplicities, respectively.

Let μ be a nondecreasing function on $\mathbf{R}$ with infinitely many points of increase such that all moments of $d\mu$ are finite. We call $d\mu$ a *measure*. The *support* of $d\mu$ is the set of points of increase of $\mu(x)$ and is denoted by $\mathrm{supp}\,(d\mu)$. If μ is absolutely continuous, $\mu' = w$ is said to be a *weight (function)*. Here and later the equality $\mu' = w$ always means $\mu' = w$, a. e. For $0 < p < +\infty$ the class of functions f which are measurable with respect to $d\mu$ supported in Δ and for which the *Stieltjes-Lebesgue integral* $\int_\Delta |f(x)|^p d\mu(x)$ exists is called $L^p_{d\mu}(\Delta)$. If $\mu(x) = x$ we write $L^p(\Delta) = L^p_{d\mu}(\Delta)$. For $m \in \mathbf{N}_0$ let $C^m(\Delta)$ denote the set of mth continuously differentiable functions on Δ; in particular $C(\Delta) := C^0(\Delta)$. Then we can define

$$\|f\|_{L^p_{d\mu}(\Delta)} = \left\{ \int_\Delta |f(x)|^p d\mu(x) \right\}^{1/p}, \quad f \in L^p_{d\mu}(\Delta) \tag{1.1.1}$$

and

$$\|f\|_\Delta = \sup_{x \in \Delta} |f(x)|, \quad f \in C(\Delta). \tag{1.1.2}$$

Of course, for $0 < p < 1$ (1.1.1) is not a norm. In the case when $\Delta = [-1,1]$ for simplicity we use the notations:

$$L^p_{d\mu} = L^p_{d\mu}[-1,1], \quad L^p = L^p[-1,1],$$

and

$$\|f\|_{d\mu,p} = \|f\|_{L^p_{d\mu}[-1,1]}, \quad \|f\| = \|f\|_{[-1,1]}.$$

We denote by $c, c_1, \dots$ positive constants independent of variables and indices, unless otherwise indicated; their value may be different at different occurrences, even in subsequent formulas. We write $a_n \sim b_n$ if $c_1 \le a_n/b_n \le c_2$ holds for every n. The notations $a(x) \sim b(x)$ and $a_n(x) \sim b_n(x)$ have similar meaning. We often use the notations

$$a_n = O(b_n), \quad a_n = o(b_n) \tag{1.1.3}$$

for $b_n > 0$, which means that a_n/b_n is bounded or tends to 0, respectively.

The symbol $\bar{z}$ denotes the conjugate complex value, $\Re(z)$ the real part, and $\Im(z)$ the imaginary part of the complex number z. The symbol $[x]$ denotes the integral part and $\{x\}$ the fractional part of the real number x. We need the symbol $(x)_+ = \max\{x, 0\}$. We need the *sign function*

$$\operatorname{sgn} x = \begin{cases} x/|x|, & x \ne 0, \\ 0, & x = 0 \end{cases} \tag{1.1.4}$$

and the *Kronecker delta*

$$\delta_{ij} = \begin{cases} 0, & i \ne j, \\ 1, & i = j. \end{cases} \tag{1.1.5}$$

The *Pochhammer symbol*, also called the *shifted factorial*, is defined for all x by

$$(x)_0 = 1, \quad (x)_{n+1} = (x+n)(x)_n, \quad n \in \mathbf{N}_0. \tag{1.1.6}$$

By the definition we have $(x)_{m+n} = (x)_m(x+m)_n$ and $(x)_n = \Gamma(x+n)/\Gamma(x)$, $m, n \in \mathbf{N}_0$, where Γ is the *gamma function*. For the *beta function* $B(x)$ we have

$$B(p,q) = \frac{\Gamma(p)\Gamma(q)}{\Gamma(p+q)} \tag{1.1.7}$$

and for the *binomial coefficient* we have

$$\binom{p}{q} = \frac{\Gamma(p+1)}{\Gamma(q+1)\Gamma(p-q+1)}. \tag{1.1.8}$$

As usual, we denote $\varphi(x) = (1-x^2)^{1/2}$,

$$E_n(f) = \min_{P \in \mathbf{P}_n} \|f - P\|,$$

and

$$\Delta_n(x) = \frac{(1-x^2)^{1/2}}{n} + \frac{1}{n^2}.$$

Denote by $\omega(f;\cdot)$ the *modulus of continuity* of f. We also need the *modulus of continuity of Ditzian-Totik* $\omega_\varphi(f;\cdot)$ [16] of f which is defined by

$$\omega_\varphi(f;t) := \sup_{0<h\le t} \|f(\cdot + h\varphi(\cdot)/2) - f(\cdot - h\varphi(\cdot)/2)\|,$$

where the expression inside $\|\cdot\|$ is taken to be zero if $x \pm h\varphi(x)/2 \notin (-1,1)$.

1.2 Orthogonal Polynomials

For a given measure $d\mu$ the corresponding system of orthonormal polynomials $\{P_n(d\mu)\}_{n=0}^{\infty}$ is defined by

$$P_n(d\mu;x) = \gamma_n(d\mu)x^n + \dots, \qquad \gamma_n(d\mu) > 0, \tag{1.2.1}$$

and

$$\int_{\mathbf{R}} P_m(d\mu;x)P_n(d\mu;x)d\mu(x) = \delta_{mn}. \tag{1.2.2}$$

For a weight w we write $P_n(w;x)$ instead of $P_n(d\mu;x)$. The zeros of $P_n(d\mu)$, which are real and distinct, and are contained in (a,b) if $\operatorname{supp}(d\mu) \subset [a,b]$, will be denoted by $x_{kn}(d\mu)$ in the increasing order: $x_{1n}(d\mu) < x_{2n}(d\mu) < \dots < x_{nn}(d\mu)$. The orthogonal polynomials $P_n(d\mu)$ satisfy the *three term recurrence relation*

$$\begin{aligned} xP_{n-1}(d\mu;x) &= \frac{\gamma_{n-1}(d\mu)}{\gamma_n(d\mu)}P_n(d\mu;x) + \alpha_{n-1}(d\mu)P_{n-1}(d\mu;x) \\ &+ \frac{\gamma_{n-2}(d\mu)}{\gamma_{n-1}(d\mu)}P_{n-2}(d\mu;x), \quad n \in \mathbf{N}_0, \end{aligned} \tag{1.2.3}$$

where $P_{-1} \equiv 0$, $\gamma_{-1} = 0$, and

$$\alpha_n(d\mu) = \int_{\mathbf{R}} xP_n(d\mu;x)^2 d\mu(x).$$

The *Christoffel function* $\lambda_n(d\mu)$ with respect to $d\mu$ is defined by

$$\lambda_n(d\mu;x) = \min_{P \in \mathbf{P}_{n-1}, P(x)=1} \int_{\mathbf{R}} P(x)^2 d\mu(x). \tag{1.2.4}$$

We have the formula

$$\lambda_n(d\mu;x)^{-1} = \sum_{k=0}^{n-1} P_k(d\mu;x)^2. \tag{1.2.5}$$

The numbers

$$\lambda_{kn}(d\mu) = \lambda_n(d\mu;x_{kn}(d\mu)) \tag{1.2.6}$$

are called *Christoffel numbers*. We have the *Markov-Stieltjes inequalities*

$$\sum_{k=1}^{i-1} \lambda_{kn}(d\mu) \le \int_{-\infty}^{x_{in}(d\mu)} d\mu(x) \le \sum_{k=1}^{i} \lambda_{kn}(d\mu) \tag{1.2.7}$$

and the *Gaussian quadrature formula*

$$\int_{\mathbf{R}} f(x)d\mu(x) = \sum_{k=1}^{n} \lambda_{kn}(d\mu)f(x_{kn}(d\mu)) \tag{1.2.8}$$

which is exact for $f \in \mathbf{P}_{2n-1}$.

For $f \in L^1_{d\mu}(\mathbf{R})$ let $\mathcal{S}_n(d\mu, f)$ denote the nth partial sum of the orthogonal Fourier series of f, which may be expressed as

$$\mathcal{S}_n(d\mu, f; x) = \int_{\mathbf{R}} f(x) K_n(d\mu; x, t) d\mu(t) \tag{1.2.9}$$

where

$$K_n(d\mu; x, t) = \sum_{k=0}^{n-1} P_k(d\mu; x) P_k(d\mu; t) \tag{1.2.10}$$

or by the *Christoffel-Darboux formula*

$$K_n(d\mu; x, t) = \frac{\gamma_{n-1}(d\mu)}{\gamma_n(d\mu)} \cdot \frac{P_{n-1}(d\mu; t) P_n(d\mu; x) - P_{n-1}(d\mu; x) P_n(d\mu; t)}{x - t}. \tag{1.2.11}$$

For a given function f the *Lagrange interpolation polynomial* $\mathcal{L}_n(d\mu, f)$ corresponding to $d\mu$ is defined to be the unique polynomial of degree at most $n-1$ which satisfies

$$\mathcal{L}_n(d\mu, f; x_{kn}(d\mu)) = f(x_{kn}(d\mu)), \quad k = 1, 2, ..., n. \tag{1.2.12}$$

If we denote by $\ell_{kn}(d\mu)$ the *fundamental polynomials of Lagrange interpolation* the polynomial $\mathcal{L}_n(d\mu, f)$ is of the form

$$\mathcal{L}_n(d\mu, f; x) = \sum_{k=1}^{n} f(x_{kn}(d\mu)) \ell_{kn}(d\mu; x). \tag{1.2.13}$$

There is a useful relation

$$\ell_{kn}(d\mu; x) = \frac{\gamma_{n-1}(d\mu)}{\gamma_n(d\mu)} \lambda_{kn}(d\mu) P_{n-1}(d\mu; x_{kn}(d\mu)) \frac{P_n(d\mu; x)}{x - x_{kn}(d\mu)}. \tag{1.2.14}$$

We also consider a measure $d\nu$ supported on the interval $[-\pi, \pi]$. We define [137, p. 268]

$$G(d\nu) = \exp\left\{\frac{1}{2\pi} \int_{-\pi}^{\pi} \ln \nu'(\theta) d\theta\right\}. \tag{1.2.15}$$

If $d\nu \in \mathbf{S}$, that is

$$\int_{-\pi}^{\pi} \ln \nu'(\theta) d\theta > -\infty, \tag{1.2.16}$$

then the *Szegő function* is defined by

$$D(z) = D(d\nu; z) = \exp\left\{\frac{1}{4\pi} \int_{-\pi}^{\pi} \frac{1 + z e^{-i\theta}}{1 - z e^{-i\theta}} \ln \nu'(\theta) d\theta\right\} \tag{1.2.17}$$

which is analytic in $|z| < 1$ [137, p. 269]. For an arbitrary continuous function $F(\theta)$ of period 2π we have [137, p. 269]

$$\lim_{r \to 1-0} \int_{-\pi}^{\pi} F(\theta) |D(d\nu; re^{i\theta})|^2 d\theta = \int_{-\pi}^{\pi} F(\theta) \nu'(\theta) d\theta. \tag{1.2.18}$$

Let for $0 < m < \infty$

$$d_n(d\nu) = \min_{b_k} \frac{1}{2\pi} \int_{-\pi}^{\pi} |z^n + b_{n-1}z^{n-1} + \cdots + b_1 z + b_0|^m d\nu(\theta), \ z = e^{i\theta}. \tag{1.2.19}$$

If $d\nu \in \mathbf{S}$, then [137, p. 376]

$$\lim_{n\to\infty} d_n(d\nu) = G(d\nu). \tag{1.2.20}$$

If $d\mu$ is a measure supported in $[-1,1]$ we define the associated measure $d\nu(\theta)$ on $[-\pi,\pi]$ by

$$\nu(\theta) = \begin{cases} \mu(\cos\theta) - \mu(1), & \theta \in [-\pi, 0], \\ \mu(1) - \mu(\cos\theta), & \theta \in [0,\pi], \end{cases} \tag{1.2.21}$$

and hence

$$d\nu(\theta) = |\sin\theta| d\mu(\cos\theta). \tag{1.2.22}$$

Then the inequality (1.2.16) is equivalent to

$$\int_{-1}^{1} \ln\mu'(x) \frac{dx}{\sqrt{1-x^2}} > -\infty. \tag{1.2.23}$$

So we also write $d\mu \in \mathbf{S}$ if condition (1.2.23) is vailid. In this case instead of (1.2.15) we put

$$G(d\mu) = \exp\left\{ \frac{1}{\pi} \int_{-1}^{1} \ln\mu'(x) \frac{dx}{\sqrt{1-x^2}} \right\}. \tag{1.2.24}$$

1.3 Jacobi Polynomials

The following results are taken from [137, Chapter IV, pp.57– 95]. The *Jacobi weight* is defined by

$$w^{(\alpha,\beta)}(x) = \begin{cases} (1-x)^\alpha (1+x)^\beta, & |x| < 1, \\ 0, & |x| \geq 1, \end{cases} \tag{1.3.1}$$

here $\alpha, \beta > -1$. The corresponding *Jacobi polynomial* $P_n^{(\alpha,\beta)}$ satisfies the differential equation

$$(1-x^2)y'' + [\beta - \alpha - (\alpha+\beta+2)x]y' + n(n+\alpha+\beta+1)y = 0 \tag{1.3.2}$$

and the normalization

$$P_n^{(\alpha,\beta)}(1) = \binom{n+\alpha}{n}. \tag{1.3.3}$$

We have

$$P_n^{(\alpha,\beta)}(x) = 2^{-n} \sum_{k=0}^{n} \binom{n+\alpha}{n-k} \binom{n+\beta}{k} (x-1)^k (x+1)^{n-k} \tag{1.3.4}$$

and

$$P_n^{(\alpha,\beta)}(-x) = (-1)^n P_n^{(\beta,\alpha)}(x). \tag{1.3.5}$$

The leading coefficient is

$$k_n(w^{(\alpha,\beta)}) = \frac{\Gamma(2n+\alpha+\beta+1)}{2^n n! \Gamma(n+\alpha+\beta+1)} \tag{1.3.6}$$

and

$$\gamma_n(w^{(\alpha,\beta)}) = \left[\frac{\Gamma(2n+\alpha+\beta+1)\Gamma(2n+\alpha+\beta+2)}{n!2^{2n+\alpha+\beta+1}\Gamma(n+\alpha+1)\Gamma(n+\beta+1)\Gamma(n+\alpha+\beta+1)}\right]^{1/2}. \tag{1.3.7}$$

There are some important special cases. $P_n = P_n^{(0,0)}$ is the nth *Legendre polynomial*. For $x = \cos\theta$, $0 \le \theta \le \pi$,

$$T_n(x) = \frac{(2n)!!}{(2n-1)!!}P_n^{(-1/2,-1/2)}(x) = \cos n\theta, \tag{1.3.8}$$

$$U_n(x) = \frac{(2n+2)!!}{2(2n+1)!!}P_n^{(1/2,1/2)}(x) = \frac{\sin(n+1)\theta}{\sin\theta}, \tag{1.3.9}$$

$$V_n(x) = \frac{(2n)!!}{(2n-1)!!}P_n^{(-1/2,1/2)}(x) = \frac{\cos[(2n+1)\theta/2]}{\cos(\theta/2)}, \tag{1.3.10}$$

$$W_n(x) = \frac{(2n)!!}{(2n-1)!!}P_n^{(1/2,-1/2)}(x) = \frac{\sin[(2n+1)\theta/2]}{\sin(\theta/2)}, \tag{1.3.11}$$

are called *Chebyshev polynomials of the first, second, third, and fourth kind*, respectively.

1.4 Laguerre and Hermite Polynomials

The following results are taken from [137, Chapter V, pp. 96–106].

The *generalized Laguerre weight* is

$$\ell^{(\alpha)}(x) = \begin{cases} x^\alpha e^{-x}, & x > 0, \\ 0, & x \le 0, \end{cases} \tag{1.4.1}$$

here $\alpha > -1$. The corresponding *generalized Laguerre polynomial* $L_n^{(\alpha)}$ satisfies the differential equation

$$xy'' + (\alpha+1-x)y' + ny = 0 \tag{1.4.2}$$

and the normalization

$$L_n^{(\alpha)}(0) = \binom{n+\alpha}{n}. \tag{1.4.3}$$

We have the representation

$$L_n^{(\alpha)}(x) = \sum_{k=0}^{n}\binom{n+\alpha}{n-k}\frac{(-x)^k}{k!}. \tag{1.4.4}$$

The leading coefficient is

$$k_n(\ell^{(\alpha)}) = \frac{(-1)^n}{n!} \tag{1.4.5}$$

and

$$\gamma_n(\ell^{(\alpha)}) = [n!\Gamma(n+\alpha+1)]^{-1/2}. \tag{1.4.6}$$

In particular, $\ell = \ell^{(0)}$ and $L_n = L_n^{(0)}$ are called the *Laguerre weight* and the *Laguerre polynomial*, respectively.

The *Hermite weight* is

$$h(x) = e^{-x^2}, \quad x \in \mathbf{R}. \tag{1.4.7}$$

The corresponding *Hermite polynomial* H_n satisfies the differential equation

$$y'' - 2xy' + 2ny = 0 \tag{1.4.8}$$

and has the representation

$$H_n(x) = \sum_{k \le n/2} \frac{(-1)^k n!}{k!(n-2k)!}(2x)^{n-2k}. \tag{1.4.9}$$

The leading coefficient is

$$k_n(h) = 2^n \tag{1.4.10}$$

and

$$\gamma_n(h) = \left(\frac{2^n}{\pi^{1/2} n!}\right)^{1/2}. \tag{1.4.11}$$

1.5 Some Frequently Used Theorems and Formulas

We list some frequently used theorems and formulas.

Theorem 1.5.1. [137, Theorem 1.21.1 (Theorem of Lukács), p. 4] *Let $P \in \mathbf{P}_n$ be nonnegative in $[-1,1]$. Then P can be represented in the form*

$$P(x) = \begin{cases} A(x)^2 + (1-x^2)B(x)^2, & n \in \mathbf{N}_2, \\ (1+x)C(x)^2 + (1-x)D(x)^2, & n \in \mathbf{N}_1. \end{cases} \tag{1.5.1}$$

Here $A(x), B(x), C(x)$, and $D(x)$ are real polynomials such that the degree of each term on the right-hand side does not exceed n.

Theorem 1.5.2. [137, Theorem 1.21.2, p. 5] *Every polynomial $P \in \mathbf{P}_n$ which is nonnegative on $\mathbf{R}$ can be represented in the form $A(x)^2 + B(x)^2$. Every polynomial $P \in \mathbf{P}_n$ which is nonnegative on $\mathbf{R}^+$ can be represented in the form $A(x)^2 + B(x)^2 + x[C(x)^2 + D(x)^2]$. Here $A(x), B(x), C(x)$, and $D(x)$ are real polynomials such that the degree of each term on the right-hand side does not exceed n.*

Theorem 1.5.3. [79] *For every $f \in C^m[-1,1]$ $(m \in \mathbf{N}_0)$ there exists a polynomial $P_n \in \mathbf{P}_n$ such that for all $x \in [-1,1]$*

$$|f^{(j)}(x) - P_n^{(j)}(x)| \le c\Delta_n(x)^{m-j}\omega(f^{(m)}; \Delta_n(x)), \quad 0 \le j \le m, \tag{1.5.2}$$

$$|P_n^{(j)}(x)| \le c\Delta_n(x)^{m-j}\omega(f^{(m)}; \Delta_n(x)), \quad j > m. \tag{1.5.3}$$

Theorem 1.5.4. [152] *For every* $f \in C^m[-1,1]$ $(m \in \mathbf{N}_0)$ *there exists a polynomial* $P_n \in \mathbf{P}_n$ *such that for all* $x \in [-1,1]$

$$|f^{(j)}(x) - P_n^{(j)}(x)| \le c\Delta_n(x)^{m-j} E_{n-m}(f^{(m)}), \quad 0 \le j \le m, \tag{1.5.4}$$

$$|P_n^{(j)}(x)| \le c\Delta_n(x)^{m-j} \omega_\varphi(f^{(m)}; 1/n), \quad j > m. \tag{1.5.5}$$

Markov inequality [135, p. 284].

$$\|P'\| \le n^2 \|P\|, \quad P \in \mathbf{P}_n. \tag{1.5.6}$$

Bernstein inequality [135, pp. 284–285].

$$\|T'\| \le n\|T\|, \quad T \in \mathbf{T}_n \tag{1.5.7}$$

or in another form,

$$|P'(x)| \le \frac{n}{\sqrt{1-x^2}} \|P\|, \quad |x| < 1, \quad P \in \mathbf{P}_n. \tag{1.5.8}$$

Chebyshev inequality [12, p. 235].

$$|P(z)| \le |T_n(z)|\, \|P\|, \quad z \in \mathbf{C} \setminus [-1,1], \quad P \in \mathbf{P}_n. \tag{1.5.9}$$

Chapter 2

Hermite Interpolation

2.1 Introduction

Let $n \in \mathbf{N}$ $(n \geq 2), m_k \in \mathbf{N}$ $(k = 1, 2, ..., n), m = \max_{1 \leq k \leq n} m_k$, and $N = \sum_{k=1}^{n} m_k - 1$. Given a system of nodes

$$a = x_0 \leq x_1 < x_2 < \cdots < x_n \leq x_{n+1} = b, \tag{2.1.1}$$

the *Hermite interpolation* $\bar{\mathcal{H}}_n(f) \in \mathbf{P}_N$ of $f \in C^{m-1}[a,b]$ is defined by

$$\bar{\mathcal{H}}_n^{(j)}(f; x_k) = f^{(j)}(x_k), \quad j = 0, 1, \ldots, m_k - 1, \quad k = 1, 2, \ldots, n \tag{2.1.2}$$

and the *Hermite-Fejér interpolation* $\mathcal{H}_n(f) \in \mathbf{P}_N$ for $f \in C[a,b]$ is defined by

$$\mathcal{H}_n^{(j)}(f; x_k) = \begin{cases} f(x_k), & j = 0, \\ 0, & j = 1, ..., m_k - 1, \end{cases} \quad k = 1, 2, \ldots. \tag{2.1.3}$$

In the next two sections we derive the *Lagrange* and *Newton formulas* for Hermite interpolation, respectively. In Section 2.4 we give basic theorems for Hermite interpolation, which will play a crucial role in the theory of Hermite interpolation and Gaussian quadrature formulas.

2.2 Lagrange Formulas

Denote by A_{kj}, $1 \leq k \leq n$, $0 \leq j \leq m_k - 1$, the *fundamental polynomials* for Hermite interpolation, i.e., the $A_{kj} \in \mathbf{P}_N$ satisfies

$$A_{kj}^{(p)}(x_q) = \delta_{jp}\delta_{kq}, \quad p = 0, 1, \ldots, m_q - 1, \quad q = 1, 2, \ldots, n. \tag{2.2.1}$$

Then $\bar{\mathcal{H}}_n(f)$ is given by

$$\bar{\mathcal{H}}_n(f; x) = \sum_{k=1}^{n} \sum_{j=0}^{m_k - 1} f^{(j)}(x_k) A_{kj}(x) \tag{2.2.2}$$

and $\mathcal{H}_n(f)$ is given by

$$\mathcal{H}_n(f;x) = \sum_{k=1}^{n} f(x_k)A_{k0}(x). \tag{2.2.3}$$

Let us give an explicit representation for A_{kj}, $1 \le k \le n$, $0 \le j \le m_k - 1$. To this end we put

$$L_k(x) = \prod_{q=1,q\neq k}^{n} \left(\frac{x-x_q}{x_k-x_q}\right)^{m_q}, \tag{2.2.4}$$

$$b_{ki} = \frac{1}{i!}\left[L_k(x)^{-1}\right]^{(i)}_{x=x_k}, \quad i = 0,1,\ldots,m_k-1, \tag{2.2.5}$$

and

$$B_{kj}(x) = \sum_{i=0}^{m_k-j-1} b_{ki}(x-x_k)^i. \tag{2.2.6}$$

The most important special case is $m_k \equiv m$. In this case we have the simpler formulas for $k = 1,2,\ldots,n$

$$L_k(x) = \ell_k(x)^m, \tag{2.2.7}$$

where

$$\ell_k(x) = \frac{\omega_n(x)}{\omega_n'(x_k)(x-x_k)}, \quad \omega_n(x) = \prod_{k=1}^{n}(x-x_k). \tag{2.2.8}$$

Then we have

Theorem 2.2.1. *The polynomial A_{kj} may be represented by*

$$A_{kj}(x) = \frac{1}{j!}(x-x_k)^j B_{kj}(x)L_k(x). \tag{2.2.9}$$

Proof. Using the interpolation condition (2.2.1) except for $q = k$ and $p \ge j$, and applying (2.2.4), we can write A_{kj} in the form

$$A_{kj}(x) = \frac{1}{j!}(x-x_k)^j L_k(x) \sum_{i=0}^{m_k-j-1} b_i(x-x_k)^i.$$

It remains to show $b_i = b_{ki}$, $i = 0,1,\ldots,m_k-j-1$. To this end we rewrite the above relation in the form

$$A_{kj}(x)L_k(x)^{-1} = \frac{1}{j!}(x-x_k)^j \sum_{i=0}^{m_k-j-1} b_i(x-x_k)^i. \tag{2.2.10}$$

In this case for $i \le m_k - j - 1$ by (2.2.1) and (2.2.5)

$$[A_{kj}(x)L_k(x)^{-1}]^{(i+j)}_{x=x_k} = \binom{i+j}{j} A^{(j)}_{kj}(x_k)[L_k(x)^{-1}]^{(i)}_{x=x_k} = \binom{i+j}{j} i!b_{ki} = \frac{(i+j)!}{j!}b_{ki}$$

and

$$\left[\frac{1}{j!}(x-x_k)^j \sum_{i=0}^{m_k-j-1} b_i(x-x_k)^i\right]^{(i+j)}_{x=x_k} = \frac{(i+j)!}{j!}b_i.$$

Hence by (2.2.10) we conclude that $b_i = b_{ki}$, $i = 0,1,\ldots,m_k-j-1$. □

The numbers b_{kj} can be calculated by a recurrence formula.

Theorem 2.2.2. *Let*

$$a_{kj} = \sum_{i \neq k} \frac{m_i}{(x_i - x_k)^j}, \quad j \geq 1. \tag{2.2.11}$$

Then

$$b_{k0} = 1, \quad b_{kj} = \frac{1}{j} \sum_{i=1}^{j} a_{ki} b_{k,j-i}, \quad j \geq 1. \tag{2.2.12}$$

Proof. We observe that

$$\frac{L_k'(x)}{L_k(x)} = \sum_{i \neq k} \frac{m_i}{x - x_i}.$$

Thus

$$a_{ki} = \frac{1}{(i-1)!} \left[-\frac{L_k'(x)}{L_k(x)} \right]_{x=x_k}^{(i-1)}.$$

By means of this formula, (2.2.5), and Newton-Leibniz formula, we get

$$\begin{aligned} b_{kj} &= \frac{1}{j!} \left[L_k(x)^{-1} \right]_{x=x_k}^{(j)} = \frac{1}{j!} \left[\left(-\frac{L_k'(x)}{L_k(x)} \right) L_k(x)^{-1} \right]_{x=x_k}^{(j-1)} \\ &= \frac{1}{j} \sum_{i=1}^{j} \frac{1}{(i-1)!} \left[-\frac{L_k'(x)}{L_k(x)} \right]_{x=x_k}^{(i-1)} \frac{1}{(j-i)!} \left[L_k(x)^{-1} \right]_{x=x_k}^{(j-i)} \\ &= \frac{1}{j} \sum_{i=1}^{j} a_{ki} b_{k,j-i}. \quad \square \end{aligned}$$

Now let us estimate the numbers ∂A_{kj} and $Z(B_{kj})$. To this end we need to introduce a definition.

Definition 2.2.1. *Let* $P \in \mathbf{P}_N$ *and* $S \subset \mathbf{R}$. *Assume that all distinct zeros of* $P^{(j)}$, $j = 0, 1, \ldots, \partial P - 1$ *are given in (2.1.1). If for* $x_k \in S$ *and* $1 \leq i < i + p \leq \partial P$ *the polynomial* P *satisfies*

$$P^{(i-1)}(x_k) \neq 0, \quad P^{(i)}(x_k) = \cdots = P^{(i+p-1)}(x_k) = 0, \quad P^{(i+p)}(x_k) \neq 0,$$

the corresponding number ρ_{ki} *is denoted by*

$$\rho_{ki} = \begin{cases} p - 1, & 2 \leq k \leq n-1 \text{ and } p \in \mathbf{N}_1, \\ p, & \text{otherwise.} \end{cases}$$

Then the number $Z''(P,S)$ *is defined by*

$$Z''(P,S) = Z'(P,S) + \sum_{x_k \in S} \sum_{i} \rho_{ki}. \tag{2.2.13}$$

We state an important result which is needed later, here $Z''(P) := Z''(P, \mathbf{R})$.

Lemma 2.2.1. [123, Theorem 2.1, p. 10] *Let* P *be a nonzero polynomial. Then*

$$Z''(P) \leq \partial P. \tag{2.2.14}$$

The following lemma provides estimates of $Z''(A_{kj})$ and $Z'(B_{kj},[A,B])$.

Lemma 2.2.2. *Let the indices k, j, $1 \le k \le n$, $0 \le j \le m_k - 1$, be fixed and let*

$$\rho_k = \begin{cases} 0, & m_k - j \in \mathbf{N}_1, \\ 1, & m_k - j \in \mathbf{N}_2. \end{cases} \tag{2.2.15}$$

Then

$$Z''(A_{kj}) \ge N - \sum_{x_k \in (x_1, x_n)} \rho_k. \tag{2.2.16}$$

Meanwhile, if $(A,B) \supset (x_1,x_n)$, then

$$Z'(B_{kj},[A,B]) \le \sum_{x_k \in (A,B)} \rho_k. \tag{2.2.17}$$

Proof. Inequality (2.2.16) follows from (2.2.13) and (2.2.15).

Let us prove (2.2.17). We observe that each zero of B_{kj} must be either a new zero or an additional multiplicity of an old zero of A_{kj}. Then by (2.2.14) and (2.2.16)

$$\begin{aligned} N &\ge \partial A_{kj} \ge Z''(A_{kj}) \ge N - \sum_{x_k \in (x_1, x_n)} \rho_k + Z'(B_{kj},[A,B]) \\ &\ge N - \sum_{x_k \in (A,B)} \rho_k + Z'(B_{kj},[A,B]) \end{aligned}$$

which gives (2.2.17). □

As a direct consequence of Lemma 2.2.2 we state

Theorem 2.2.3. *We have*

$$\partial A_{kj} \ge N - 1. \tag{2.2.18}$$

Furthermore, if $k = 1$, or $k = n$, or $m_k - j \in \mathbf{N}_1$ then

$$\partial A_{kj} = N. \tag{2.2.19}$$

Proof. By (2.2.15) and (2.2.16) we see $Z''(A_{kj}) \ge N - 1$ and further, if $k = 1$, or $k = n$, or $m_k - j \in \mathbf{N}_1$ we obtain $Z''(A_{kj}) \ge N$. Remembering $\partial A_{kj} \le N$, it remains to apply Lemma 2.2.1. □

Theorem 2.2.4. *If the number $m_k - j$ is odd then*

$$B_{kj}(x) > 0, \qquad x \in \mathbf{R}, \quad b_{k,m_k-j-1} > 0. \tag{2.2.20}$$

If the number $m_k - j$ is even then

$$Z'(B_{kj}) = \begin{cases} 1, & b_{k,m_k-j-1} \ne 0, \\ 0, & b_{k,m_k-j-1} = 0. \end{cases} \tag{2.2.21}$$

Proof. If $m_k - j \in \mathbf{N}_1$ then by (2.2.15) and (2.2.17) we have $Z(B_{kj}) = 0$. Recalling $B_{kj}(x_k) = 1$, relation (2.2.20) follows.

Let $m_k - j \in \mathbf{N}_2$. By (2.2.15) and (2.2.17) we have $Z'(B_{kj}) \le 1$. If $b_{k,m_k-j-1} \ne 0$ then the number $\partial B_{kj} = m_k - j - 1$ is odd and hence $Z'(B_{kj}) \ge 1$. So $Z'(B_{kj}) = 1$. If $b_{k,m_k-j-1} = 0$ then by (2.2.18) the number $\partial B_{kj} = m_k - j - 2$ is even and hence $Z'(B_{kj}) = 0$. □

For $k = 1$ or $k = n$ we can obtain further results which will be needed later.

Theorem 2.2.5. *Denote $d_1 = x_2 - x_1$ and $d_n = x_n - x_{n-1}$. Then*

$$\begin{cases} b_{1j} > \frac{b_{1,j-1}}{d_1} > 0, & j = 1,2,\dots,m_1 - 1, \\ (-1)^j b_{nj} > \frac{(-1)^{j-1} b_{n,j-1}}{d_n} > 0, & j = 1,2,\dots,m_n - 1. \end{cases} \tag{2.2.22}$$

Moreover, if the number $m_1 - j$ is even, then the polynomial B_{1j} has a unique zero ξ such that

$$x_1 - d_1 < \xi < x_1 \tag{2.2.23}$$

and

$$B_{1j}(x) \begin{cases} \ge 0, & x \ge \xi, \\ < 0, & x < \xi; \end{cases} \tag{2.2.24}$$

if the number $m_n - j$ is even, then the polynomial B_{nj} has a unique zero η such that

$$x_n < \eta < x_n + d_n \tag{2.2.25}$$

and

$$B_{nj}(x) \begin{cases} \le 0, & x \ge \eta, \\ > 0, & x < \eta. \end{cases} \tag{2.2.26}$$

Proof. It is enough to give the proof for $k = 1$. First let us prove

$$b_{1,m_1-j-1} > 0, \quad j = 0,1,\dots,m_1 - 1. \tag{2.2.27}$$

This inequality has been given by (2.2.20) if $m_1 - j$ is odd; we have only to prove (2.2.27) for the case when $m_1 - j$ is even. In this case by (2.2.19) we see $b_{1,m_1-j-1} \ne 0$ and by (2.2.17) we conclude

$$Z(B_{1j}, [x_1, +\infty)) = 0. \tag{2.2.28}$$

Since $B_{1j}(x_1) = 1$, $b_{1,m_1-j-1} > 0$, which proves (2.2.27).

Next, put

$$L_1^*(x) = L_1(x) \left(\frac{x - x_2}{x_1 - x_2} \right)^{-1}.$$

Thus by (2.2.27)

$$b_{1j}^* = \frac{1}{j!} \left[L_1^*(x)^{-1} \right]_{x=x_1}^{(j)} = b_{1j} + \frac{1}{x_1 - x_2} b_{1,j-1} > 0, \quad j \ge 1,$$

which gives the first line in (2.2.22).

Finally, by (2.2.21) and (2.2.27) we see $Z'(B_{1j}) = 1$. Then the polynomial B_{1j} has a unique zero, which is denoted by ξ. Relation (2.2.28) implies inequality $\xi < x_1$ and relation (2.2.24). Further, using (2.2.22) we have that for $x \leq x_1 - d_1$

$$\begin{aligned} B_{1j}(x) &= \sum_{\nu=0}^{m_1-j-1} b_{1\nu}(x-x_1)^\nu = \sum_{\nu=0}^{(m_1-j-2)/2} [b_{1,2\nu+1}(x-x_1)+b_{1,2\nu}](x-x_1)^{2\nu} \\ &\leq - \sum_{\nu=0}^{(m_1-j-2)/2} [b_{1,2\nu+1}d_1 - b_{1,2\nu}](x-x_1)^{2\nu} < 0. \end{aligned}$$

Then we conclude that $\xi > x_1 - d_1$. This proves (2.2.23). □

2.3 Newton Formulas

We introduce the notation $f[x_1^{m_1},\dots,x_n^{m_n}]$ for the *divided difference* of f at the points (2.1.1), here $x_k^{m_k}$ means that the point x_k is repeated m_k times.

Definition 2.3.1. [58, p. 373] *The divided difference $f[x_1^{m_1},\dots,x_n^{m_n}]$ for the function $f \in C^{m-1}[a,b]$ is defined to be the leading coefficient of the unique polynomial $\bar{\mathcal{H}}_n(f) \in \mathbf{P}_N$, which satisfies the condition (2.1.2).*

Theorem 2.3.1. *Let $\xi_0,\dots,\xi_N$ be given, not necessarily distinct. For $f \in C^m[a,b]$ we have*

$$f(x) = H(x) + f[\xi_0,\dots,\xi_N,x]\prod_{j=0}^{N}(x-\xi_j), \tag{2.3.1}$$

where the polynomial H may be represented in the form

$$H(x) = \sum_{i=0}^{N} f[\xi_0,\dots,\xi_i]\prod_{j=0}^{i-1}(x-\xi_j). \tag{2.3.2}$$

Proof. Denote by $\bar{\mathcal{H}}(\xi_0,\dots,\xi_i;\cdot) \in \mathbf{P}_{i-1}$ $(i \geq 1)$ the Hermite interpolation of f at the points $\xi_0,\dots,\xi_i$. Then by Definition 2.3.1

$$\bar{\mathcal{H}}(\xi_0,\dots,\xi_i;y) - \bar{\mathcal{H}}(\xi_0,\dots,\xi_{i-1};y) = f[\xi_0,\dots,\xi_i]\prod_{j=0}^{i-1}(y-\xi_j).$$

For a fixed point $x \in [a,b]$ by induction we get

$$\bar{\mathcal{H}}(\xi_0,\dots,\xi_N,x;y) = \sum_{i=0}^{N} f[\xi_0,\dots,\xi_i]\prod_{j=0}^{i-1}(y-\xi_j) + f[\xi_0,\dots,\xi_N,x]\prod_{j=0}^{N}(y-\xi_j).$$

Let $x \notin \{\xi_0,\dots,\xi_N\}$. Inserting $y = x$ into the above relation and noticing by definition $\bar{\mathcal{H}}(\xi_0,\dots,\xi_N,x;x) = f(x)$, we obtain (2.3.1) with (2.3.2). By continuity relation (2.3.1) with (2.3.2) holds for every $x \in [a,b]$. □

Clearly, H is just a Newton formula of Hermite interpolation for f at the points $\xi_0,\dots,\xi_N$. By means of different permutation for the points $\xi_0,\dots,\xi_N$ we can obtain different Newton formulas of Hermite interpolation for f for different applications.

Theorem 2.3.2. *We have*

$$\bar{\mathcal{H}}_n(f;x) = \sum_{k=1}^{n}\sum_{j=0}^{m_k-1} f[x_1^{m_1},\ldots,x_{k-1}^{m_{k-1}},x_k^{j+1}]\prod_{i=1}^{k-1}(x-x_i)^{m_i}(x-x_k)^j \tag{2.3.3}$$

and

$$\begin{aligned}&\bar{\mathcal{H}}_n(f;x)\\ &= \sum_{j=0}^{m-1}\sum_{m_k\geq j+1} f[x_1^{\min\{j+1,m_1\}},\ldots,x_k^{\min\{j+1,m_k\}},x_{k+1}^{\min\{j,m_{k+1}\}},\ldots,x_n^{\min\{j,m_n\}}]\\ &\quad\times\prod_{i=1}^{k-1}(x-x_i)^{\min\{j+1,m_i\}}\prod_{i=k}^{n}(x-x_i)^{\min\{j,m_i\}}.\end{aligned} \tag{2.3.4}$$

Proof. To derive (2.3.3) we choose ξ_i's as x_k's in the order

$$\overbrace{x_1,\ldots,x_1}^{m_1},\overbrace{x_2,\ldots,x_2}^{m_2},\ldots,\overbrace{x_n,\ldots,x_n}^{m_n},$$

i.e., we put $\xi_i = x_k$ if the indices i and k satisfy

$$\sum_{j=1}^{k-1} m_j \leq i < \sum_{j=1}^{k} m_j, \quad 1\leq k\leq n.$$

Then relation (2.3.2) becomes (2.3.3).

To derive (2.3.4) we choose ξ_i's as x_k's in the m ordered groups corresponding to $j = 0,1,\ldots,m-1$

$$x_{j_1},x_{j_2},\ldots,x_{j_{n_j}}, \quad 1\leq j_1<j_2<\cdots<j_{n_j}\leq n,$$

where the index j_k appears if and only if $m_{j_k}\geq j+1$. In this case relation (2.3.2) yields (2.3.4). □

We state an interesting special case when $m_1=\cdots=m_n=m$.

Corollary 2.3.1. *If $m_1=\cdots=m_n=m$ then*

$$\bar{\mathcal{H}}_n(f;x) = \sum_{k=1}^{n}\sum_{j=0}^{m-1} f[x_1^{m},\ldots,x_{k-1}^{m},x_k^{j+1}]\prod_{i=1}^{k-1}(x-x_i)^{m}(x-x_k)^j, \tag{2.3.5}$$

$$\bar{\mathcal{H}}_n(f;x) = \sum_{k=1}^{n}\sum_{j=0}^{m-1} f[x_1^{j+1},\ldots,x_k^{j+1},x_{k+1}^{j},\ldots,x_n^{j}]\omega_n(x)^j\prod_{i=1}^{k-1}(x-x_i), \tag{2.3.6}$$

and

$$\bar{\mathcal{H}}_n(f;x) = \sum_{k=1}^{n}\sum_{j=0}^{m-1} f[x_1^{j},\ldots,x_n^{j},x_k]\omega_n(x)^j\ell_k(x). \tag{2.3.7}$$

Proof. The relation (2.3.5) or (2.3.6) is a direct consequence of (2.3.3) or (2.3.4), respectively. To derive (2.3.7) we observe that for a fixed index j the expression

$$\begin{aligned}&\sum_{k=1}^{n} f[x_1^{j+1},\ldots,x_k^{j+1},x_{k+1}^{j},\ldots,x_n^{j}]\prod_{i=1}^{k-1}(x-x_i)\\ =&\sum_{k=1}^{n} f[x_1^{j},\ldots,x_n^{j},x_1,\ldots,x_k]\prod_{i=1}^{k-1}(x-x_i)\end{aligned}$$

or

$$\sum_{k=1}^{n} f[x_1^{j},\ldots,x_n^{j},x_k]\ell_k(x)$$

is the Newton or the Lagrange interpolation of the function $f[x_1^j,\ldots,x_n^j,x]$, respectively. By uniqueness they are equal. Hence (2.3.7) follows from (2.3.6). □

In order to derive Gaussian quadrature formulas we need other Hermite interpolation formulas.

Theorem 2.3.3. *We have*

$$\begin{aligned}&\bar{\mathcal{H}}_n(f;x)\\ =&\sum_{k=1}^{n}\sum_{j=0}^{m_k-2} f[x_1^{m_1-1},\ldots,x_{k-1}^{m_{k-1}-1},x_k^{j+1}]\prod_{i=1}^{k-1}(x-x_i)^{m_i-1}(x-x_k)^j\\ &+\sum_{k=1}^{n} f[x_1^{m_1},\ldots,x_k^{m_k},x_{k+1}^{m_{k+1}-1},\ldots,x_n^{m_n-1}]\prod_{i=1}^{n}(x-x_i)^{m_i-1}\prod_{i=1}^{k-1}(x-x_i)\end{aligned}\tag{2.3.8}$$

and

$$\begin{aligned}&\bar{\mathcal{H}}_n(f;x)\\ =&\sum_{j=0}^{m-2}\sum_{m_k\ge j+2} f[x_1^{\min\{j+1,m_1-1\}},\ldots,x_k^{\min\{j+1,m_k-1\}},\\ &x_{k+1}^{\min\{j,m_{k+1}-1\}},\ldots,x_n^{\min\{j,m_n-1\}}]\\ &\times\prod_{i=1}^{k-1}(x-x_i)^{\min\{j+1,m_i-1\}}\prod_{i=k}^{n}(x-x_i)^{\min\{j,m_i-1\}}\\ &+\sum_{k=1}^{n} f[x_1^{m_1},\ldots,x_k^{m_k},x_{k+1}^{m_{k+1}-1},\ldots,x_n^{m_n-1}]\prod_{i=1}^{n}(x-x_i)^{m_i-1}\prod_{i=1}^{k-1}(x-x_i).\end{aligned}\tag{2.3.9}$$

Proof. To derive (2.3.8) choose ξ_i's as x_k's in the order

$$\overbrace{x_1,\ldots,x_1}^{m_1-1},\overbrace{x_2,\ldots,x_2}^{m_2-1},\ldots,\overbrace{x_n,\ldots,x_n}^{m_n-1},x_1,x_2,\ldots,x_n.$$

Then we obtain (2.3.8).

To derive (2.3.4) choose ξ_i, $0 \le i \le N-n$ as x_k's in the $m-1$ ordered groups corresponding to $j = 0, 1, \ldots, m-2$

$$x_{j_1}, x_{j_2}, \ldots, x_{j_{n_j}}, \quad 1 \le j_1 < j_2 < \cdots < j_{n_j} \le n,$$

where the index j_k appears if and only if $m_{j_k} \ge j+2$. Then we choose $\xi_{N-n+k} = x_k$, $k = 1, 2, \ldots, n$. In this case relation (2.3.2) yields (2.3.9). □

As an immediate consequence of Theorem 2.3.3 we state an interesting special case when $m_1 = \cdots = m_n = m$.

Corollary 2.3.2. *If $m_1 = \cdots = m_n = m$ then*

$$\begin{aligned}\bar{\mathcal{H}}_n(f;x) &= \sum_{k=1}^{n} \sum_{j=0}^{m-2} f[x_1^{m-1}, \ldots, x_{k-1}^{m-1}, x_k^{j+1}] \prod_{i=1}^{k-1} (x-x_i)^{m-1} (x-x_k)^j \\ &\quad + \sum_{k=1}^{n} f[x_1^m, \ldots, x_k^m, x_{k+1}^{m-1}, \ldots, x_n^{m-1}] \omega_n(x)^{m-1} \prod_{i=1}^{k-1} (x-x_i).\end{aligned} \tag{2.3.10}$$

The following theorem gives an explicit expression of divided differences in term of the function and its derivatives.

Theorem 2.3.4. *Let $f \in C^{m-1}[a,b]$. Then*

$$f[x_1^{m_1}, \ldots, x_n^{m_n}] = \sum_{k=1}^{n} \sum_{j=0}^{m_k-1} \frac{1}{j!} b_{k,m_k-j-1} f^{(j)}(x_k) \prod_{i \ne k} (x_k - x_i)^{-m_i} \tag{2.3.11}$$

and

$$f[x_1^{m_1}, \ldots, x_n^{m_n}] = \sum_{k=1}^{n} \frac{1}{(m_k-1)!} \left[f(x) \prod_{i \ne k} (x-x_i)^{-m_i} \right]_{x=x_k}^{(m_k-1)}. \tag{2.3.12}$$

In particular, if $m_1 = \cdots = m_n = m$ then

$$f[x_1^m, \ldots, x_n^m] = \sum_{k=1}^{n} \sum_{j=0}^{m-1} \frac{1}{j!} b_{k,m-j-1} \omega_n'(x_k)^{-m} f^{(j)}(x_k) \tag{2.3.13}$$

and

$$f[x_1^m, \ldots, x_n^m] = \sum_{k=1}^{n} \frac{1}{(m-1)!} \left[f(x) \left(\frac{\omega_n(x)}{x-x_k} \right)^{-m} \right]_{x=x_k}^{(m-1)}. \tag{2.3.14}$$

Proof. By Definition 2.3.1 we have to determine the leading coefficient of the polynomial $\bar{\mathcal{H}}_n(f)$. The formulas (2.2.4), (2.2.6), and (2.2.9) show that the leading coefficient of A_{kj} is

$$\frac{1}{j!} b_{k,m_k-j-1} \prod_{i \ne k} (x_k - x_i)^{-m_i}$$

which coupled with (2.2.2) gives (2.3.11). Applying Newton-Leibniz rule and using (2.2.4) and (2.2.5) from (2.3.11) we can easily get (2.3.12). The formulas (2.3.13) and (2.3.14) follow directly from (2.3.11) and (2.3.12), respectively. □

Let us conclude this section with an identity which will be needed in Section 6.5.

Theorem 2.3.5. *Let $f \in C^m[a,b]$. Then*

$$\sum_{k=1}^{n} m_k f[x_1^{m_1},\ldots,x_n^{m_n},x_k] = f'[x_1^{m_1},\ldots,x_n^{m_n}]. \tag{2.3.15}$$

Proof. Applying (2.3.12) and using the symbol $g_k(x) = \prod_{i\neq k}(x-x_i)^{-m_i}$

$$\begin{aligned}
&\sum_{k=1}^{n} m_k f[x_1^{m_1},\ldots,x_n^{m_n},x_k] \\
=\;& \sum_{k=1}^{n} \frac{1}{(m_k-1)!}[f(x)g_k(x)]_{x=x_k}^{(m_k)} + \sum_{k=1}^{n}\sum_{i\neq k} \frac{m_k}{(m_i-1)!}\left[\frac{f(x)g_i(x)}{x-x_k}\right]_{x=x_i}^{(m_i-1)} \\
=\;& \sum_{k=1}^{n} \frac{1}{(m_k-1)!}[f'(x)g_k(x)]_{x=x_k}^{(m_k-1)} + \sum_{k=1}^{n} \frac{1}{(m_k-1)!}[f(x)g_k'(x)]_{x=x_k}^{(m_k-1)} \\
&+ \sum_{k=1}^{n}\sum_{i\neq k} \frac{m_k}{(m_i-1)!}\left[\frac{f(x)g_i(x)}{x-x_k}\right]_{x=x_i}^{(m_i-1)} \\
=\;& S_1+S_2+S_3.
\end{aligned} \tag{2.3.16}$$

By (2.3.12) $S_1 = f'[x_1^{m_1},\ldots,x_n^{m_n}]$. Meanwhile, we observe that

$$g_k'(x) = -g_k(x)\sum_{i\neq k}\frac{m_i}{x-x_i}$$

and hence

$$S_2 = -\sum_{k=1}^{n}\sum_{i\neq k}\frac{m_i}{(m_k-1)!}\left[-\frac{f(x)g_k(x)}{x-x_i}\right]_{x=x_k}^{(m_k-1)}.$$

Using a simple identity about double summation

$$\sum_{k=1}^{n}\sum_{i=1,i\neq k}^{n} a_{ki} = \sum_{k=1}^{n}\sum_{i=1,i\neq k}^{n} a_{ik}$$

we obtain $S_2 = -S_3$. Inserting S_1, S_2, and S_3 into (2.3.16) we obtain (2.3.15). □

2.4 Basic Theorems

Let

$$d_k = \max\{|x_k - x_{k-1}|, |x_k - x_{k+1}|\}, \quad k = 1,2,\ldots,n, \tag{2.4.1}$$

and

$$I = \begin{cases} [a,+\infty), & k=1, \\ \mathbf{R}, & 2\le k\le n-1, \\ (-\infty,b], & k=n. \end{cases} \tag{2.4.2}$$

For a polynomial P and a set $S \subset \mathbf{R}$ the symbol $Z^o(P,S)$ means the number of zeros of odd multiplicity for P in S.

Lemma 2.4.1. *Let an index k, $2 \le k \le n-1$ $(n \ge 4)$, be fixed and $0 \le j < i < m_k$, where both $m_k - i$ and $m_k - j$ are odd. Assume that a polynomial G satisfies*

$$\partial G = N - 2 \tag{2.4.3}$$

and

$$G^{(p)}(x_q) = 0,$$
$$p = \begin{cases} 0, 1, \ldots, m_q - 1, & q \neq k, \\ 0, 1, \ldots, j-1, j+1, \ldots, i-1, i+1, \ldots, m_k - 1, & q = k. \end{cases} \tag{2.4.4}$$

Then

$$|G^{(j)}(x_k)| \le \frac{j!}{i!} d_k^{i-j} |G^{(i)}(x_k)|. \tag{2.4.5}$$

Proof. By (2.4.4) the Taylor expansion of G at $x = x_k$ is of the form

$$\begin{cases} G(x) = c_0(x-x_k)^j + c_1(x-x_k)^i + (x-x_k)^{m_k} D_1(x), \\ c_0 = \frac{G^{(j)}(x_k)}{j!}, \quad c_1 = \frac{G^{(i)}(x_k)}{i!}, \quad D_1 \in \mathbf{P}_{N-2}. \end{cases} \tag{2.4.6}$$

We observe that $c_0 \neq 0$, for otherwise by (2.4.4) and (2.4.3)

$$Z''(G) = \sum_{q=1}^{n} m_q - 1 = N > \partial G$$

would occur, a contradiction to (2.2.14). Now we introduce the polynomials D and F:

$$D(x) = \frac{G(x)}{c_0(x-x_k)^j} = 1 + \frac{c_1}{c_0}(x-x_k)^{i-j} + \frac{1}{c_0}(x-x_k)^{m_k-j} D_1(x), \tag{2.4.7}$$

$$D'(x) = (x-x_k)^{i-j-1} F(x), \tag{2.4.8}$$

where

$$F(x) = (i-j)\frac{c_1}{c_0} + (x-x_k)^{m_k-i} D_2(x), \quad D_2 \in \mathbf{P}_{N-2}. \tag{2.4.9}$$

Denote $\Delta = (x_{k-1}, x_{k+1})$ and let $\Delta' = (y_1, y_2) \supset \Delta$ be the largest open interval such that $Z(F, \Delta') = Z(F, \Delta)$. Of course, we have

$$\begin{cases} F(y_1) = 0, & y_1 > -\infty, \\ F(y_2) = 0, & y_2 < +\infty. \end{cases} \tag{2.4.10}$$

The first preliminary step of the proof is to show

$$Z'(F, \Delta') = Z'(F, \Delta) = 0 \tag{2.4.11}$$

and

$$Z'(F', \Delta') \le m_k - i. \tag{2.4.12}$$

In fact, by (2.4.6) and (2.4.7) we see that

$$D^{(p)}(x_q) = 0, \quad p = \begin{cases} 0,1,\ldots,m_q-1, & q \neq k, \\ 1,\ldots,i-j-1,i-j+1,\ldots,m_k-j-1, & q = k. \end{cases} \tag{2.4.13}$$

Applying Rolle theorem we obtain $s \geq n-3$ zeros $z_1,\ldots,z_s$ of D' outside Δ from $n-3$ intervals $(x_1,x_2),\ldots,(x_{k-2},x_{k-1}),(x_{k+1},x_{k+2}),\ldots,(x_{n-1},x_n)$ constructed by the zeros of D:

$$D'(z_q) = 0, \quad q = 1,\ldots,s. \tag{2.4.14}$$

Relations (2.4.8) and (2.4.13) yield

$$Z''(D',\Delta) \geq (i-j-1) + Z'(F,\Delta) + (m_k - i - 1) = Z'(F,\Delta) + m_k - j - 2. \tag{2.4.15}$$

Meanwhile, by (2.4.13) and (2.4.14) we have

$$Z''(D',\mathbf{R}\setminus\Delta') = Z'(D',\mathbf{R}\setminus\Delta) \geq \sum_{q\neq k}(m_q-1) + s = N - n - m_k + s + 2. \tag{2.4.16}$$

It follows by Lemma 2.2.1 from (2.4.3) and (2.4.7) that

$$N - j - 3 = \partial D' \geq Z''(D') = Z''(D',\Delta') + Z''(D',\mathbf{R}\setminus\Delta').$$

Using (2.4.15) and (2.4.16), recalling that $s \geq n-3$, we get (2.4.11) and

$$s = n-3. \tag{2.4.17}$$

Thus inequality (2.4.16) becomes

$$Z'(F,\mathbf{R}\setminus\Delta') \geq N - m_k - 1$$

and hence by Rolle theorem $Z'(F',\mathbf{R}\setminus\Delta') \geq N - m_k - 3$. Since $\partial F' = N - i - 3$, by Lemma 2.2.1 we have

$$Z'(F',\Delta') \leq Z''(F') - Z'(F',\mathbf{R}\setminus\Delta') \leq \partial F' - Z'(F',\mathbf{R}\setminus\Delta') \leq m_k - i,$$

which proves (2.4.12).

The second preliminary step of the proof is to choose an interval $\delta_k = [x_{k-1},x_k]$ or $\delta_k = [x_k,x_{k+1}]$ so that the function $|F|$ is monotone on δ_k and is attained its maximum on δ_k at x_k. Since by (2.4.9)

$$F^{(p)}(x_k) = 0, \quad p = 1,2,\ldots,m_k - i - 1, \tag{2.4.18}$$

according to (2.4.12) we see that $Z^o(F',\Delta') \leq 1$. Noticing that $n \geq 4$, by (2.4.10) we conclude either $F(y_1) = 0$ or $F(y_2) = 0$. Suppose without loss of generality that $F(y_1) = 0$. If $Z^o(F',\Delta') = 0$, then the function $|F|$ is increasing on the interval Δ' and in this case we put $\delta_k = [x_{k-1},x_k]$. If $Z^o(F',\Delta') = 1$, then the polynomial F' has a unique zero y of odd multiplicity on the interval Δ' and the point y must be a maximum point of the function $|F|$ on the interval Δ'. In this case we put

$$\delta_k = \begin{cases} [x_{k-1},x_k], & y \geq x_k, \\ [x_k,x_{k+1}], & y < x_k. \end{cases}$$

Then in both the cases the function $|F|$ is monotone on δ_k and is attained its maximum on δ_k at x_k. So by (2.4.9) we have

$$|F(x)| \le |F(x_k)| = (i-j)\left|\frac{c_1}{c_0}\right|, \quad x \in \delta_k.$$

Now for the proof proper, with the help of this inequality it follows from (2.4.8) and (2.4.6) that

$$\left|\int_{\delta_k} D'(x)\,dx\right| = \left|\int_{\delta_k} (x-x_k)^{i-j-1}F(x)\,dx\right| \le (i-j)\left|\frac{c_1}{c_0}\right| \int_{\delta_k} |x-x_k|^{i-j-1}\,dx$$
$$= \left|\frac{c_1}{c_0}\right| |\delta_k|^{i-j} \le \left|\frac{c_1}{c_0}\right| d_k^{i-j} = \frac{j!d_k^{i-j}|G^{(i)}(x_k)|}{i!|G^{(j)}(x_k)|}.$$

Meanwhile, according to (2.4.4) and (2.4.7) we see $D(x_{k-1}) = D(x_{k+1}) = 0$ and hence

$$\left|\int_{\delta_k} D'(x)\,dx\right| = 1.$$

Therefore

$$\frac{j!d_k^{i-j}|G^{(i)}(x_k)|}{i!|G^{(j)}(x_k)|} = 1$$

which is equivalent to (2.4.5). □

With the help of Lemma 2.4.1 we may now state the first main result of this section.

Theorem 2.4.1. *For a fixed index k, $1 \le k \le n$ $(n \ge 4)$, let the number $m_k - j$ be odd and $j < i < m_k$. Then with $c = 1$*

$$B_{kj}(x) \ge c d_k^{j-i}|x-x_k|^{i-j}|B_{ki}(x)|, \quad x \in I, \tag{2.4.19}$$

$$|A_{kj}(x)| \ge c\frac{i!}{j!}d_k^{j-i}|A_{ki}(x)|, \quad x \in I, \tag{2.4.20}$$

and

$$b_{k,m_k-j-1} \ge c d_k^{j-i}|b_{k,m_k-i-1}|. \tag{2.4.21}$$

Moreover, both the order d_k^{j-i} and the constant $c = 1$ in (2.4.19)–(2.4.21) are the best possible.

Proof. First we prove (2.4.19). To this end we distinguish two cases when $m_k - i \in \mathbf{N}_1$ and $m_k - i \in \mathbf{N}_2$.

Case 1. $m_k - i \in \mathbf{N}_1$. Denote

$$x_0' = x_1 - d_1, \quad x_{n+1}' = x_n + d_n,$$

and

$$I' = \begin{cases} [x_0', +\infty), & k = 1, \\ \mathbf{R}, & 2 \le k \le n-1, \\ (-\infty, x_{n+1}'], & k = n. \end{cases}$$

Since $x_1 = a$ or $x_n = b$ may occur, here we need to use I' instead of I and put

$$c = c_{kji} = \sup\{d : B_{kj}(x) \geq d(x-x_k)^{i-j}B_{ki}(x), \quad x \in I'\}.$$

Clearly $I \subset I'$ and

$$C(x) = B_{kj}(x) - c(x-x_k)^{i-j}B_{ki}(x) \geq 0, \quad x \in I'.$$

By means of (2.2.9) we consider the expression consisting of A_{kj} and A_{ki}:

$$G(x) = C(x)L_k(x)\frac{(x-x_k)^j}{j!} = A_{kj}(x) - \frac{ci!}{j!}A_{ki}(x). \tag{2.4.22}$$

To apply Lemma 2.4.1 the first step is to find more zeros of polynomial G than those explicitly contained in (2.4.22). To this end we have to discuss zeros of polynomial C.

Claim 1. $Z'(C) \leq 2$.

Suppose to the contrary that the polynomial C has three zeros, say, $\alpha_1 = \alpha_2 < \alpha_3$: $C(\alpha_1) = C'(\alpha_1) = C(\alpha_3) = 0$. Hence $G(\alpha_1) = G'(\alpha_1) = G(\alpha_3) = 0$. We observe that by (2.4.22) each zero of C may not be x_k and must be either a new zero or an additional multiplicity of an old zero of L_k and hence of $A_{kj} - (ci!/j!)A_{ki}$. This would lead to

$$Z''(G) \geq Z''(A_{kj} - (ci!/j!)A_{ki}) + 3 \geq (N-2) + 3 = N+1 > \partial G,$$

a contradiction to Lemma 2.2.1. This contradiction proves Claim 1.

Claim 2. One and only one of the following three cases may occur.

Case A. $Z'(C) = 2$, that is, the polynomial C with $\partial C = \partial B_{kj}$ has two zeros $\alpha_1 \leq \alpha_2$, which satisfy $I' \cap \{\alpha_1, \alpha_2\} \neq \emptyset$ and

$$\begin{cases} C(\alpha_1) = C'(\alpha_1) = 0, & \alpha_1 = \alpha_2, \\ C(\alpha_1) = C(\alpha_2) = 0, \quad \alpha_1 = x'_{n+1} \text{ or } \alpha_2 = x'_0, & \alpha_1 < \alpha_2. \end{cases} \tag{2.4.23}$$

The case when $\alpha_1 < \alpha_2$ may occur for $k = 1$ and $\alpha_2 = x'_0$ or for $k = n$ and $\alpha_1 = x'_{n+1}$ only.

Case B. $Z'(C) = 1$, that is, the polynomial C with $\partial C = \partial B_{kj} - 1$ has a unique simple zero $\alpha_1 \in \{x'_0, x'_{n+1}\}$; it may occur for $k = 1$ and $\alpha_1 = x'_0$ or for $k = n$ and $\alpha_1 = x'_{n+1}$ only.

Case C. $Z'(C) = 0$, that is, the polynomial C with $\partial C = \partial B_{kj} - 2$ has no real zeros and this may occur for those k, $2 \leq k \leq n-1$, only.

In fact, it is easy to see that at least one zero of polynomial C (if any) belongs to the interval I'; meanwhile, if $C(\alpha) = 0$ and $\alpha \in (I')^o$ then the zero α must be of even multiplicity.

If $Z'(C) = 2$, denoting by α_1, α_2, $\alpha_1 \leq \alpha_2$ the zeros of C, then $I' \cap \{\alpha_1, \alpha_2\} \neq \emptyset$ and (2.4.23) holds. Meanwhile, the case when $\alpha_1 < \alpha_2$ may occur for $k = 1$ or $k = n$ only. Further, we must have $\alpha_2 = x'_0$ for $k = 1$, for otherwise by the same argument as that of Claim 1 it would lead to a contradiction. Similarly, we have $\alpha_1 = x'_0$ for $k = n$. This proves Case A.

If $Z'(C) = 1$, then polynomial C has a unique simple zero $\alpha_1 \in \{x'_0, x'_{n+1}\}$ and, recalling that $\partial B_{kj} = m_k - j - 1$ is even by (2.2.19), $\partial C \leq \partial B_{kj} - 1$. Clearly, it may occur for $k = 1$ or $k = n$ only. On the other hand, we must have $\partial C \geq \partial B_{kj} - 1$ and $\alpha_1 = x'_0$ for $k = 1$ or $\alpha_1 = x'_{n+1}$ for $k = n$, for otherwise by the same argument as that of Claim 1 it would lead to a contradiction. This proves Case B.

If $Z'(C)=0$, then $\partial C \le \partial B_{kj}-2$ and we also must have $\partial C \ge \partial B_{kj}-2$ by the same argument as above. Further we have $2 \le k \le n-1$, for otherwise it leads to $Z''(G) \ge (N-2)+1 = N-1 > N-2 = \partial G$, a contradiction to Lemma 2.2.1.

This completes the proof of Claim 2.

According to Claim 2 for Case A

$$\begin{cases} G(\alpha_1)=G'(\alpha_1)=0, & & \alpha_1=\alpha_2, \\ G(\alpha_1)=G(\alpha_2)=0, & \begin{cases} \alpha_2 = x'_0, & k=1, \\ \alpha_1 = x'_{n+1}, & k=n, \end{cases} & \alpha_1<\alpha_2, \end{cases} \tag{2.4.24}$$

and for Case B

$$G(\alpha_1)=0, \quad \alpha_1 = \begin{cases} x'_0, & k=1, \\ x'_{n+1}, & k=n. \end{cases} \tag{2.4.25}$$

We do not add an interpolation condition and an interpolation node for Case C. We add two interpolation conditions given in (2.4.24) for Case A or one interpolation condition given in (2.4.25) for Case B. Then we put

$$\{x_1^*,\ldots,x_{n^*}^*\} = \begin{cases} \{x_1,\ldots,x_n\}\cup\{\alpha_1,\alpha_2\}, & \text{for Case A,} \\ \{x_1,\ldots,x_n\}\cup\{\alpha_1\}, & \text{for Case B} \end{cases}$$

with $x_1^* < \cdots < x_{n^*}^*$. Each of the additional zeros may appear as a new zero not contained in $\{x_1,\ldots,x_n\}$ or as an additional multiplicity of an old zero already known. Clearly $\partial G = N-\rho$, here for convenience we put $\rho = 0,1,2$ for Cases A,B,C, respectively.

To apply Lemma 2.4.1 the second step is to show that for each node x_k $(1\le k\le n)$ satisfies

$$x_2^* \le x_k \le x_{n^*-1}^*. \tag{2.4.26}$$

Clearly, it is enough to show (2.4.26) for $k=1$ and $k=n$. To this end it suffices to prove that for $\rho \le 1$

$$\begin{cases} x'_0 \le \alpha_{2-\rho} < x_1, & k=1, \\ x'_{n+1} \ge \alpha_1 > x_n, & k=n. \end{cases} \tag{2.4.27}$$

In fact, for Case A with $\alpha_1<\alpha_2$ and Case B relation (2.4.27) has been given by (2.4.24) and (2.4.25), respectively. For Case A with $\alpha_1=\alpha_2$ we must also have (2.4.27), for otherwise it would imply that $Z''(G) \ge N+1 > \partial G$, a contradiction to Lemma 2.2.1. This contradiction proves (2.4.27).

By definition there is an index i_k such that $x_{i_k}^* = x_k$, for which by Claim 2 relation (2.4.26) holds.

We can now apply Lemma 2.4.1 to get

$$\begin{aligned} 1 &= |G^{(j)}(x_k)| = |G^{(j)}(x_{i_k}^*)| \le \frac{j!(d_{i_k}^*)^{i-j}}{i!}|G^{(i)}(x_{i_k}^*)| \\ &= \frac{j!(d_{i_k}^*)^{i-j}}{i!}|G^{(i)}(x_k)| = \frac{j!(d_{i_k}^*)^{i-j}}{i!}\cdot\frac{ci!}{j!} = c(d_{i_k}^*)^{i-j}, \end{aligned} \tag{2.4.28}$$

here we also use (2.4.22).

If we can show

$$d^*_{i_k} = \max\{|x^*_{i_k} - x^*_{i_k-1}|, |x^*_{i_k} - x^*_{i_k+1}|\} \le d_k, \tag{2.4.29}$$

then relation (2.4.28) implies $c \ge (d^*_{i_k})^{j-i} \ge d_k^{j-i}$, which proves (2.4.19). Let us prove (2.4.29).

Clearly, (2.4.29) holds for $2 \le k \le n-1$, since $\{x^*_1, \dots, x^*_{n^*}\} \supset \{x_1, \dots, x_n\}$. Let $k = 1$. Using (2.4.27), and recalling that $x^*_{i_1} = x_1$ and $x'_0 = x_1 - d_1$, we obtain

$$d^*_{i_1} \le \max\{|x_1 - x_2|, |x_1 - \alpha_{2-\rho}|\} \le \max\{|x_1 - x_2|, |x_1 - x'_0|\} \le d_1.$$

Similarly, we can prove (2.4.29) for $k = n$.

Case 2. $m_k - i \in \mathbf{N}_2$. In this case we separate two subcases when $i = j+1$ and $i \ge j+3$.

Case 2.1. $i = j+1$. In this case put

$$L^*_k(x) = L_k(x)\left(\frac{x - x_q}{x_k - x_q}\right)^{-1}, \quad q \ne k.$$

We use formula (2.2.5) to calculate the corresponding numbers

$$b^*_{k\nu} = \frac{1}{\nu!}\left[L^*_k(x)^{-1}\right]^{(\nu)}_{x=x_k} = b_{k\nu} + \frac{1}{x_k - x_q} b_{k,\nu-1}, \quad \nu \ge 1. \tag{2.4.30}$$

Then by (2.2.20) we construct the corresponding polynomial

$$B^*_{kj}(x) = \sum_{\nu=0}^{m_k-j-1} b^*_{k\nu}(x - x_k)^\nu = B_{kj}(x) + \frac{x - x_k}{x_k - x_q} B_{k,j+1}(x) > 0. \tag{2.4.31}$$

For $2 \le k \le n-1$ substituting $q = k \pm 1$ into (2.4.31) yields (2.4.19) with $c = 1$.

Let us prove (2.4.19) for the case when $k = 1$. In this case relation (2.4.31) with $q = 2$ gives only

$$B_{1j}(x) > -\frac{x - x_1}{x_1 - x_2} B_{1,j+1}(x). \tag{2.4.32}$$

We have to show

$$B_{1j}(x) > \frac{x - x_1}{x_1 - x_2} B_{1,j+1}(x), \tag{2.4.33}$$

since it coupled with (2.4.32) may give

$$B_{1j}(x) > \left|\frac{x - x_1}{x_1 - x_2} B_{1,j+1}(x)\right| \ge \left|\frac{x - x_1}{d_1} B_{1,j+1,1}(x)\right|. \tag{2.4.34}$$

To prove (2.4.33) we notice that according to Theorem 2.2.5 the polynomial $B_{1,j+1}$ has a unique zero $\xi \in (x_1 - d_1, x_1)$ and satisfies that $B_{1,j+1}(x) \ge 0$ for $x \ge \xi$ and $B_{1,j+1}(x) < 0$ otherwise. Thus the term of the right-hand side in (2.4.33) is negative for $x < \xi$ or $x > x_1$ and hence inequality (2.4.33) holds for these x; for $x \in [\xi, x_1]$ by means of (2.2.22) we obtain

$$B_{1j}(x) = b_{1,m_1-j-1}(x - x_1)^{m_1-j-1} + B_{1,j+1}(x) \ge B_{1,j+1}(x) \ge \left|\frac{x - x_1}{x_1 - x_2}\right| B_{1,j+1}(x).$$

This proves (2.4.33) and hence proves (2.4.19) for $k = 1$. Similarly we can prove (2.4.19) for $k = n$.

Case 2.2. $i \geq j+3$. In this case from the conclusion of Case 1 and by induction we have

$$B_{kj}(x) \geq \left(\frac{x-x_k}{d_k}\right)^{i-j-1} B_{k,i-1}(x) \geq \left|\frac{x-x_k}{d_k}\right|^{i-j} |B_{ki}(x)|.$$

This completes the proof of (2.4.19).

With the help of (2.2.9) and (2.4.19)

$$\begin{aligned} |A_{kj}(x)| &= \left|\frac{1}{j!}(x-x_k)^j B_{kj}(x) L_k(x)\right| \geq \frac{i!}{j!} d_k^{j-i} \left|\frac{1}{i!}(x-x_k)^i B_{ki}(x) L_k(x)\right| \\ &= \frac{i!}{j!} d_k^{j-i} |A_{ki}(x)|. \end{aligned}$$

This proves (2.4.20). Comparing the leading coefficients of both the sides of (2.4.19) yields (2.4.21).

To prove the last conclusion of the theorem it is enough to show this for (2.4.21), because (2.4.19) implies (2.4.20) and (2.4.21).

To show that the order is the best possible let $m_k \equiv m$ and let the system of nodes (2.1.1) be defined as follows, in which the index k, $1 \leq k \leq n$, is fixed:

$$x_{k-1} = -\frac{1}{n^2}, \quad x_k = 0, \quad x_{k+1} = \frac{1}{2n^2}, \quad |x_q| \geq \frac{1}{2}, \quad |q-k| \geq 2.$$

Assume that both i and j are even. By a simple calculation we get $d_k = 1/n^2$,

$$\left|\sum_{|q-k|\geq 2} \frac{m}{x_q^j}\right| \leq m2^j(n-2)$$

and

$$\begin{aligned} a_{kj} &= \sum_{q\neq k} \frac{m}{(x_q-x_k)^j} = \sum_{q\neq k} \frac{m}{x_q^j} \\ &= m(2n^2)^j + m(-n^2)^j + \sum_{|q-k|\geq 2} \frac{m}{x_q^j} \sim n^{2j} \sim d_k^{-j}. \end{aligned}$$

Thus by (2.2.12) we obtain

$$b_{kj} \sim d_k^{-j} \sim d_k^{i-j} b_{ki}, \quad i > j \geq 0.$$

This proves that the order of (2.4.21) is the best possible.

To show that the constant $c = 1$ of (2.4.21) is the best possible. Let us consider the system of nodes (2.1.1) ($n = 2p-1,\ p \geq 3$):

$$x_p = 0, \quad x_{p+1} = \frac{1}{p}, \quad x_{p+k} \geq \frac{1}{2}, \quad k \geq 2; \quad x_{p-k} = -x_{p+k}, \quad k = 1, 2, \ldots, p-1.$$

Besides we choose

$$m_k = \begin{cases} 3, & k = p, \\ 1, & k \neq p. \end{cases}$$

Then by (2.2.11) and (2.2.12) $a_{p1} = 0,\ b_{p2} = (a_{p1}^2 + a_{p2})/2 = a_{p2}/2, d_p = 1/p$, and by (2.4.21)

$$\begin{aligned} c &\le \frac{d_p^2 b_{p2}}{b_{p0}} = d_p^2 b_{p2} = \frac{d_p^2 a_{p2}}{2} = \frac{1}{2p^2}\sum_{q\neq p}\frac{m_q}{(x_q - x_p)^2} \\ &= \frac{1}{p^2}\sum_{q=1}^{p-1}\frac{1}{x_q^2} = 1 + \frac{1}{p^2}\sum_{q=1}^{p-2}\frac{1}{x_q^2} \le 1 + \frac{4(p-2)}{p^2} \longrightarrow 1, \end{aligned}$$

as $p \longrightarrow \infty$. This proves our conclusion. □

We observe that one can not apply Theorem 2.4.1 to the whole real axis **R**. The only reason is that when $k = 1$ or $k = n$ this theorem does not hold. To derive an analogue of Theorem 2.4.1 which holds for the whole real axis **R** we will modify the definitions of d_1 and d_n. Let

$$\begin{cases} \overline{d}_1 = \begin{cases} |x_1 - x_2|, & m_2 > 1, \\ |x_1 - x_3|, & m_2 = 1, \end{cases} \\ \overline{d}_n = \begin{cases} |x_n - x_{n-1}|, & m_{n-1} > 1, \\ |x_n - x_{n-2}|, & m_{n-1} = 1, \end{cases} \\ \overline{d}_k = \max\{|x_k - x_{k-1}|, |x_k - x_{k+1}|\}, \quad k = 2, \ldots, n-1. \end{cases} \tag{2.4.35}$$

The second main result of this section is as follows.

Theorem 2.4.2. *For a fixed index k, $1 \le k \le n$ ($n \ge 4$), if the number $m_k - j$ is odd and $j < i < m_k$, then with $c = 1$*

$$B_{kj}(x) \ge c\overline{d}_k^{\,j-i}|x - x_k|^{i-j}|B_{ki}(x)|, \quad x \in \mathbf{R}, \tag{2.4.36}$$

$$|A_{kj}(x)| \ge c\frac{i!}{j!}\overline{d}_k^{\,j-i}|A_{ki}(x)|, \quad x \in \mathbf{R}, \tag{2.4.37}$$

and

$$b_{k,m_k-j-1} \ge c\overline{d}_k^{\,j-i}|b_{k,m_k-i-1}|. \tag{2.4.38}$$

Moreover, both the order $\overline{d}_k^{\,j-i}$ and the constant $c = 1$ in (2.4.36)–(2.4.38) are the best possible.

Proof. It suffices to show the theorem for $k = 1$ and $k = n$; we do this for $k = 1$ only (leaving the case for $k = n$ to the reader).

We distinguish two cases.

Case 1. $i = j + 1$. In this case inequality (2.4.36) follows directly from (2.4.34).

Case 2. $i = j + 2$. Inequality (2.4.19) with $k = 1$ reads

$$B_{1j}(x) \ge \left|\frac{x - x_1}{d_1}\right|^2 |B_{1,j+2}(x)|, \quad x \ge x_0. \tag{2.4.39}$$

If we choose $x_0 = x_1 - \overline{d}_1$, then $d_1 = \overline{d}_1$ and inequality (2.4.39) becomes

$$B_{1j}(x) \ge \left|\frac{x - x_1}{d_1}\right|^2 |B_{1,j+2}(x)| = \left|\frac{x - x_1}{\overline{d}_1}\right|^2 |B_{1,j+2}(x)|, \quad x \ge x_1 - \overline{d}_1.$$

So it suffices to show that

$$B_{1j}(x) \geq \left|\frac{x-x_1}{\overline{d}_1}\right|^2 |B_{1,j+2}(x)|, \quad x < x_1 - \overline{d}_1. \tag{2.4.40}$$

To this end we use (2.4.30) with $k = 1$ and $q = 2$ to obtain

$$B^*_{1,j+1}(x) = B_{1,j+1}(x) + \frac{x-x_1}{x_1-x_2} B_{1,j+2}(x). \tag{2.4.41}$$

Since $m_1 - j - 1$ is even, according to (2.2.24) we conclude

$$B^*_{1,j+1}(x) < 0, \quad B_{1,j+1}(x) < 0, \quad x < x_1 - \overline{d}_1. \tag{2.4.42}$$

Recalling that $B_{1,j+2}(x) > 0$, $x \in \mathbf{R}$, relations (2.4.41) and (2.4.42) imply

$$\begin{aligned} |B_{1,j+1}(x)| &= -B_{1,j+1}(x) = \frac{x-x_1}{x_1-x_2} B_{1,j+2}(x) - B^*_{1,j+1}(x) \\ &> \frac{x-x_1}{x_1-x_2} B_{1,j+2}(x) \geq \left|\frac{x-x_1}{\overline{d}_1}\right| B_{1,j+2}(x), \quad x < x_1 - \overline{d}_1, \end{aligned}$$

which by (2.4.36) with $i = j+1$ gives (2.4.40) (see Case 1). This proves that inequality (2.4.36) holds for $i = j+1$ and $i = j+2$. By induction (2.4.36) holds for every i with $j < i \leq m_1 - 1$.

The other statements of the theorem may be obtained from Theorem 2.4.1 □

2.5 Notes

2.5.1. The formula (2.3.7) for Hermite interpolation is due to Bojanov [8].

2.5.2. The first result on the basic theorems of Hermite interpolation on an arbitrary system of nodes was given by Szabados [133]. In his important paper he proved (2.4.36) with $c = c(m)$ for $m_k \equiv m$ and $i = m - 1$; he also developed a nice approach to these theorems.

2.5.3. Using a modification of Szabados' approach Shi [109] proved Theorem 2.4.1.

2.5.4. Theorem 2.4.2 is due to Zhou [161].

2.5.5. An extension of Theorems 2.4.1 to Birkhoff interpolation may be found in [123, pp. 139–156].

Chapter 3

General Properties of Power Orthogonal Polynomials

3.1 Definitions and Preliminaries

The smallest interval containing supp$(d\mu)$ is denoted by $\Delta(d\mu) = \langle a,b\rangle$, where

$$\langle a,b\rangle = \begin{cases} [a,b], & -\infty < a < b < +\infty, \\ (-\infty,b], & a=-\infty,\ b<+\infty, \\ [a,+\infty), & a>-\infty,\ b=+\infty, \\ (-\infty,+\infty), & a=-\infty,\ b=+\infty. \end{cases}$$

In what follows for $n \in \mathbf{N}$ assume that $m_1, m_2, \ldots, m_n \geq 1$ are *arbitrary real numbers*, unless otherwise indicated. Put

$$\mathbf{m} = (m_1, m_2, \ldots, m_n),$$

$$\mathbf{Y} = \{\mathbf{y} = (y_1, y_2, \ldots, y_n) :\ a < y_1 < y_2 < \cdots < y_n < b\},$$

$$\Omega_n(x) = \Omega_n(\mathbf{y};x) = \Omega_n(\mathbf{m},\mathbf{y};x) = \prod_{k=1}^{n} |x-y_k|^{m_k}, \quad \mathbf{y} \in \overline{\mathbf{Y}},$$

and

$$\Phi_n(\mathbf{y}) = \Phi_n(d\mu,\mathbf{m};\mathbf{y}) = \int_a^b \Omega_n(\mathbf{m},\mathbf{y};x)\,d\mu(x),$$

where $\overline{\mathbf{Y}}$ is the *closure* of the set $\mathbf{Y}$. If a vector $\mathbf{x} = \mathbf{x}(d\mu,\mathbf{m}) \in \mathbf{Y}$ is a solution of the extremal problem

$$\Phi_n(\mathbf{x}) = \inf_{\mathbf{y}\in\overline{\mathbf{Y}}} \Phi_n(\mathbf{y}), \tag{3.1.1}$$

the corresponding polynomial

$$\omega_n(\mathbf{x};x) = \omega_n(d\mu,\mathbf{m};x) = (x-x_1)(x-x_2)\cdots(x-x_n)$$

is said to be the nth *power orthogonal polynomial* or the nth $\mathbf{m}$ *orthogonal polynomial* with respect to $d\mu$. Its zeros are denoted by $x_k = x_k(d\mu,\mathbf{m})$.

Two cases will be particularly interesting. One of them is the case when $m_k \equiv m$; in this case we agree to replace $\mathbf{m}$ by m, say, $\omega_n(d\mu,\mathbf{m};x) = \omega_n(d\mu,m;x)$ and call it the nth *m orthogonal polynomial* with respect to $d\mu$. In this case we also need the nth *power orthonormal polynomial* or the nth *m orthonormal polynomial* with respect to $d\mu$

$$P_n(\mathbf{x};x) = P_n(d\mu,m;x) = \gamma_n(d\mu,m)\omega_n(d\mu,m;x), \tag{3.1.2}$$

where

$$\gamma_n(d\mu,m) = \Phi_n(d\mu,m;\mathbf{x})^{-1/m}. \tag{3.1.3}$$

Also, the zeros of $P_n(d\mu,m)$ are denoted by $x_k = x_{kn}(d\mu,m)$, $k = 1,2,\ldots,n$.

The other is the case when all the m_k are positive integers. In this case the solution $\mathbf{x}$ of (3.1.1) admits the *generalized Gaussian quadrature formula* (see Chapter 4).

As we know, the case when $m_k \equiv 2$ is the classical orthogonal polynomials $P_n(d\mu) = P_n(d\mu,2)$; it has a long history of study and a classical theory. Orthogonal polynomials are the object of the treatise by Szegő [137]. In 1930, Bernstein [5] showed that the monic Chebyshev polynomial $2^{1-n}T_n(x)$ is the solution of (3.1.1) with respect to the *Chebyshev weight* $v(x) = (1-x^2)^{-1/2}$ on the interval $[-1,1]$ for every vector $\mathbf{m} = (m,\ldots,m)$. Power orthogonal polynomials with even numbers m_k's were investigated by Turán [140] for the case $m_1 = \cdots = m_n$, and by Popoviciu [77] and Tschakaloff [139] for the general case, respectively. Later these polynomials were investigated mainly by Bojanov [6]–[9] and Italian mathematicians, e.g., Ossicini [69, 70], Ghizzetti and Ossicini [28]–[32], Guerra [38, 39], Ossicini and Rosati [72]–[74], Gori [33], Gori and Lo Cascio [34].

3.2 Existence, Uniqueness, and Characterizations

Put

$$\varphi_k(\mathbf{y}) = \frac{\partial\Phi_n(\mathbf{y})}{\partial y_k}, \quad k = 1,2,\ldots,n. \tag{3.2.1}$$

The system of equations

$$\varphi_k(\mathbf{y}) = 0, \quad k = 1,2,\ldots,n, \tag{3.2.2}$$

is called the *normal system of equations*.

Lemma 3.2.1. *Let $d\mu$ be a measure on* $\mathbf{R}$. *Then*

$$\varphi_k(\mathbf{y}) = -m_k \int_{\mathbf{R}} \frac{\Omega_n(\mathbf{y};x)}{x-y_k} d\mu(x), \quad k = 1,2,\ldots,n \tag{3.2.3}$$

and

$$\frac{\partial\varphi_k(\mathbf{y})}{\partial y_j} = \begin{cases} m_k m_j \int_{\mathbf{R}} \frac{\Omega_n(\mathbf{y};x)}{(x-y_k)(x-y_j)} d\mu(x) = \frac{m_k\varphi_j(\mathbf{y}) - m_j\varphi_k(\mathbf{y})}{y_k - y_j}, & j \neq k, \\ m_k(m_k-1) \int_{\mathbf{R}} \frac{\Omega_n(\mathbf{y};x)}{(x-y_k)^2} d\mu(x), & j = k, \quad m_k > 1, \\ 2\prod_{i=1,i\neq k}^n |y_k - y_i|^{m_i} \mu'(y_k), & j = k, \quad m_k = 1. \end{cases} \tag{3.2.4}$$

Here the relation (3.2.4) with $j = k$ holds provided the terms of its right-hand side exist.

Proof. By an easy calculation and the identity

$$\frac{1}{(x-y_k)(x-y_j)} = \frac{1}{y_k - y_j}\left(\frac{1}{x-y_k} - \frac{1}{x-y_j}\right)$$

we can obtain (3.2.3) and (3.2.4). □

Lemma 3.2.2. *Suppose that $d\mu$ is a measure on $\langle a,b\rangle$, and further that $\mu \in C\langle a,b\rangle$ if $\min_{1\le k\le n} m_k = 1$. If a vector $\mathbf{x} \in \mathbf{Y}$ satisfies (3.1.1), then*

(a) the vector $\mathbf{x}$ is a solution of the system of normal equations:

$$\varphi_k(\mathbf{x}) = 0, \quad k = 1,2,\ldots,n; \tag{3.2.5}$$

(b) the vector $\mathbf{x}$ satisfies the orthogonality relation:

$$\int_a^b \frac{\Omega_n(\mathbf{x};x)}{\omega_n(\mathbf{x};x)} q(x)d\mu(x) = 0, \quad \forall q \in \mathbf{P}_{n-1}; \tag{3.2.6}$$

(c) we have

$$\frac{\partial \varphi_k(\mathbf{x})}{\partial y_j} = 0, \quad k \neq j, \quad k,j = 1,2,\ldots,n. \tag{3.2.7}$$

Moreover, Statements (a) and (b) are equivalent.

Proof. (a) Since the function $\Phi_n(\mathbf{y})$ attains its minimum at $\mathbf{y} = \mathbf{x}$,

$$\varphi_k(\mathbf{x}) = \left.\frac{\partial \Phi_n(\mathbf{y})}{\partial y_k}\right|_{\mathbf{y}=\mathbf{x}} = 0. \quad k = 1,2,\ldots,n.$$

(a)$\Longleftrightarrow$ (b) Using (3.2.3) we rewrite (3.2.5) as

$$\int_{\mathbf{R}} \frac{\Omega_n(\mathbf{x};x)}{\omega_n(\mathbf{x};x)} \cdot \frac{\omega_n(\mathbf{x};x)}{x - x_k} d\mu(x) = 0, \quad k = 1,2,\ldots,n. \tag{3.2.8}$$

Since the system of polynomials

$$\frac{\omega_n(\mathbf{x};x)}{x - x_k}, \quad k = 1,2,\ldots,n$$

spans the set $\mathbf{P}_{n-1}$. This shows that relation (3.2.8) is equivalent to (3.2.6).

(c) For $k \neq j$ relation (3.2.7) follows directly from (3.2.4) and (3.2.5). □

Theorem 3.2.1. *Suppose that $d\mu$ is a measure on $\langle a,b\rangle$, and further that $\mu \in C\langle a,b\rangle$ if $\min_{1\le k\le n} m_k = 1$. Then there exists a vector $\mathbf{x} \in \mathbf{Y}$ satisfying (3.1.1) and Statements (a) and (b) of Lemma 3.2.2 are equivalent.*

Proof. We see that the function $\Phi_n(\mathbf{y})$ is continuous and tends to ∞ as $y_1 \to -\infty$ or $y_n \to +\infty$. So the function $\Phi_n(\mathbf{y})$ attains its minimum at $\mathbf{y} = \mathbf{x} \in \overline{\mathbf{Y}}$. It is enough to show $\mathbf{x} \in \mathbf{Y}$.

We will use an idea of Tschakaloff given in [139]. Suppose to the contrary that $x_k = x_{k+1} = \cdots = x_j = c$ and $x_{k-1} < x_k$ if $k > 1$, and $x_j < x_{j+1}$ if $j < n$. Let us consider the function

$$F(h) = \int_a^b \Omega_n(\mathbf{x};x)|x-c|^{-m_k-m_j}|x-c+m_jh|^{m_k}|x-c-m_kh|^{m_j}d\mu(x), \quad h \geq 0,$$

for which $F(0) = \Phi_n(\mathbf{x})$. By differentiation

$$\begin{aligned} F'(h) = m_km_j \textstyle\int_a^b \Omega_n(\mathbf{x};x)|x-c|^{-m_k-m_j}|x-c+m_jh|^{m_k-1}|x-c-m_kh|^{m_j-1} \\ \times \left\{|x-c-m_kh|\text{sgn}\,(x-c+m_jh) - |x-c+m_jh|\text{sgn}\,(x-c-m_kh)\right\}d\mu(x) \end{aligned}$$

and

$$\begin{aligned}
F''(h) &= m_k m_j(m_k+m_j)\left\{-\int_a^{c-m_jh}\Omega_n(\mathbf{x};x)|x-c|^{-m_k-m_j}\right.\\
&\times\frac{d}{dh}\left[h(c-m_jh-x)^{m_k-1}(c+m_kh-x)^{m_j-1}\right]d\mu(x)\\
&+\int_{c-m_jh}^{c+m_kh}\Omega_n(\mathbf{x};x)|x-c|^{-m_k-m_j}\\
&\times\frac{d}{dh}\left[h(x-c+m_jh)^{m_k-1}(c+m_kh-x)^{m_j-1}\right]d\mu(x)\\
&-\int_{c+m_kh}^{b}\Omega_n(\mathbf{x};x)|x-c|^{-m_k-m_j}\\
&\left.\times\frac{d}{dh}\left[h(x-c+m_jh)^{m_k-1}(x-c-m_kh)^{m_j-1}\right]d\mu(x)\right\}.
\end{aligned}$$

Clearly, $F'(0)=0$ and

$$F''(0)=-m_k m_j(m_k+m_j)\int_a^b\Omega_n(\mathbf{x};x)|x-c|^{-2}d\mu(x)<0.$$

Hence $F(h)<F(0)=\Phi_n(\mathbf{x})$ for h small enough. Meanwhile by definition $F(h)=\Phi_n(\mathbf{x}_h)$, where

$$\mathbf{x}_h=(x_1,\ldots,x_{k-1},c-m_jh,c,\ldots,c,c+m_kh,x_{j+1},\ldots,x_n),$$

and hence $\Phi_n(\mathbf{x}_h)<\Phi_n(\mathbf{x})$. This is a contradiction to the assumption that $\mathbf{x}$ minimizes $\Phi_n(\mathbf{y})$.

It remains to show $a<x_1$ if $a>-\infty$ or $x_n<b$ if $b<+\infty$. It is enough to show $a<x_1$ if $a>-\infty$. Suppose to the contrary that $a\ge x_1$. In this case by (3.2.3)

$$\varphi_1(\mathbf{x})=\int_a^b\left\{-m_1\frac{\Omega_n(\mathbf{x};x)}{x-x_1}\right\}d\mu(x)<0,$$

since the integrand is not positive on $\langle a,b\rangle$. This contradicts (3.2.5).

The second conclusion of the theorem follows from Lemma 3.2.2 □

Remark 3.2.1. The following example shows that the condition $\mu\in C\langle a,b\rangle$ can not be dropped for the existence in the case when $\min_{1\le k\le n}m_k=1$ in general.

Let $m_1=\cdots=m_{N+1}=1$, $a=-1$, $b=1$, and

$$\mu(x)=\begin{cases}x-1, & x\in[-1,0],\\ x+1, & x\in(0,1].\end{cases}$$

In this case for any system of nodes

$$-1=x_0<x_1<\cdots<x_{N+1}<x_{N+2}=1$$

we can choose an index p, $0\le p\le N+1$, so that $x_p\le 0<x_{p+1}$. So we see

$$\begin{aligned}
&\left|\sum_{k=0}^{N+1}(-1)^k\int_{x_k}^{x_{k+1}}d\mu(x)\right|\\
\ge\ &\int_{x_p}^{x_{p+1}}d\mu(x)-\left[\int_{-1}^{x_p}d\mu(x)+\int_{x_{p+1}}^{1}d\mu(x)\right|>2-2=0.
\end{aligned}$$

This shows that the orthogonality relation (3.2.6) is violated. Therefore the extremal problem (3.1.1) has no solutions.

Theorem 3.2.2. *Let $d\mu$ be a measure on $\langle a,b\rangle$ satisfying $\mu \in C\langle a,b\rangle$. Assume that if $\min_{1\le k\le n} m_k = 1$ then $\mu \in C^1\langle a,b\rangle$ and $\mu'(x) > 0$, $x \in (a,b)$. Then there exists a unique vector $\mathbf{x} \in \mathbf{Y}$ satisfying (3.1.1) and the following statements are equivalent.*

(a) A vector $\mathbf{x}$ satisfies (3.1.1).

(b) A vector $\mathbf{x}$ is a solution of the system of normal equations (3.2.2).

(c) A vector $\mathbf{x}$ satisfies the orthogonality relations (3.2.6).

Proof. The existence of solutions of (3.1.1) has been given by Theorem 3.2.1.

(a)$\Longrightarrow$(b) Apply Lemma 3.2.2.

Now let us prove uniqueness. The proof uses a modification of ideas of Bojanov in [6] and Shi in [107].

It is particularly simple to prove the uniqueness for the case when $m_1 = \cdots = m_n = m$: in this case the extremal problem (3.1.1) is equivalent to the problem of best $L^m_{d\mu}$ approximation to the function x^n from the subspace $\mathbf{P}_{n-1}$ and the uniqueness follows from Corollaries 3.5 and 2.4 in [128, p. 111 and p. 236].

For a general case we may assume that there is an index r, $1 \le r \le n$, such that $m_r > 1$. By Lemma 3.2.2 in order to prove the uniqueness it suffices to show that the system of normal equations (3.2.2) has a unique solution in $\mathbf{Y}$. For exactness such a problem is called the $(m_1,\ldots,m_n;d\mu)$ problem. The proof proceeds by induction on n. The uniqueness for $n=1$ is true according to the above conclusion. Now suppose, as an induction hypothesis, that the uniqueness of the $(m_1,\ldots,m_{n-1};d\mu)$ problem is true for every system of real numbers $m_1,\ldots,m_{n-1} \ge 1$ $(n \ge 2)$ and for every measure $d\mu$ on $\langle a,b\rangle$ satisfying the assumptions of the theorem. Denote by

$$x_k = x_k(\xi) := x_k(d\mu_\xi),\ k \in \mathbf{K}_r := \{1,\ldots,r-1,r+1,\ldots,n\},$$

the unique solution of the $(m_1,\ldots,m_{r-1},m_{r+1},\ldots,m_n;d\mu_\xi)$ problem which by Theorem 3.2.1 satisfies

$$a = x_0(\xi) < x_1(\xi) < \cdots < x_{r-1}(\xi) < x_{r+1}(\xi) < \cdots < x_n(\xi) < x_{n+1}(\xi) = b, \tag{3.2.9}$$

where

$$d\mu_\xi(x) = |x-\xi|^{m_r}\,d\mu(x), \quad \xi \in \langle a,b\rangle.$$

If $r=1$ then we choose $\xi_1 = a$. If $r>1$ then by Theorem 3.2.1 there is a solution $\mathbf{y} = (y_r,y_1,\ldots,y_{r-1},y_{r+1},\ldots,y_n)$ for the $(m_r,m_1,\ldots,m_{r-1},m_{r+1},\ldots,m_n;d\mu)$ problem, which satisfies

$$a < y_r < y_1 < \cdots < y_{r-1} < y_{r+1} < \cdots < y_n < b. \tag{3.2.10}$$

Further, by the implication (a)$\Longrightarrow$(b) we see

$$\varphi_k(\mathbf{y}) = 0, \quad k = 1,2,\ldots,n. \tag{3.2.11}$$

We rewrite the $n-1$ equations in (3.2.11) with $k \in \mathbf{K}_r$ as

$$\begin{aligned}
\varphi_k(\mathbf{y}) &= -m_k \int_a^b \frac{\Omega_n(\mathbf{y};x)}{|x-y_r|^{m_r}(x-y_k)}|x-y_r|^{m_r}\,d\mu(x) \\
&= -m_k \int_a^b \frac{\Omega_n(\mathbf{y};x)}{|x-y_r|^{m_r}(x-y_k)}\,d\mu_{y_r}(x) = 0.
\end{aligned}$$

By the induction hypothesis the corresponding $(m_1,\ldots,m_{r-1},m_{r+1},\ldots,m_n;d\mu_{y_r})$ problem has a unique solution. Hence according to the above relations and the implication (a)$\Longrightarrow$(b), the vector $(y_1,\ldots,y_{r-1},y_{r+1},\ldots,y_n)$ is just its solution, that is, $y_k = x_k(y_r)$, $k\in \mathbf{K}_r$, which, coupled with (3.2.10), gives

$$x_{r-1}(y_r) = y_{r-1} > y_r. \tag{3.2.12}$$

Similarly, if we consider the $(m_1,\ldots,m_{r-1},m_{r+1},\ldots,m_n,m_r;d\mu)$ problem, we can obtain its solution $\mathbf{z} = (z_1,\ldots,z_{r-1},z_{r+1},\ldots,z_n,z_r)$ and then conclude $x_{r-1}(z_r) = z_{r-1} < z_r$, which, together with (3.2.12), implies that there must exist a point $\xi_1 \in \langle a,b\rangle$ such that

$$x_{r-1}(\xi_1) = \xi_1, \tag{3.2.13}$$

since according to Theorem 3.4.1 below the function $x_{r-1} = x_{r-1}(\xi)$ is continuous with respect to ξ.

We claim that the solution of (3.2.13) in $\langle a,b\rangle$ is unique. In fact, suppose to the contrary that there would be another point $\xi_2 \in \langle a,b\rangle$, $\xi_2 \neq \xi_1$, such that $x_{r-1}(\xi_2) = \xi_2$. This implies that the $(m_1,\ldots,m_{r-1}+m_r,m_{r+1},\ldots,m_n;d\mu)$ problem would have the two solutions $\{x_k(\xi_1) : k\in \mathbf{K}_r\}$ and $\{x_k(\xi_2) : k\in \mathbf{K}_r\}$, contradicting the induction hypothesis. This contradiction proves our claim.

Similarly, if $r = n$ then we choose $\xi_2 = b$. If $r < n$ then we may conclude that there must exist a unique point $\xi_2 \in \langle a,b\rangle$ such that $x_{r+1}(\xi_2) = \xi_2$. We observe that

$$\xi_1 \neq \xi_2, \tag{3.2.14}$$

for otherwise $x_{r-1}(\xi_1) = \xi_1 = \xi_2 = x_{r+1}(\xi_2) = x_{r+1}(\xi_1)$ would occur, contradicting (3.2.9). This contradiction proves (3.2.14).

Thus the inequalities $x_{r-1}(\xi) < \xi < x_{r+1}(\xi)$ are true if and only if $\min\{\xi_1,\xi_2\} < \xi < \max\{\xi_1,\xi_2\}$. Denote $x_r = \xi$ and

$$\varphi(\xi) := \varphi_r(x_1(\xi),\ldots,x_{r-1}(\xi),\xi,x_{r+1}(\xi),\ldots,x_n(\xi)) := -m_r \int_a^b \frac{\Omega_n(\mathbf{x}(\xi);x)}{x-\xi}\,d\mu(x).$$

Clearly

$$\varphi'(\xi) = \frac{\partial\varphi_r}{\partial x_r} + \sum_{k\in\mathbf{K}_r} \frac{\partial\varphi_r}{\partial x_k} x_k'(\xi). \tag{3.2.15}$$

If $\varphi(\xi) = 0$, then it follows by the implication (a)$\Longrightarrow$(b) from (3.2.2) that the vector $\mathbf{x}(\xi) = (x_1(\xi),\ldots,x_{r-1}(\xi),\xi,x_{r+1}(\xi),\ldots,x_n(\xi))$ is a solution of (3.2.2). Applying (3.2.7) we have

$$\frac{\partial\varphi_r}{\partial x_k} = 0, \quad k\in\mathbf{K}_r.$$

Hence the relation (3.2.15) becomes

$$\varphi'(\xi) = \frac{\partial\varphi_r}{\partial x_r}.$$

Since $m_r > 1$ by assumption, using (3.2.4) we obtain

$$\varphi'(\xi) = \frac{\partial\varphi_r}{\partial x_r} = m_r(m_r-1)\int_a^b \frac{\Omega_n(\mathbf{x}(\xi);x)}{(x-\xi)^2}\,d\mu(x) > 0,$$

whenever $\varphi(\xi) = 0$, $\min\{\xi_1, \xi_2\} < \xi < \max\{\xi_1, \xi_2\}$. This means that there is at most one point ξ satisfying $\varphi(\xi) = 0$. This prove the uniqueness.

(b)$\Longrightarrow$(a) As proved before, this follows directly from the implication (a)$\Longrightarrow$ (b) and the uniqueness of solutions of the system of normal equations (3.2.2).

(b)$\Longleftrightarrow$(c) Use Lemma 3.2.2. □

Remark 3.2.2. The condition of Theorem 3.2.2 that $\mu \in C\langle a,b\rangle$ may be dropped if $\min_{1\le k\le n} m_k \ge 2$.

Remark 3.2.3. In general the condition of Theorem 3.2.2 that $\mu \in C^1\langle a,b\rangle$ and $\mu'(x) > 0$, $x \in (a,b)$ in case $\min_{1\le k\le n} m_k = 1$ can not be dropped. For example, let $m_1 = 1$, $m_2 = 2$, $a = -1$, $b = 1$, and

$$\mu'(x) = \begin{cases} 20, & x \in [-1,-1/2), \\ 0, & x \in [-1/2,-1/4), \\ 416, & x \in [-1/4,0), \\ 11, & x \in [0,1]. \end{cases}$$

In this case by a calculation it is easy to check that every vector $(x_1, 0)$ with $-1/2 \le x_1 \le -1/4$ is a solution of (3.1.1).

3.3 Elementary Properties of the Zeros

Lemma 3.3.1. *Let $\partial P = n$, $n-1 \le \partial Q \le n$, $P \ne cQ$, and $p \ge -1$. Then the following statements are equivalent:*

(a) the function

$$f(c,d,p;x) = c|P(x)|^p P(x) + d|Q(x)|^p Q(x) \tag{3.3.1}$$

has at least $n-1$ sign changes for every nonzero pair $\{c,d\}$;

(b) both P and Q have simple real zeros only, and the zeros of P and Q mutually separate each other.

Proof. We separate two cases when $p = 0$ and $p \ne 0$.

Case 1. $p = 0$.

(a)$\Longrightarrow$(b) First we prove some claims.

Claim 1. The polynomial

$$f(c,d,0;x) = cP(x) + dQ(x)$$

has simple real zeros only for every nonzero pair $\{c,d\}$.

In fact, this claim directly follows from that the polynomial f has at least $n-1$ sign changes and its degree is n or $n-1$.

Claim 2. The polynomials P and Q have no common zeros.

In fact, suppose not and assume that $P(y) = Q(y) = 0$ for some number $y \in \mathbf{R}$. Then the polynomial

$$f(Q'(y), -P'(y), 0; x) = Q'(y)P(x) - P'(y)Q(x)$$

would have a zero y of multiplicity ≥ 2, a contradiction to Claim 1.

Claim 3. Let x_k, $k=1,2,\ldots,n$, in (2.1.1) be the zeros of P. Then each interval (x_k,x_{k+1}), $k=0,1,\ldots,n$, can not contain more than one zero of Q.

In fact, suppose to the contrary that the interval (x_j,x_{j+1}), $0\le j\le n$, contains two zeros of Q. Then we may choose a pair $\{c,d\}$ so that the polynomial $f(c,d,0;x)$ has a zero y of multiplicity ≥ 2, a contradiction to Claim 1. This contradiction proves Claim 3.

By interchanging the role of P and Q we can prove that each interval (y_k,y_{k+1}), $k=0,1,\ldots,\partial Q$, can not contain more than one zero of P, where y_k, $k=1,2,\ldots,\partial Q$, are the zeros of Q and

$$-\infty=y_0<y_1<\cdots<y_{\partial Q}<y_{\partial Q+1}=+\infty.$$

Hence we conclude that between two consecutive zeros of P (or Q) we have exactly one zero of Q (or P).

(b)$\Longrightarrow$(a) This implication is trivial for the case when $d=0$. We treat the case when $d\ne 0$. Assume that P has the zeros (2.1.1). Since

$$f(c,d,0;x_k)=dQ(x_k),\quad k=1,2,\ldots,n,$$

and

$$Q(x_k)Q(x_{k+1})<0,\quad k=1,2,\ldots,n-1,$$

the polynomial $f(c,d,0;x)$ has at least $n-1$ sign changes.

Case 2. $p\ne 0$. In this case we claim that the inequality

$$F(x)=[P(x)+Q(x)][|P(x)|^pP(x)+|Q(x)|^pQ(x)]\ge 0$$

holds for every point x. In fact, for a fixed point x, if $|P(x)|>|Q(x)|$ or $P(x)=Q(x)$

$$\text{sgn }[P(x)+Q(x)]=\text{sgn }[|P(x)|^pP(x)+|Q(x)|^pQ(x)]=\text{sgn }[P(x)]$$

and hence $F(x)\ge 0$; similarly, if $|P(x)|<|Q(x)|$ or $P(x)=-Q(x)$ then $F(x)\ge 0$. This proves our claim.

With the help of this claim if we write

$$\begin{aligned} f(c,d,p;\cdot) &= \left|(\text{sgn } c)|c|^{1/(p+1)}P\right|^p\left[(\text{sgn } c)|c|^{1/(p+1)}P\right] \\ &\quad +\left|(\text{sgn } d)|d|^{1/(p+1)}Q\right|^p\left[(\text{sgn } d)|d|^{1/(p+1)}Q\right] \end{aligned}$$

we can conclude that the function $f(c,d,p;\cdot)$ with $p\ne 0$ has at least $n-1$ sign changes for every nonzero pair $\{c,d\}$ if and only if the polynomial

$$f((\text{sgn } c)|c|^{1/(p+1)},(\text{sgn } d)|d|^{1/(p+1)},0;\cdot)$$

has at least $n-1$ sign changes for every nonzero pair $\{c,d\}$, which is equivalent to that the polynomial $f(c,d,0;\cdot)=cP+dQ$ has at least $n-1$ sign changes for every nonzero pair $\{c,d\}$. It remains to apply the conclusion of Case 1. □

The following theorems deal with the zeros of $P_n(d\mu,m)$ and are extensions of [137, Theorems 3.3.2, 3.3.4, and 3.3.5, pp. 45–46], respectively.

Theorem 3.3.1. *Let the assumptions of Theorem 3.2.2 prevail and let $m\ge 1$. Then the zeros of $P_n(d\mu,m)$ and $P_{n+1}(d\mu,m)$ mutually separate each other.*

Proof. By Theorem 3.2.2 we have

$$\int_a^b |P_k(d\mu,m;x)|^{m-2}P_k(d\mu,m;x)q(x)d\mu(x)=0, \quad \forall q\in \mathbf{P}_{k-1}, \quad k=n,n+1.$$

Thus for $P=P_{n+1}(d\mu,m)$ and $Q=P_n(d\mu,m)$ we have

$$\int_a^b f(c,d,m-2;x)q(x)d\mu(x)=0, \quad \forall q\in \mathbf{P}_{n-1}.$$

This shows that the function $f(c,d,m-2;\cdot)$ has at least n sign changes for every nonzero pair $\{c,d\}$. Applying Lemma 3.3.1 proves our theorem. □

Theorem 3.3.2. *Let the assumptions of Theorem 3.2.2 prevail and let c be an arbitrary real constant. Then the polynomial*

$$P_{n+1}(d\mu,m)-cP_n(d\mu,m) \tag{3.3.2}$$

has $n+1$ distinct real zeros. If $c>0$ (or $c<0$), these zeros lie in (a,b), with the exception of the greatest (or least) zero which lies in $\langle a,b\rangle$ only for $c\le P_{n+1}(d\mu,m;b)/P_n(d\mu,m;b)$ (or $c\ge P_{n+1}(d\mu,m;a)/P_n(d\mu,m;a)$).

Proof. The function $P_{n+1}(d\mu,m;x)/P_n(d\mu,m;x)$ increases from $-\infty$ to $+\infty$ in the intervals $(x_{kn}(d\mu,m),x_{k+1,n}(d\mu,m))$, $k=0,1,\ldots,n$. This completes the proof. □

Lemma 3.3.2. *If $P\in\widehat{\mathbf{P}}_n$ then with the notation of (2.2.8)*

$$\sum_{k=1}^{n}\frac{P(x_k)}{\omega_n'(x_k)}=1. \tag{3.3.3}$$

Proof. Since $\omega_n(x)-xP(x)$ is a polynomial of degree at most $n-1$, by (2.2.8) we have

$$\omega_n(x)-xP(x)=\sum_{k=1}^{n}[\omega_n(x_k)-x_kP(x_k)]\ell_k(x)=-\sum_{k=1}^{n}x_kP(x_k)\ell_k(x).$$

Again by (2.2.8) we see

$$\begin{aligned}\omega_n(x) &= xP(x)-\sum_{k=1}^{n}x_kP(x_k)\ell_k(x)=x\sum_{k=1}^{n}P(x_k)\ell_k(x)-\sum_{k=1}^{n}x_kP(x_k)\ell_k(x)\\ &= \sum_{k=1}^{n}P(x_k)(x-x_k)\ell_k(x)=\omega_n(x)\sum_{k=1}^{n}\frac{P(x_k)}{\omega_n'(x_k)}.\end{aligned}$$

We obtain (3.3.3). □

Theorem 3.3.3. *The following decomposition into partial fractions holds:*

$$\frac{P_n(d\mu,m;x)}{P_{n+1}(d\mu,m;x)}=\sum_{k=1}^{n+1}\frac{r_k}{x-x_{k,n+1}(d\mu,m)} \tag{3.3.4}$$

and

$$r_k>0, \quad k=1,2,\ldots,n+1; \quad \sum_{k=1}^{n+1}r_k=\frac{\gamma_n(d\mu,m)}{\gamma_{n+1}(d\mu,m)}. \tag{3.3.5}$$

Proof. Clearly, we have

$$r_k = \frac{P_n(d\mu,m;x_{k,n+1}(d\mu,m))}{P'_{n+1}(d\mu,m;x_{k,n+1}(d\mu,m))} > 0.$$

The last relation in (3.3.5) follows from Lemma 3.3.2. □

Let us conclude this section with a conjecture.

Conjecture 3.3.1. *If $p > m \geq 1$ then on each interval $(x_k(d\mu,m), x_{k+1}(d\mu,m))$, $k = 0,1,\dots,n$, contains at most one zero of $P_n(d\mu,p)$.*

3.4 Continuity of Power Orthogonal Polynomials with Respect to $d\mu$ and m

Theorem 3.4.1. *Let $d\mu_\xi$ be a measure on* **R** *satisfying*

$$\sup_{c\leq\xi\leq d} [\mu_\xi(+\infty) - \mu_\xi(-\infty)] < +\infty \tag{3.4.1}$$

and depending continuously on a parameter $\xi \in [c,d]$ in the sense that if ξ approaches ξ_0 then the sequence $\mu_\xi(x)$ tends to $\mu_{\xi_0}(x)$ for any fixed x. Assume that the solution $\mathbf{x}(\xi) := \mathbf{x}(d\mu_\xi)$ of the system of normal equations (3.2.2) with $d\mu = d\mu_\xi$ is always unique for every $\xi \in [c,d]$. Then the solution $\mathbf{x}(\xi)$ depends continuously on ξ in $[c,d]$.

Proof. Let $\xi_p, \xi \in [c,d]$ satisfy

$$\lim_{p\to\infty} \xi_p = \xi, \quad \lim_{p\to\infty} \mathbf{x}(\xi_p) = \mathbf{x}. \tag{3.4.2}$$

According to the definition of $\mathbf{x}(\xi_p)$ we have

$$\int_{\mathbf{R}} \frac{\Omega_n(\mathbf{x}(\xi_p);x)}{x - x_k(\xi_p)} d\mu_{\xi_p}(x) = 0, \quad k = 1,2,\dots,n.$$

By (3.4.1) and (3.4.2) the above equations lead to

$$\int_{\mathbf{R}} \frac{\Omega_n(\mathbf{x};x)}{x - x_k} d\mu_\xi(x) = 0, \quad k = 1,2,\dots,n.$$

By the assumption of uniqueness we conclude $\mathbf{x}(\xi) = \mathbf{x}$. Thus (3.4.2) means

$$\lim_{\xi_p\to\xi} \mathbf{x}(\xi_p) = \mathbf{x}(\xi). \quad \square$$

Theorem 3.4.2. *Let the assumptions of Theorem 3.2.2 prevail. Then the nth power orthogonal polynomial $\omega_n(\mathbf{x};x) = \omega_n(\mathbf{x}(\mathbf{m});x)$ with respect to $d\mu$ depends continuously on the index vector* **m**.

Proof. Let $\mathbf{x}(\mathbf{m}) = (x_1(\mathbf{m}), x_2(\mathbf{m}), \ldots, x_n(\mathbf{m}))$ and $\mathbf{x}(\mathbf{m}^*)$ ($\mathbf{m}^* = (m_1^*, m_2^*, \ldots, m_n^*)$) denote the vectors of zeros of the unique nth $\mathbf{m}$ and $\mathbf{m}^*$ orthogonal polynomial with respect to $d\mu$, respectively. It is enough to show that $\lim_{\mathbf{m}\to\mathbf{m}^*} \mathbf{x}(\mathbf{m}) = \mathbf{x}(\mathbf{m}^*)$.

First we observe that the set $\{\mathbf{x}(\mathbf{m}) : \|\mathbf{m}-\mathbf{m}^*\| := \sum_{k=1}^n |m_k - m_k^*| \le 1\}$ is bounded. In fact, if we choose a number $d > 0$ so large that $\int_{-d}^d d\mu(x) > 0$, then for $y_1 \le -d-1$ we see that

$$\begin{aligned}
&\inf_{\|\mathbf{m}-\mathbf{m}^*\|\le 1} \int_a^b \prod_{k=1}^n |x-y_k|^{m_k} d\mu(x) \\
\ge\ & \inf_{\|\mathbf{m}-\mathbf{m}^*\|\le 1} \int_{-d}^d \prod_{k=1}^n |x-y_k|^{m_k} d\mu(x) \\
\ge\ & \inf_{\|\mathbf{m}-\mathbf{m}^*\|\le 1} |y_1+d|^{m_1} \int_{-d}^d \prod_{k=2}^n |x-y_k|^{m_k} d\mu(x) \\
\ge\ & |y_1+d| \inf_{\|\mathbf{m}-\mathbf{m}^*\|\le 1} \inf_{y_2\le\cdots\le y_n} \int_{-d}^d \prod_{k=2}^n |x-y_k|^{m_k} d\mu(x).
\end{aligned}$$

Since

$$\inf_{\|\mathbf{m}-\mathbf{m}^*\|\le 1} \inf_{y_2\le\cdots\le y_n} \int_{-d}^d \prod_{k=2}^n |x-y_k|^{m_k} d\mu(x) > 0,$$

we conclude that

$$\lim_{y_1\to-\infty} \inf_{\|\mathbf{m}-\mathbf{m}^*\|\le 1} \int_a^b \prod_{k=1}^n |x-y_k|^{m_k} d\mu(x) = \infty,$$

which shows that $\inf_{\|\mathbf{m}-\mathbf{m}^*\|\le 1} x_1(\mathbf{m}) \ge c_1 > -\infty$. Similarly, we can show that $\sup_{\|\mathbf{m}-\mathbf{m}^*\|\le 1} x_n(\mathbf{m}) \le c_2 < +\infty$. This proves boundedness of the set $\{\mathbf{x}(\mathbf{m}) : \|\mathbf{m}-\mathbf{m}^*\| \le 1\}$.

Suppose now that $\mathbf{x}^*$ is an arbitrary limit point of $\mathbf{x}(\mathbf{m})$ when $\mathbf{m}$ approaches $\mathbf{m}^*$, say,

$$\lim_{\mathbf{m}^{(i)}\to\mathbf{m}^*} \mathbf{x}(\mathbf{m}^{(i)}) = \mathbf{x}^*, \tag{3.4.3}$$

where $\mathbf{m}^{(i)} = (m_1^{(i)}, m_2^{(i)}, \ldots, m_n^{(i)})$, $i = 1, 2, \ldots$ and $\mathbf{x}^* = (x_1^*, x_2^*, \ldots, x_n^*)$. By the definition of $\mathbf{x}(\mathbf{m}^{(i)})$ applying Theorem 3.2.2 for each i we have

$$-m_k^{(i)} \int_a^b \frac{\prod_{\nu=1}^n |x - x_\nu(\mathbf{m}^{(i)})|^{m_\nu^{(i)}}}{x - x_k(\mathbf{m}^{(i)})} d\mu(x) = 0, \quad k = 1, 2, \ldots, n.$$

As $\mathbf{m}^{(i)} \to \mathbf{m}^*$ by (3.4.3) the above equations lead to

$$-m_k^* \int_a^b \frac{\prod_{\nu=1}^n |x - x_\nu^*|^{m_\nu^*}}{x - x_k^*} d\mu(x) = 0, \quad k = 1, 2, \ldots, n,$$

which according to Theorem 3.2.2 means that $\mathbf{x}^* = \mathbf{x}(\mathbf{m}^*)$. Since $\mathbf{x}^*$ is an arbitrary limit point of $\mathbf{x}(\mathbf{m})$ when $\mathbf{m}$ approaches $\mathbf{m}^*$, we obtain that $\lim_{\mathbf{m}\to\mathbf{m}^*} \mathbf{x}(\mathbf{m}) = \mathbf{x}(\mathbf{m}^*)$. This completes the proof. □

3.5 Notes

3.5.1. The existence of a solution of the extremal problem (3.1.1) with even multiplicities $\{m_k\}$ was proved by Turán [140] for the case $m_1 = \cdots = m_n$, and by Popoviciu [77] and Tschakaloff [139] for arbitrary multiplicities, respectively. Later the existence was verified by Bojanov [7] for arbitrary numbers $m_k \geq 1$.

3.5.2. The uniqueness of a solution of the extremal problem (3.1.1) with even integers $\{m_k\}$ was proved much later by Ghizzetti and Ossicini [32]. This result was extended by Bojanov, Braess and Dyn [10] to the case with arbitrary positive integers $\{m_k\}$ under the assumption:

$$\mu \in C^1[a,b], \quad \mu' > 0. \tag{3.5.1}$$

Under the assumption (3.5.1) the uniqueness of the extremal problem (3.3.1) was verified by Bojanov [6, 7] for the case $m_k = qq_k$, where $q \geq 1$ is an arbitrary real number and q_k's are positive integers using an induction method, and for the case when $m_k \geq 1$ are arbitrary real numbers using a topological method. The topological method was first used by Barrow [4]. Shi [107] used a modification of ideas of Bojanov in [6] to prove the uniqueness for an arbitrary measure with a bounded support when $\min_{1 \leq k \leq n} m_k > 1$, dropping the assumption (3.5.1). In a recent paper Shi and Xu [127] extended the results in [107] to more general case, dropping the assumption of a bounded support.

3.5.3. Theorems 3.4.1 and 3.4.2 may be found in [127].

Chapter 4

Gaussian Quadrature Formulas

4.1 Introduction and Christoffel Numbers

In this chapter we investigate the *Gaussian* (or *Gauss-Turán*) *quadrature formulas* (see (4.1.3) below). Throughout this chapter let all the m_k be positive integers. Given a system of nodes (2.1.1) let the polynomials A_{kj} be defined in (2.2.9). Denote

$$S_n(x) = S_n(\mathbf{x},\mathbf{m};x) = \mathrm{sgn}\prod_{k=1}^{n}(x-x_k)^{m_k} \tag{4.1.1}$$

and

$$\lambda_{kj} = \lambda_{kjn}(d\mu,\mathbf{x},\mathbf{m}) = \int_a^b A_{kj}(x)S_n(x)d\mu(x),$$
$$j = 0,1,\ldots,m_k-1, \quad k = 1,2,\ldots,n. \tag{4.1.2}$$

Theorem 4.1.1. *Suppose that $d\mu$ is a measure on $\langle a,b\rangle$, and further that $\mu \in C\langle a,b\rangle$ if $\min_{1\le k\le n} m_k = 1$. Then*

(d) there exists a vector $\mathbf{x}\in\mathbf{Y}$ such that the (generalized) Gaussian quadrature formula

$$\int_a^b f(x)S_n(x)d\mu(x) = \sum_{k=1}^{n}\sum_{j=0}^{m_k-2}\lambda_{kj}f^{(j)}(x_k) \tag{4.1.3}$$

holds for all $f\in\mathbf{P}_N$, where the function $S_n = S_n(d\mu,\mathbf{m})$ is given by (4.1.1) and the Christoffel numbers $\lambda_{kj} = \lambda_{kjn}(d\mu,\mathbf{m})$ are given by (4.1.2).

Meanwhile, Statement (d) is equivalent to each of Statements (b) and (c) of Theorem 3.2.2.

Furthermore, if the assumptions of Theorem 3.2.2 prevails, then there exists a unique vector $\mathbf{x}\in\mathbf{Y}$ satisfying (4.1.3) and Statement (d) and Statements (a)–(c) of Theorem 3.2.2 are equivalent.

Proof. By Theorems 3.2.1 and 3.2.2 it is enough to prove that Statement (d) is equivalent to Statement (b) of Theorem 3.2.2. To this end we use (2.2.2) for $f\in\mathbf{P}_N$

$$f(x) = \sum_{k=1}^{n}\sum_{j=0}^{m_k-1} f^{(j)}(x_k)A_{kj}(x).$$

This yields

$$\int_a^b f(x)S_n(x)d\mu(x) = \sum_{k=1}^{n}\sum_{j=0}^{m_k-1} \lambda_{kj} f^{(j)}(x_k), \tag{4.1.4}$$

where the numbers λ_{kj} are given by (4.1.2). Comparing (4.1.4) with (4.1.3) we see that these two formulas are equivalent if and only if

$$\lambda_{k,m_k-1} = 0, \quad k = 1,2,\ldots,n. \tag{4.1.5}$$

This is the case, because by (2.2.4), (2.2.6), and (2.2.9) we have

$$\begin{aligned} A_{k,m_k-1}(x) &= \frac{1}{(m_k-1)!}(x-x_k)^{m_k-1}L_k(x) \\ &= \frac{S_n(x)\Omega_n(x)}{(m_k-1)![\prod_{i\neq k}(x_k-x_i)^{m_i}](x-x_k)}, \quad k = 1,2,\ldots,n, \end{aligned}$$

and hence relation (4.1.5) is equivalent to Statement (b) of Theorem 3.2.2. □

Theorem 4.1.2. *Let the assumptions of Theorem 4.1.1 prevail and let relation (4.1.3) hold for all $f \in \mathbf{P}_N$. If $m_k - j \in \mathbf{N}_2$ and $1 \leq k \leq n$, then*

$$sgn\ \lambda_{kj} = S_n(\mathbf{x}; x_k + 0). \tag{4.1.6}$$

Proof. By (4.1.2), (2.2.9), and (2.2.6) for $0 \leq j \leq m_k - 1$

$$\begin{aligned} \lambda_{kj} &= \frac{1}{j!}\int_a^b (x-x_k)^j B_{kj}(x)L_k(x)S_n(x)d\mu(x) \\ &= \frac{1}{j!}\int_a^b (x-x_k)^j B_{k,j+1}(x)L_k(x)S_n(x)d\mu(x) \\ &\quad + \frac{1}{j!}b_{k,m_k-j-1}\int_a^b (x-x_k)^{m_k-1}L_k(x)S_n(x)d\mu(x). \end{aligned}$$

Applying (4.1.3) the last term in the above relation is zero. Thus

$$\lambda_{kj} = \frac{1}{j!}\int_a^b (x-x_k)^j B_{k,j+1}(x)L_k(x)S_n(x)d\mu(x). \tag{4.1.7}$$

It is easy to see that

$$S_n(x_k+0) = \text{sgn} \prod_{q=1,q\neq k}^{n} (x_k - x_q)^{m_q}.$$

Since $m_k - j \in \mathbf{N}_2$, by Theorem 2.2.4 we have $B_{k,j+1}(x) > 0$, $x \in \mathbf{R}$ and hence

$$S_n(x_k+0)\lambda_{kj} = \frac{1}{j!}\int_a^b B_{k,j+1}(x)|(x-x_k)^j L_k(x)|d\mu(x) > 0. \tag{4.1.8}$$

We obtain (4.1.6). □

Theorem 4.1.3. *Let the assumptions of Theorem 4.1.2 prevail. If $m_k - j \in \mathbf{N}_2$, $j < i < m_k$, and $1 \leq k \leq n$, then*

$$|\lambda_{ki}| \leq \frac{j!}{i!} S_n(x_k+0) d_k^{i-j} \lambda_{kj} \tag{4.1.9}$$

and

$$|\lambda_{ki}| \leq \frac{j!}{i!} S_n(x_k+0) \overline{d}_k^{i-j} \lambda_{kj}. \tag{4.1.10}$$

Proof. By (4.1.8) and (2.4.19)

$$\begin{aligned} S_n(x_k+0)\lambda_{kj} &= \frac{1}{j!}\int_a^b B_{k,j+1}(x)|(x-x_k)^j L_k(x)|d\mu(x) \\ &\geq \frac{1}{j!} d_k^{j-i} \left| \int_a^b B_{k,i+1}(x)|(x-x_k)^i L_k(x) S_n(x) \right| d\mu(x) \\ &= \frac{i!}{j!} d_k^{j-i} |\lambda_{ki}|. \end{aligned}$$

We obtain (4.1.9).

Similarly, by (4.1.8) and (2.4.36) we obtain (4.1.10). □

4.2 A General Quadrature Formula

Denote by $\mathbf{F}(d\mu, \langle a,b \rangle)$ the set of all Riemann-Stieltjes integrable functions with respect to $d\mu$ on $\langle a,b \rangle$ and $\mathbf{F}(d\mu) = \mathbf{F}(d\mu, \mathbf{R})$.

Definition 4.2.1. [24, p. 62] *$d\mu \in \mathbf{E}$ means that if a measure $d\nu$ satisfies*

$$\int_{\mathbf{R}} x^n d\nu(x) = \int_{\mathbf{R}} x^n d\mu(x), \quad n = 0, 1, \ldots,$$

then

$$\nu(x) - \nu(-\infty) = \mu(x) - \mu(-\infty),$$

apart from denumerably many points of discontinuity.

We state two lemmas without proof.

Lemma 4.2.1. [24, Theorem II.2.2, p. 64] *If a measure $d\mu$ is supported in a finite interval $[a,b]$, then $d\mu \in \mathbf{E}$.*

Lemma 4.2.2. [24, Theorem II.5.2, p. 80] *If for a number $c > 0$ the relation*

$$\int_{\mathbf{R}} e^{c|x|} d\mu(x) = M < \infty \tag{4.2.1}$$

is satisfied, then $d\mu \in \mathbf{E}$.

Let $\mathbf{x}_n = (x_{1n}, x_{2n}, \ldots, x_{nn})$ satisfy

$$a = x_{0n} \leq x_{1n} < x_{2n} < \cdots < x_{nn} \leq x_{n+1,n} = b, \quad n = 1, 2, \ldots. \tag{4.2.2}$$

Put for each n

$$I_{kn} = \langle x_{kn}, x_{k+1,n} \rangle, \quad y_{kn} \in I_{kn}, \quad k = 0,1,\ldots,n,$$

and

$$\rho_n = \max_{0\leq k\leq n} \min\left\{ |I_{kn}|, \int_{I_{kn}} d\mu(x) \right\}.$$

Definition 4.2.2. [24, p. 88] *Let*

$$c_{kn} \geq 0, \quad k = 1,2,\ldots,n, \quad n = 1,2,\ldots,$$

and

$$Q_n(\mathbf{x}_n; f) = \sum_{k=1}^{n} c_{kn} f(x_{kn}).$$

A positive quadrature procedure $\{Q_n\}$ *is said to belong to the measure* $d\mu$ *if the relation*

$$\lim_{n\to\infty} Q_n(\mathbf{x}_n; f) = \int_a^b f(x)d\mu(x) \tag{4.2.3}$$

holds for every polynomial f.

The following results are interesting.

Theorem 4.2.1. *Let* $d\mu$ *be a measure supported on a finite interval* $[a,b]$. *If* $f \in \mathbf{F}(d\mu,[a,b])$ *then*

$$\lim_{\rho_n\to 0} \sum_{k=0}^{n} \int_{I_{kn}} f(y_{kn})d\mu(x) = \int_a^b f(x)d\mu(x) \tag{4.2.4}$$

apart from any choice of I_{kn} *and* y_{kn}.

Conversely, if

$$\lim_{\rho_n\to 0} \sum_{k=0}^{n} \int_{I_{kn}} f(y_{kn})d\mu(x) = I$$

apart from any choice of I_{kn} *and* y_{kn}, *then* $f \in \mathbf{F}(d\mu,[a,b])$ *and*

$$I = \int_a^b f(x)d\mu(x).$$

Proof. Assume that $f \in \mathbf{F}(d\mu,[a,b])$. By definition for an arbitrary number $\varepsilon > 0$ there exists a number $\delta > 0$ such that if

$$\max_{0\leq k\leq n} \{|I'_{kn}|\} \leq \delta$$

then

$$\left| \sum_{k=0}^{n} \int_{I'_{kn}} f(y'_{kn})d\mu(x) - \int_a^b f(x)d\mu(x) \right| \leq \varepsilon/2, \tag{4.2.5}$$

where

$$I'_{kn} = \langle x'_{kn}, x'_{k+1,n} \rangle, \quad y'_{kn} \in I'_{kn}, \quad k = 0,1,\ldots,n,$$

and

$$x'_{0n} = a \leq x'_{1n} < x_{2n} < \cdots < x'_{nn} \leq x'_{n+1,n} = b.$$

Now we choose

$$0 < \rho_n \leq \frac{\varepsilon\delta}{4\|f\|_{[a,b]}(b-a+\delta)+\varepsilon} \tag{4.2.6}$$

and write

$$\sum_{k=0}^{n}\int_{I_{kn}} f(y_{kn})d\mu(x) = \sum_{|I_{kn}|\leq\delta}\int_{I_{kn}} f(y_{kn})d\mu(x) + \sum_{|I_{kn}|>\delta}\int_{I_{kn}} f(y_{kn})d\mu(x). \tag{4.2.7}$$

Then we divide each interval I_{kn} with $|I_{kn}| > \delta$ into disjoint subintervals

$$I_{kn} = \cup_{j=1}^{i_{kn}} I_{jkn}$$

such that $|I_{jkn}| \leq \delta$ for $j = 1,2,\ldots,i_{kn}$, and choose a point $y_{jkn} \in I_{jkn}$ for $j = 1,2,\ldots,i_{kn}$. Thus by (4.2.5)

$$\left|\sum_{|I_{kn}|\leq\delta}\int_{I_{kn}} f(y_{kn})d\mu(x) + \sum_{|I_{kn}|>\delta}\sum_{j=1}^{i_{kn}}\int_{I_{jkn}} f(y_{jkn})d\mu(x) - \int_a^b f(x)d\mu(x)\right| \leq \varepsilon/2. \tag{4.2.8}$$

On the other hand, by (4.2.6) we see $\rho_n \leq \delta$. Hence we conclude that if $|I_{kn}| > \delta$ then

$$\int_{I_{kn}} d\mu(x) \leq \rho_n,$$

since $\min\left\{|I_{kn}|, \int_{I_{kn}} d\mu(x)\right\} \leq \rho_n$. Thus according to (4.2.6)

$$\begin{aligned}&\left|\sum_{|I_{kn}|>\delta}\int_{I_{kn}} f(y_{kn})d\mu(x) - \sum_{|I_{kn}|>\delta}\sum_{j=1}^{i_{kn}}\int_{I_{jkn}} f(y_{jkn})d\mu(x)\right| \\ \leq\ & 2\|f\|_{[a,b]}\sum_{|I_{kn}|>\delta}\int_{I_{kn}} d\mu(x) \leq 2\rho_n\|f\|_{[a,b]}\left(\frac{b-a}{\delta}+1\right) \leq \varepsilon/2,\end{aligned}$$

which, coupled with (4.2.7) and (4.2.8), gives

$$\left|\sum_{k=0}^{n}\int_{I_{kn}} f(y_{kn})d\mu(x) - \int_a^b f(x)d\mu(x)\right| \leq \varepsilon.$$

Then relation (4.2.4) follows.

Conversely, we observe that for any choice of I_{kn} by definition

$$\rho_n \leq \max_{0\leq k\leq n}|I_{kn}|.$$

Then we can conclude that $f \in \mathbf{F}(d\mu,[a,b])$ and

$$I = \int_a^b f(x)d\mu(x). \quad \square$$

Remark 4.2.1. If $a = -\infty$ or $b = +\infty$ the theorem may be violated, since the value $|f(y_{n+1,n})|$ or $|f(y_{0n})|$ may be sufficiently large.

To prove our next theorem we state the following result without proof.

Theorem 4.2.2. [24, Theorem III.1.2, p. 89] *Let $d\mu \in \mathbf{E}$ and let a positive quadrature procedure $\{Q_n\}$ belong to the measure $d\mu$. If $f \in \mathbf{F}(d\mu)$ satisfies*

$$|f(x)| \leq A + B|x|^s, \quad x \in \mathbf{R}, \quad A, B, s > 0, \tag{4.2.9}$$

then relation (4.2.3) holds.

Theorem 4.2.3. *Let $d\mu \in \mathbf{E}$ and let a positive quadrature procedure $\{Q_n\}$ belong to the measure $d\mu$. If*

$$\int_c^d d\mu(x) > 0$$

then for sufficiently large n the interval $\langle c, d\rangle$ contains at least one point of x_{kn}'s.

Proof. Suppose to the contrary that the interval $\langle c, d\rangle$ contains no any point of x_{kn}'s for n large enough. Now put

$$f(x) = \begin{cases} 1, & x \in \langle c, d\rangle, \\ 0, & x \notin \langle c, d\rangle. \end{cases}$$

Applying Theorem 4.2.2 and noticing that the interval $\langle c, d\rangle$ contains no any point of x_{kn}'s, for n large enough we have

$$0 = \lim_{n\to\infty} Q_n(\mathbf{x}_n; f) = \int_a^b f(x)d\mu(x) = \int_c^d d\mu(x) > 0,$$

a contradiction. □

Theorem 4.2.4. *Let $d\mu \in \mathbf{E}$ and let a positive quadrature procedure $\{Q_n\}$ belong to the measure $d\mu$.*

(a) $\mu \in C\langle a, b\rangle$ if and only if

$$\lim_{n\to\infty} \max_{0\leq k\leq n} \int_{I_{kn}} d\mu(x) = 0. \tag{4.2.10}$$

(b) If μ is strictly increasing on a finite interval $[c, d]$ then

$$\lim_{n\to\infty} \max_{I_{kn}\subset[c,d]} |I_{kn}| = 0. \tag{4.2.11}$$

(c) If $\mu \in C\langle a, b\rangle$ or μ is strictly increasing on $\langle a, b\rangle$ then

$$\lim_{n\to\infty} \rho_n = 0. \tag{4.2.12}$$

Proof. (a) Let for each n

$$\int_{I_{k_n,n}} d\mu(x) = \max_{0\leq k\leq n} \int_{I_{kn}} d\mu(x), \quad 0 \leq k_n \leq n.$$

Assume, passing to a subsequence if necessary, that

$$\lim_{n\to\infty} I_{k_n,n} = \langle C, D\rangle. \tag{4.2.13}$$

We must show

$$\int_C^D d\mu(x) = 0.$$

Suppose to the contrary that

$$\int_C^D d\mu(x) > 0.$$

Choose two numbers c,d so that $C < c < d < D$ and $\int_c^d d\mu(x) > 0$. By Theorem 4.2.3 for sufficiently large n the interval $[c,d]$ contains at least one point of x_{kn}'s. But by (4.2.13) for n large enough we have $x_{k_n,n} < c < d < x_{k_n+1,n}$, which shows that for n large enough the interval $[c,d]$ contains no any point of x_{kn}'s. This is a contradiction to Theorem 4.2.3. This contradiction proves (4.2.10).

Conversely, if μ is discontinuous at, say $x = y$, and $y \in I_{kn}$ then $\int_{I_{kn}} d\mu(x) \geq \mu(y+0) - \mu(y-0) > 0$, which shows that (4.2.10) is violated.

(b) Let for each n

$$|I_{k_n,n}| = \max_{I_{kn} \subset [c,d]} |I_{kn}|, \quad I_{k_n,n} \subset [c,d].$$

Assume, passing to a subsequence if necessary, that (4.2.13) holds. We must show $C = D$. Suppose to the contrary that $C < D$. Clearly by the assumption we have $\int_C^D d\mu(x) > 0$. By the same argument as that of Statement (a) we again get a contradiction to Theorem 4.2.3. This contradiction proves (4.2.11).

(c) Let for each n

$$\min\left\{|I_{k_n,n}|, \int_{I_{k_n,n}} d\mu(x)\right\} = \max_{0\leq k\leq n} \min\left\{|I_{kn}|, \int_{I_{kn}} d\mu(x)\right\}, \quad 0 \leq k_n \leq n.$$

Assume, passing to a subsequence if necessary, that (4.2.13) holds. We must show

$$\min\left\{D - C, \int_C^D d\mu(x)\right\} = 0.$$

Suppose to the contrary that

$$\min\left\{D - C, \int_C^D d\mu(x)\right\} > 0,$$

that is,

$$D - C > 0, \quad \int_C^D d\mu(x) > 0.$$

Choose two numbers c,d so that $C < c < d < D$ and $\int_c^d d\mu(x) > 0$. By the same argument as that of Statement (a) we can get a contradiction. This contradiction proves Statement (c). □

Remark 4.2.2. If $\mu \notin C\langle a,b\rangle$ and μ is not strictly increasing then Statement (c) may be not true. For example,

$$\mu(x) = \begin{cases} x, & 1/2 \leq |x| \leq 1, \\ 0, & |x| \leq 1/2, \\ -1, & x < -1, \\ 1, & x > 1. \end{cases}$$

In this case the Gaussian quadrature procedure $\{Q_n(\mathbf{x}_n(d\mu))\}$ belongs to the measure $d\mu$. But $P_n(d\mu;x)$ has at most one zero in the interval $(-1/2,1/2)$ (see, say, [24, Theorem I.2.4, p. 18]). Thus $\rho_n \geq 1/2$.

Theorem 4.2.5. *Let $d\mu \in \mathbf{E}$ and let a positive quadrature procedure $\{Q_n\}$ belong to the measure $d\mu$. Assume that $\mu \in C\langle a,b\rangle$ or μ is strictly increasing on $\langle a,b\rangle$. Then for any finite interval $[c,d]$*

$$\lim_{n\to\infty} \sum_{I_{kn}\subset[c,d]} \int_{I_{kn}} f(y_{kn})d\mu(x) = \int_c^d f(x)d\mu(x).$$

Proof. Apply Theorems 4.2.1 and 4.2.4 to the function

$$g(x) = \begin{cases} f(x), & x \in [c,d], \\ 0, & x \notin [c,d]. \end{cases} \quad \square$$

4.3 Erdős-Turán Type Inequalities

Let a system of nodes (2.1.1) be given and let the numbers $s_r = s_r(\mathbf{m})$ be defined by

$$\begin{cases} s_r = S_n(x_r - 0) = (-1)^{m_r+m_{r+1}+\cdots+m_n}, & r = 1,2,\ldots,n, \\ s_0 = s_1, \quad s_{n+1} = s_{n+2} = 1. \end{cases} \tag{4.3.1}$$

Lemma 4.3.1. *Given a point $\xi \in \langle a,b\rangle$ let an $(N+1)$th continuously differentiable function g satisfy*

$$g^{(j)}(x) > 0, \quad x \in \langle a,\xi\rangle, \quad j = 0,1,\ldots,N+1, \tag{4.3.2}$$

and let $P \in \mathbf{P}_N$, $P \neq 0$. Then

$$Z'(g-P,\langle a,\xi\rangle) + Z'(P,\langle \xi,b\rangle) \leq N+1. \tag{4.3.3}$$

Proof. We distinguish three cases.

Case 1. $Z'(g-P,\langle a,\xi\rangle) \leq 1$. In this case by (2.2.14) we have $Z'(P,\langle \xi,b\rangle) \leq \partial P \leq N$. Then inequality (4.3.3) follows.

Case 2. $Z'(P,\langle \xi,b\rangle) = 0$. In this case if inequality (4.3.3) is not true, then inequality $Z'(g-P,\langle a,\xi\rangle) \geq N+2$ would occur. Applying Rolle theorem $N+1$ times would yield

$$Z'(g^{(N+1)},\langle a,\xi\rangle) = Z'(g^{(N+1)} - P^{(N+1)},\langle a,\xi\rangle) \geq 1,$$

a contradiction to (4.3.2). This contradiction proves (4.3.3).

Case 3. $Z'(g-P,\langle a,\xi\rangle) \geq 2$ and $Z'(P,\langle \xi,b\rangle) \geq 1$. We shall prove (4.3.3) by induction with respect to N. The inequality (4.3.3) is obviously true for $N=0$, because the equality $N=0$ implies $Z'(P,\langle \xi,b\rangle) = 0$, which has been proved in Case 2. Suppose now, as an induction hypothesis, that inequality (4.3.3) is true for $N = r-1$, $r \geq 1$. Let y_1 be the greatest zero of $g-P$ in $\langle a,\xi\rangle$ and y_2 the least zero of P in $\langle \xi,b\rangle$. We separate two cases.

Case 3.1. The point y_1 is a simple zero of $g-P$. Since $Z'(g-P,\langle a,\xi\rangle) \geq 2$, there exists a greatest but one zero y_0 of $g-P$. By Rolle theorem we can find a point $z_1 \in (y_0,y_1)$ such that $g'(z_1) - P'(z_1) = 0$. This gives $P'(z_1) = g'(z_1) > 0$. On the other hand, by the assumptions we see that $P(y_1) = g(y_1) > 0 = P(y_2)$, which shows that there exists a point $z_2 \in (y_1, y_2)$ such that $P'(z_2) < 0$. Then there is a point $z \in (z_1, z_2)$ such that $P'(z) = 0$. Using the induction hypothesis we have

$$\begin{aligned} r &\geq Z'(g' - P', \langle a, z_1]) + Z'(P', [z_1, b\rangle) \\ &= Z'(g' - P', \langle a, z_1]) + Z'(P', [z_1, z_2)) + Z'(P', [z_2, b\rangle) \\ &\geq Z'(g - P, \langle a, \xi]) - 1 + 1 + Z'(P, [\xi, b\rangle) - 1 \\ &= Z'(g - P, \langle a, \xi]) + Z'(P, [\xi, b\rangle) - 1, \end{aligned}$$

which yields

$$Z'(g - P, \langle a, \xi]) + Z'(P, [\xi, b\rangle) \leq r + 1.$$

This proves inequality (4.3.3) for $N = r$. By induction inequality (4.3.3) holds.

Case 3.2. The point y_1 is not a simple zero of $g - P$. Then $Z'(g' - P', \langle a, y_1]) \geq Z'(g - P, \langle a, \xi]) - 1$. From the inequalities $P'(y_1) = g'(y_1) > 0$ and $P'(z_2) < 0$ we may find a point $z' \in (y_1, z_2)$ such that $P'(z') = 0$. Again using the induction hypothesis we obtain

$$\begin{aligned} r &\geq Z'(g' - P', \langle a, y_1]) + Z'(P', [y_1, b\rangle) \\ &= Z'(g' - P', \langle a, y_1]) + Z'(P', [y_1, z_2)) + Z'(P', [z_2, b\rangle) \\ &\geq Z'(g - P, \langle a, \xi]) - 1 + 1 + Z'(P, [\xi, b\rangle) - 1 \\ &= Z'(g - P, \langle a, \xi]) + Z'(P, [\xi, b\rangle) - 1. \end{aligned}$$

This gives

$$Z'(g - P, \langle a, \xi]) + Z'(P, [\xi, b\rangle) \leq r + 1$$

and proves inequality (4.3.3) for $N = r$. By induction inequality (4.3.3) is true. □

Using Lemma 4.3.1 we can establish the following Erdős-Turán type inequalities.

Theorem 4.3.1. *Let an $(N+1)$th continuously differentiable function g satisfy*

$$g^{(j)}(x) \geq 0, \quad x \in \langle a, x_r \rangle, \quad j = 0, 1, \ldots, N+1. \tag{4.3.4}$$

Then for $0 \leq r \leq n+1$

$$s_r S_n(x) \left[g(x) - \sum_{k=1}^{r-1} \sum_{j=0}^{m_k - 1} g^{(j)}(x_k) A_{kj}(x) \right] \geq 0, \quad x \in \langle a, x_r \rangle, \tag{4.3.5}$$

$$s_r S_n(x) \sum_{k=1}^{r-1} \sum_{j=0}^{m_k - 1} g^{(j)}(x_k) A_{kj}(x) \geq 0, \quad x \in \langle x_r, b \rangle, \tag{4.3.6}$$

$$s_{r+1} S_n(x) \left[g(x) - \sum_{k=1}^{r} \sum_{j=0}^{m_k - 1} g^{(j)}(x_k) A_{kj}(x) \right] \geq 0, \quad x \in \langle a, x_r \rangle, \tag{4.3.7}$$

and

$$s_{r+1} S_n(x) \sum_{k=1}^{r} \sum_{j=0}^{m_k - 1} g^{(j)}(x_k) A_{kj}(x) \geq 0, \quad x \in \langle x_r, b \rangle. \tag{4.3.8}$$

Here we agree $\sum_{k=1}^{n+1} = \sum_{k=1}^{n}$.

Proof. It suffices to show (4.3.5)–(4.3.8) for the case when $a > -\infty$ and $b < +\infty$. In fact, if, say, $a = -\infty$ and $b < +\infty$, we can choose an arbitrary number $A, -\infty < A < x_1$. The inequality (4.3.4) implies that

$$g^{(j)}(x) \geq 0, \quad x \in [A, x_r], \quad j = 0, 1, \ldots, N+1.$$

If the inequality

$$s_r S_n(x) \left[g(x) - \sum_{k=1}^{r-1} \sum_{j=0}^{m_k - 1} g^{(j)}(x_k) A_{kj}(x) \right] \geq 0, \quad x \in [A, x_r],$$

holds for every A then inequality (4.3.5) holds. So does for inequality (4.3.7).

Now let $a > -\infty$ and $b < +\infty$. By (4.1.1) and (4.3.1) we observe that

$$S_n(x) = s_r, \quad x \in (x_{r-1}, x_r), \quad 1 \leq r \leq n+1. \tag{4.3.9}$$

On the other hand, it is enough to show (4.3.5)–(4.3.8) for g satisfying

$$g^{(j)}(x) > 0, \quad x \in [a, x_r], \quad j = 0, 1, \ldots, N+1, \tag{4.3.10}$$

instead of (4.3.4) (because we intend to apply Lemma 4.3.1). In fact, if g satisfies (4.3.4) only, then we consider the function $g_\varepsilon(x) = g(x) + \varepsilon e^x$, $\varepsilon > 0$, which already satisfies (4.3.10). Applying the above conclusion for the function g_ε and letting $\varepsilon \longrightarrow 0$ yields (4.3.5)–(4.3.8).

Let us prove (4.3.5) and (4.3.6) first. In this case for simplicity put

$$P(x) = \sum_{k=1}^{r-1} \sum_{j=0}^{m_k - 1} g^{(j)}(x_k) A_{kj}(x).$$

Then inequalities (4.3.5) and (4.3.6) become

$$s_r S_n(x) \left[g(x) - P(x) \right] \geq 0, \quad x \in \langle a, x_r \rangle, \tag{4.3.11}$$

and

$$s_r S_n(x) P(x) \geq 0, \quad x \in \langle x_r, b \rangle, \tag{4.3.12}$$

respectively.

To prove (4.3.11) and (4.3.12) we distinguish two cases.

Case 1. $r \leq 1$. In this case $P = 0$. Then inequality (4.3.12) is trivial and inequality (4.3.11) becomes

$$s_r S_n(x) g(x) \geq 0, \quad x \in [a, x_r],$$

which by (4.3.9) and (4.3.10) is true.

Case 2. $r \geq 2$. If we introduce the function

$$G_r(x) = \begin{cases} g(x) - P(x), & x < x_r, \\ P(x), & x \geq x_r, \end{cases}$$

then inequalities (4.3.11) and (4.3.12) may be written as one inequality

$$s_r S_n(x) G_r(x) \geq 0, \quad x \in [a,b], \tag{4.3.13}$$

and we have to prove (4.3.13). Since the proof is long, we break it into a series of claims.

Claim 1. The function G_r satisfies the condition:

$$G_r^{(j)}(x_k) = 0, \quad j = 0,1,\ldots,m_k - 1, \quad k = 1,2,\ldots,n.$$

This is easy to check by the definition of G_r.

Claim 2. We have

$$Z(G_r, (x_{r-1}, x_r)) = Z(P, (x_{r-1}, x_r)) = 0. \tag{4.3.14}$$

In fact, by Lemma 4.3.1 and Claim 1 we see

$$\begin{aligned} N+1 &\geq Z(g - P, [a, x_r]) + Z(P, [x_r, b]) \\ &= Z(G_r, [a, x_{r-1}]) + Z(G_r, (x_{r-1}, x_r)) + Z(G_r, [x_r, b]) \\ &\geq N + 1 + Z(G_r, (x_{r-1}, x_r)) \end{aligned}$$

and hence $Z(G_r, (x_{r-1}, x_r)) = 0$. Similarly, by Lemma 4.3.1 and Claim 1 we obtain

$$\begin{aligned} N+1 &\geq Z(g - P, [a, x_{r-1}]) + Z(P, [x_{r-1}, b]) \\ &= Z(G_r, [a, x_{r-1}]) + Z(P, [x_{r-1}, x_r)) + Z(G_r, [x_r, b]) \\ &\geq N + 1 + Z(P, (x_{r-1}, x_r)) \end{aligned}$$

and hence $Z(P, (x_{r-1}, x_r)) = 0$. This proves Claim 2.

Claim 3. We have for $(2 \leq) r \leq n$

$$G_r(x) > 0, \quad P(x) > 0, \quad x \in (x_{r-1}, x_r). \tag{4.3.15}$$

In fact, in case $r \leq n$ since $P(x_{r-1}) = g(x_{r-1}) > 0$ and $P(x_r) = 0$, by (4.3.10) and (4.3.14) we obtain (4.3.15).

Claim 4. The function $S_n G_r$ does not change sign on $[a,b]$.

In fact, suppose to the contrary that the function $S_n G_r$ would change sign at $z \in (a,b)$. If $z \notin \{x_1, \ldots, x_n\}$ then $G_r(z) = 0$. If $z = x_k$ holds for some k, $1 \leq k \leq n$, then the inequality

$$[S_n(x_k - \delta) G_r(x_k - \delta)][S_n(x_k + \delta) G_r(x_k + \delta)] < 0$$

would hold for all small $\delta > 0$, which by (4.1.1) implies that the inequality

$$(-1)^{m_k} G_r(x_k - \delta) G_r(x_k + \delta) < 0 \tag{4.3.16}$$

would hold for all small $\delta > 0$. If we agree to write for simplicity

$$G_r^{(j)}(x) = \begin{cases} g^{(j)}(x) - P^{(j)}(x), & x < x_r, \\ P^{(j)}(x), & x \geq x_r, \end{cases}$$

then we can conclude

$$G_r^{(m_k)}(x_k) = 0. \tag{4.3.17}$$

Indeed, if $k \neq r$ then by Claim 1 and (4.3.16) inequality (4.3.17) is true; if $k = r$ then by (4.3.15) inequality (4.3.16) leads to

$$(-1)^{m_r} P(x_r - \delta) P(x_r + \delta) < 0$$

and hence (4.3.17) with $k = r$ follows. This proves (4.3.17). By (4.3.17) we have $Z'(G_r, [a,b]) \geq \sum_{k=1}^n m_k + 1 = N+2$, contradicting Lemma 4.3.1. This contradiction proves our claim.

We are in a position to prove (4.3.13). To this end according to Claim 4 it is enough to determine the sign of the function $S_n G_r$ at its one nonzero point. We separate two cases when $r \leq n$ and $r = n+1$.

Case 2.1. $r \leq n$. In this case by means of (4.3.15) and (4.3.9)

$$\text{sgn}\,[S_n(x) G_r(x)] = S_n(x) = s_r, \quad x \in (x_{r-1}, x_r),$$

which gives (4.3.13).

Case 2.2. $r = n+1$. In this case we have $Z'(G_{n+1}, [a,b]) = Z'(g - P, [a,b]) = N+1$. By Rolle theorem for each j, $j = 0, 1, \ldots, N$, there must exist the largest zero ξ_j of $G_{n+1}^{(j)}$ such that

$$G_{n+1}^{(j)}(\xi_j) = 0, \quad Z(G_{n+1}^{(j)}, (\xi_j, b]) = 0, \quad j = 0, 1, \ldots, N, \tag{4.3.18}$$

and

$$x_n = \xi_0 = \cdots = \xi_{m_n - 1} > \xi_{m_n} > \cdots > \xi_N.$$

Recalling that $G_{n+1}^{(N+1)}(x) = g^{(N+1)}(x) > 0$ in $[a,b]$ by (4.3.10) with $j = N+1$ and noticing (4.3.18) with $j = N$, we conclude $G_{n+1}^{(N)}(x) > 0$ for $x > \xi_N$. By induction we further obtain

$$G_{n+1}^{(j)}(x) > 0, \quad x > \xi_j, \quad j = N, N-1, \ldots, 1, 0.$$

In particular for $j = 0$ we have that $G_{n+1}(x) > 0$ for $x \in (x_n, b)$, which implies that the equality

$$\text{sgn}\,[S_n(x) G_{n+1}(x)] = 1 = s_{n+1}$$

holds for $x \in (x_n, b)$. Then inequality (4.3.13) is valid.

This completes the proof of (4.3.5) and (4.3.6).

Let us prove (4.3.7) and (4.3.8). In this case for simplicity we need to put

$$P(x) = \sum_{k=1}^{r} \sum_{j=0}^{m_k - 1} g^{(j)}(x_k) A_{kj}(x)$$

and

$$G_r(x) = \begin{cases} g(x) - P(x), & x \leq x_r, \\ P(x), & x > x_r. \end{cases} \tag{4.3.19}$$

Again we distinguish two cases.

Case I. $r = 0$. In this case inequalities (4.3.7) and (4.3.8) are trivial, because $P = 0$.

Case II. $r \geq 1$. In this case by the same argument as above we again conclude that the function $S_n G_r$ does not change sign on $[a,b]$. But in this case we have to prove

$$s_{r+1} S_n(x) G_r(x) \geq 0, \quad x \in [a,b], \tag{4.3.20}$$

instead of (4.3.13). If $r \leq n$, then by the same argument as before we can obtain that $G_r(x) = P(x) > 0$ for $x \in (x_r, x_{r+1})$ and hence by (4.3.9)

$$\operatorname{sgn}\,[S_n(x)G_r(x)] = S_n(x) = s_{r+1}, \quad x \in (x_r, x_{r+1}),$$

which yields (4.3.20). If $r = n+1$, then inequality (4.3.8) is trivial and inequality (4.3.7) is equivalent to (4.3.5). With the help of (4.3.19) and (4.3.20) we directly get (4.3.7) and (4.3.8) under the assumption (4.3.10). □

Theorem 4.3.1 is very general and useful. We state some useful corollaries.

Corollary 4.3.1. *Let $0 \leq r < q \leq n+1$ and let an $(N+1)$th continuously differentiable function g satisfy (4.3.4) with $r = q$. If*

$$s_r = s_{q+1} \tag{4.3.21}$$

then

$$s_r S_n(x) \left[g(x) - \sum_{k=r}^{q} \sum_{j=0}^{m_k-1} g^{(j)}(x_k) A_{kj}(x) \right] \geq 0, \quad x \in \langle x_r, x_q \rangle. \tag{4.3.22}$$

Proof. Clearly by (4.3.1) equality (4.3.21) is equivalent to that the sum $\sum_{k=r}^{q} m_k$ is even. The inequality (4.3.7) with $r = q$ reads

$$s_{q+1} S_n(x) \left[g(x) - \sum_{k=1}^{q} \sum_{j=0}^{m_k-1} g^{(j)}(x_k) A_{kj}(x) \right] \geq 0, \quad x \in \langle a, x_q \rangle.$$

This, coupled with (4.3.6) and (4.3.21), gives that for $x \in \langle x_r, x_q \rangle$

$$\begin{aligned} 0 \quad &\leq \quad s_{q+1} S_n(x) \left[g(x) - \sum_{k=1}^{q} \sum_{j=0}^{m_k-1} g^{(j)}(x_k) A_{kj}(x) \right] \\ &\qquad + s_r S_n(x) \sum_{k=1}^{r-1} \sum_{j=0}^{m_k-1} g^{(j)}(x_k) A_{kj}(x) \\ &= \quad s_r S_n(x) \left[g(x) - \sum_{k=r}^{q} \sum_{j=0}^{m_k-1} g^{(j)}(x_k) A_{kj}(x) \right]. \end{aligned}$$

We obtain (4.3.22). □

Corollary 4.3.2. *Let $0 \leq r \leq n+1$ and let an $(N+1)$th continuously differentiable function g satisfy (4.3.4) with replacing r by $r+1$. If the sum $m_r + m_{r+1}$ is even, then*

$$(-1)^{m_r} \left[g(x) - \sum_{k=r}^{r+1} \sum_{j=0}^{m_k-1} g^{(j)}(x_k) A_{kj}(x) \right] \geq 0, \quad x \in \langle x_r, x_{r+1} \rangle. \tag{4.3.23}$$

In particular, if both m_r and m_{r+1} are odd, then

$$\sum_{k=r}^{r+1}\sum_{j=0}^{m_k-1} g^{(j)}(x_k)A_{kj}(x) \geq g(x), \quad x \in \langle x_r, x_{r+1}\rangle. \tag{4.3.24}$$

Proof. Since the sum $m_r + m_{r+1}$ is even, the equality (4.3.21) with $q = r+1$ is true and hence inequality (4.3.22) with $q = r+1$ gives

$$s_r S_n(x)\left[g(x) - \sum_{k=r}^{r+1}\sum_{j=0}^{m_k-1} g^{(j)}(x_k)A_{kj}(x)\right] \geq 0, \quad x \in \langle x_r, x_{r+1}\rangle. \tag{4.3.25}$$

But by (4.3.9) for $x \in (x_r, x_{r+1})$ we have $s_r S_n(x) = s_r s_{r+1} = (-1)^{m_r}$ and hence inequality (4.3.25) becomes (4.3.23). The inequality (4.3.24) is an immediate consequence of (4.3.23) if both m_r and m_{r+1} are odd. □

Remark 4.3.1. For Lagrange interpolation ($m_1 = \cdots = m_n = 1$) the inequality (4.3.24) with $g = 1$ becomes the classical Erdős-Turán inequality [22, Lemma IV]

$$\ell_k(x) + \ell_{k+1}(x) \geq 1, \quad x \in [x_k, x_{k+1}], \quad 0 \leq k \leq n \ (\ell_0 = \ell_{n+1} = 0). \tag{4.3.26}$$

Corollary 4.3.3. *Let $0 \leq r < q \leq n+1$ and let an $(N+1)$th continuously differentiable function g satisfy (4.3.4) with $r = q$. If each number m_k is even for $k = r, r+1, \ldots, q$, then*

$$\sum_{k=r}^{q}\sum_{j=0}^{m_k-1} g^{(j)}(x_k)A_{kj}(x) \leq g(x), \quad x \in \langle x_r, x_q\rangle. \tag{4.3.27}$$

Proof. Since the number m_k is even for $k = r, r+1, \ldots, q$, by (4.3.1) we have $s_r = s_{r+1} = \cdots = s_{q+1}$. Thus by (4.3.9) we see

$$s_r S_n(x) \geq 0, \quad x \in \langle x_r, x_q\rangle.$$

It remains to apply Corollary 4.3.1. □

Corollary 4.3.4. *Let $0 \leq r \leq n+1$ and let an $(N+1)$th continuously differentiable function g satisfy (4.3.4). If the number m_r is odd, then*

$$s_{r+1}S_n(x)\,[sgn\ (x - x_r)] \sum_{j=1}^{m_r-1} g^{(j)}(x_r)A_{rj}(x) \geq 0, \quad x \in \langle a, b\rangle. \tag{4.3.28}$$

Proof. If the number m_r is odd, then $s_r = -s_{r+1}$. By (4.3.5) and (4.3.7) the inequality

$$s_{r+1}S_n(x)\,[\mathrm{sgn}\ (x - x_r)] \sum_{j=0}^{m_r-1} g^{(j)}(x_r)A_{rj}(x) \geq 0 \tag{4.3.29}$$

holds for $x \in \langle a, x_r\rangle$; besides, inequalities (4.3.6) and (4.3.8) imply that inequality (4.3.29) holds for $x \in \langle x_r, b\rangle$. □

4.4 The Generalized Markov-Stieltjes Inequalities

As an important application of the Erdős-Turán type inequalities, in this section we will derive the generalized Markov-Stieltjes inequalities, which also have many applications: estimations for the Christoffel numbers (Section 4.5), location of the zeros of power orthogonal polynomials, etc..

Theorem 4.4.1. *Let the assumptions of Theorem 4.1.2 prevail. Let $1 \le r \le n+1$ and let an $(N+1)$th continuously differentiable function g satisfy (4.3.4). Then with the notations of Theorem 4.1.1*

$$s_r \int_a^{x_{r-1}} g(x)S_n(x)d\mu(x) \le s_r \sum_{k=1}^{r-1} \sum_{j=0}^{m_k-2} \lambda_{kj} g^{(j)}(x_k) \le s_r \int_a^{x_r} g(x)S_n(x)d\mu(x), \tag{4.4.1}$$

if the integrals exist.

Proof. We prove the second inequality only, because the first may be proved similarly.

If $r = n+1$, integrating (4.3.5) from a to b and recalling (4.1.5), we obtain (4.4.1). Now let $1 \le r \le n$ and let

$$m_k^* = \begin{cases} m_r - 1, & k = r, \\ m_k, & k \ne r, \quad 1 \le k \le n. \end{cases}$$

It is easy to see that $s_r(\mathbf{m}^*) = -s_r(\mathbf{m}) = -s_r$ and

$$S_n(\mathbf{x},\mathbf{m}^*;x) = [\text{sgn } (x-x_r)]S_n(\mathbf{x},\mathbf{m};x) = [\text{sgn } (x-x_r)]S_n(x).$$

Denote by $A_{kj}^* \in \mathbf{P}_{N-1}$ the fundamental polynomials for Hermite interpolation with multiplicities m_k^* on the vector of nodes $\mathbf{x}$. Applying Theorem 4.3.1 to A_{kj}^* inequalities (4.3.5) and (4.3.6) read

$$(-s_r)[\text{sgn } (x-x_r)]S_n(x) \left[g(x) - \sum_{k=1}^{r-1} \sum_{j=0}^{m_k^*-1} g^{(j)}(x_k)A_{kj}^*(x) \right] \ge 0, \quad x \in \langle a, x_r \rangle, \tag{4.4.2}$$

and

$$(-s_r)[\text{sgn } (x-x_r)]S_n(x) \sum_{k=1}^{r-1} \sum_{j=0}^{m_k^*-1} g^{(j)}(x_k)A_{kj}^*(x) \ge 0, \quad x \in \langle x_r, b \rangle, \tag{4.4.3}$$

respectively. Integrating (4.4.2) from a to x_r and (4.4.3) from x_r to b, we obtain

$$s_r \int_a^{x_r} g(x)S_n(x)d\mu(x) \ge s_r \sum_{k=1}^{r-1} \sum_{j=0}^{m_k^*-1} g^{(j)}(x_k) \int_a^{x_r} A_{kj}^*(x)S_n(x)d\mu(x)$$

and

$$0 \ge s_r \sum_{k=1}^{r-1} \sum_{j=0}^{m_k^*-1} g^{(j)}(x_k) \int_{x_r}^{b} A_{kj}^*(x)S_n(x)d\mu(x),$$

respectively. Summing these two inequalities yields

$$s_r \int_a^{x_r} g(x)S_n(x)d\mu(x) \ge s_r \sum_{k=1}^{r-1} \sum_{j=0}^{m_k^*-1} g^{(j)}(x_k) \int_a^{b} A_{kj}^*(x)S_n(x)d\mu(x). \tag{4.4.4}$$

Substituting A^*_{kj} into (4.1.3), we get

$$\int_a^b A^*_{kj}(x)S_n(x)d\mu(x) = \lambda_{kj}, \quad j=0,1,\ldots,m_k-2, \quad k=1,2,\ldots,n,$$

and

$$\int_a^b A^*_{k,m^*_k-1}(x)S_n(x)d\mu(x) = 0, \ k \neq r.$$

Then the second inequality of (4.4.1) follows from (4.4.4). □

For the convenience of use we formulate a special case of Theorem 4.4.1 when $g=1$.

Corollary 4.4.1. *Let the assumptions of Theorem 4.1.2 prevail and let* $1 \le r \le n+1$. *Then with the notations of Theorem 4.1.1*

$$s_r \int_a^{x_{r-1}} S_n(x)d\mu(x) \le s_r \sum_{k=1}^{r-1} \lambda_{k0} \le s_r \int_a^{x_r} S_n(x)d\mu(x). \tag{4.4.5}$$

Corollary 4.4.2. *Let the assumptions of Theorem 4.1.2 prevail. Let* $1 \le r \le q \le n$ *and let an* $(N+1)$*th continuously differentiable function g satisfy (4.3.4) with* $r = q+1$. *Then with the notations of Theorem 4.1.1*

$$\begin{aligned} &\left| \sum_{k=r}^{q} \sum_{j=0}^{m_k-2} \lambda_{kj} g^{(j)}(x_k) - \int_{x_r}^{x_q} g(x)S_n(x)d\mu(x) \right| \\ \le &\int_{x_{r-1}}^{x_r} g(x)d\mu(x) + \int_{x_q}^{x_{q+1}} g(x)d\mu(x), \end{aligned} \tag{4.4.6}$$

if the integrals exist.

Proof. Subtracting from each term of (4.4.1) its last one and recalling (4.3.9), we obtain

$$-\int_{x_{r-1}}^{x_r} g(x)d\mu(x) \le s_r \left[\sum_{k=1}^{r-1} \sum_{j=0}^{m_k-2} \lambda_{kj} g^{(j)}(x_k) - \int_a^{x_r} g(x)S_n(x)d\mu(x) \right] \le 0. \tag{4.4.7}$$

Similarly, subtracting from each term of (4.4.1) its first one and recalling (4.3.9) (after replacing r by $q+1$), we get

$$0 \le s_{q+1} \left[\sum_{k=1}^{q} \sum_{j=0}^{m_k-2} \lambda_{kj} g^{(j)}(x_k) - \int_a^{x_q} g(x)S_n(x)d\mu(x) \right] \le \int_{x_q}^{x_{q+1}} g(x)d\mu(x). \tag{4.4.8}$$

Then inequality (4.4.6) follows from (4.4.7) and (4.4.8). □

As an application of Theorems 4.1.3 and 4.4.1 we formulate a useful result, which is needed later.

Theorem 4.4.2. *Let* $-\infty < a < b < +\infty$ *and let the assumptions of Theorem 4.1.2 prevail. We have the estimation*

$$|\lambda_{kj}| \le \frac{22}{j!} d_k^j \int_{x_{k-1}}^{x_{k+1}} d\mu(x), \quad 0 \le j \le m_k - 2, \quad 1 \le k \le n. \tag{4.4.9}$$

Proof. Applying (4.4.6) with $r = q = k$ to $g(x) = 1$ and $g(x) = x - a$, we have

$$|\lambda_{k0}| \le \int_{x_{k-1}}^{x_{k+1}} d\mu(x) \tag{4.4.10}$$

and

$$|(x_k - a)\lambda_{k0} + \lambda_{k1}| \le \int_{x_{k-1}}^{x_{k+1}} (x-a)d\mu(x), \tag{4.4.11}$$

respectively. It follows from (4.4.11) that

$$|\lambda_{k1}| \le (x_k - a)|\lambda_{k0}| + \int_{x_{k-1}}^{x_{k+1}} (x-a)d\mu(x). \tag{4.4.12}$$

Meanwhile according to (4.1.9) we obtain

$$|\lambda_{kj}| \le \begin{cases} \frac{1}{j!} d_k^j |\lambda_{k0}|, & m_k \in \mathbf{N}_2, \\ \frac{1}{j!} d_k^{j-1} |\lambda_{k1}|, & m_k \in \mathbf{N}_1. \end{cases} \tag{4.4.13}$$

Hence by (4.4.10) it suffices to show (4.4.9) for the case when $j = 1$ and $m_k \in \mathbf{N}_1$. To this end we distinguish two cases.

Case 1. $d_k > (x_k - a)/2$. In this case since

$$x_{k+1} - a = x_k - a + x_{k+1} - x_k \le x_k - a + d_k, \tag{4.4.14}$$

by (4.4.12) and (4.4.10) we have

$$\begin{aligned} |\lambda_{k1}| &\le (x_k - a)|\lambda_{k0}| + (x_{k+1} - a) \int_{x_{k-1}}^{x_{k+1}} d\mu(x) \\ &\le (x_k - a)|\lambda_{k0}| + (x_k - a + d_k) \int_{x_{k-1}}^{x_{k+1}} d\mu(x) \\ &\le 2d_k|\lambda_{k0}| + 3d_k \int_{x_{k-1}}^{x_{k+1}} d\mu(x) \\ &\le 5d_k \int_{x_{k-1}}^{x_{k+1}} d\mu(x). \end{aligned}$$

Case 2. $d_k \le (x_k - a)/2$. In this case let $M = [(x_k - a)/(2d_k)]$. Clearly

$$1 \le M \le \frac{x_k - a}{2d_k} < M + 1.$$

This implies

$$Md_k \le \frac{x_k - a}{2} \tag{4.4.15}$$

and

$$Md_k \ge \frac{1}{2}(M+1)d_k > \frac{x_k - a}{4}. \tag{4.4.16}$$

Further, applying (4.4.6) with $r = q = k$ to $g(x) = (x-a)^M$ gives

$$\left| \sum_{j=0}^{m_k - 2} \lambda_{kj} g^{(j)}(x_k) \right| \le \int_{x_{k-1}}^{x_{k+1}} (x-a)^M d\mu(x). \tag{4.4.17}$$

By means of (4.4.13) and (4.4.15)

$$\begin{aligned}\left|\sum_{j=2}^{m_k-2}\lambda_{kj}g^{(j)}(x_k)\right| &\le \sum_{j=2}^{\min\{m_k-2,M\}} M^j(x_k-a)^{M-j}|\lambda_{kj}|\\ &\le M(x_k-a)^{M-1}|\lambda_{k1}|\sum_{j=2}^{\min\{m_k-2,M\}}\frac{1}{j!}\left(\frac{Md_k}{x_k-a}\right)^{j-1}\\ &\le M(x_k-a)^{M-1}|\lambda_{k1}|\sum_{j=2}^{\min\{m_k-2,M\}}\frac{1}{j!2^{j-1}}\\ &\le \frac{1}{2}M(x_k-a)^{M-1}|\lambda_{k1}|,\end{aligned}$$

from which inequality (4.4.17) yields

$$\begin{aligned}\int_{x_{k-1}}^{x_{k+1}}(x-a)^M d\mu(x) &\ge |\lambda_{k1}g'(x_k)|-|\lambda_{k0}g(x_k)|-\left|\sum_{j=2}^{m_k-2}\lambda_{kj}g^{(j)}(x_k)\right|\\ &\ge \frac{1}{2}M(x_k-a)^{M-1}|\lambda_{k1}|-(x_k-a)^M|\lambda_{k0}|. \qquad (4.4.18)\end{aligned}$$

Thus by (4.4.18), (4.4.10), and (4.4.14)

$$\begin{aligned}|\lambda_{k1}| &\le \frac{2(x_k-a)}{M}\left[|\lambda_{k0}|+\frac{1}{(x_k-a)^M}\int_{x_{k-1}}^{x_{k+1}}(x-a)^M d\mu(x)\right]\\ &\le \frac{2(x_k-a)}{M}\left[1+\left(\frac{x_{k+1}-a}{x_k-a}\right)^M\right]\int_{x_{k-1}}^{x_{k+1}}d\mu(x)\\ &\le \frac{2(x_k-a)}{M}\left[1+\left(1+\frac{d_k}{x_k-a}\right)^M\right]\int_{x_{k-1}}^{x_{k+1}}d\mu(x).\end{aligned}$$

Then using a well known inequality $1+x\le e^x$, and applying (4.4.15) and (4.4.16), we obtain

$$\begin{aligned}|\lambda_{k1}| &\le \frac{2(x_k-a)}{M}\left[1+e^{Md_k/(x_k-a)}\right]\int_{x_{k-1}}^{x_{k+1}}d\mu(x)\\ &\le 8(1+e^{1/2})d_k\int_{x_{k-1}}^{x_{k+1}}d\mu(x)\le 22d_k\int_{x_{k-1}}^{x_{k+1}}d\mu(x). \quad \square\end{aligned}$$

Here we derive a theorem, which is valid for general Hermitian quadrature formulas and needs a weaker condition (4.4.19) than (4.3.4).

Theorem 4.4.3. *Let $d\mu$ be a measure supported on $\langle a,b\rangle$ and let an $(N+1)$th continuously differentiable function g satisfy*

$$g(x)\ge 0,\quad g^{(N+1)}(x)\ge 0,\quad x\in\langle a,b\rangle. \qquad (4.4.19)$$

Then the inequality

$$\sum_{k=1}^{n}\sum_{j=0}^{m_k-1}\lambda_{kj}g^{(j)}(x_k)\le\int_a^b g(x)S_n(x)d\mu(x) \qquad (4.4.20)$$

holds, if the integral on the right-hand side exists, where λ_{kj} are given by (4.1.2).

In particular, if the vector $\mathbf{x}$ *is the solution of the extremal problem (3.1.1), then inequality (4.4.20) with (4.1.5) holds.*

Proof. Following the idea of [24, p. 92] let $\varepsilon > 0$ be arbitrary and let

$$g_\varepsilon(x) = g(x) + \varepsilon x^{N+1}.$$

Assume that $P \in \mathbf{P}_N$ is the Hermite interpolation polynomial of g_ε at nodes $x_1, \ldots, x_n$ with the corresponding multiplicity $m_1, \ldots, m_n$. For simplicity write $G = g_\varepsilon - P$. Then

$$G^{(j)}(x_k) = 0, \quad j = 0, 1, \ldots, m_k - 1, \quad k = 1, 2, \ldots, n. \tag{4.4.21}$$

We claim that the function $S_n G$ does not change sign in (a,b). In fact, suppose to the contrary that the function $S_n G$ would change sign at $z \in (a,b)$. If $z \notin \{x_1, \ldots, x_n\}$ then $G(z) = 0$. If $z = x_r$ holds for some r, $1 \le r \le n$, then the inequality

$$[S_n(x_r - \delta)G(x_r - \delta)][S_n(x_r + \delta)G(x_r + \delta)] < 0$$

would hold for all small $\delta > 0$, which by (4.1.1) implies that the inequality

$$(-1)^{m_r} G(x_r - \delta)G(x_r + \delta) < 0 \tag{4.4.22}$$

would hold for all small $\delta > 0$. Using (4.4.21) we obtain $G^{(m_r)}(x_r) = 0$. Again using (4.4.21) we conclude $Z'(G) \ge \sum_{k=1}^{n} m_k + 1 = N + 2$. Applying Rolle theorem $N+1$ times we get $Z'(g_\varepsilon^{(N+1)}) = Z'(G^{(N+1)}) \ge 1$, a contradiction to (4.4.19). This proves our claim.

Moreover, by interpolation theory we observe that

$$G(x) = (x - x_1)^{m_1} \cdots (x - x_n)^{m_n} G^{(N+1)}(\xi)/(N+1)!, \quad x, \xi \in (a,b)$$

and hence $S_n(x)G(x) \ge 0$, $\quad x \in \langle a,b \rangle$. Integrating this inequality from a to b and letting $\varepsilon \to 0$ yields (4.4.20) with (4.1.2). Further, if relation (3.1.1) is true then by Theorem 4.1.1 relation (4.1.5) holds. □

4.5 Convergence of Gaussian Quadrature Formulas on Finite Intervals

In this section we discuss convergence of Gausas quadrature formula on a finite interval $[a,b]$.

Let $n \in \mathbf{N}$ $(n \ge 2)$. Assume that $m_{kn} \in \mathbf{N}$, $1 \le k \le n$. Denote $\mathbf{m}_n = (m_{1n}, \ldots, m_{nn})$, $M_n = \max_{1 \le k \le n} m_{kn}$, $m = \sup_n M_n$, and $N_n = \sum_{k=1}^{n} m_{kn} - 1$. Assume that $M_n \ge 2$ and $m < +\infty$.

If the assumptions of Theorem 4.1.1 is valid then for each n there exists a vector of nodes $\mathbf{x}_n$ satisfying (4.2.2) for which the generalized Gaussian quadrature formula

$$\int_a^b f(x) S_n(x) d\mu(x) = \sum_{k=1}^{n} \sum_{j=0}^{m_{kn}-2} \lambda_{kjn} f^{(j)}(x_{kn}) \tag{4.5.1}$$

is exact for all $f \in \mathbf{P}_{N_n}$, where

$$S_n(x) = S_n(\mathbf{x}_n, \mathbf{m}_n; x) = S_n(d\mu, \mathbf{m}_n; x) \tag{4.5.2}$$

and

$$\lambda_{kjn} = \lambda_{kjn}(d\mu, \mathbf{m}_n), \quad j = 0, 1, \ldots, m_{kn} - 2, \quad k = 1, 2, \ldots, n. \tag{4.5.3}$$

Let the sequence of nonnegative integers $\{r_{kn}\}$ satisfy

$$r_{kn} \le m_{kn} - 2, \quad 1 \le k \le n. \tag{4.5.4}$$

Put

$$r = \sup_{1 \le k \le n,\ n \ge 2} r_{kn}$$

and consider the *truncated generalized Gaussian quadrature formula*

$$Q_{nr}(d\mu; f) = Q_{nr}(d\mu, \mathbf{m}_n; f) = \sum_{k=1}^{n} \sum_{j=0}^{r_{kn}} \lambda_{kjn} f^{(j)}(x_{kn}), \quad f \in C^r[-1, 1]. \tag{4.5.5}$$

We shall establish the conditions of convergence and the rate of convergence for $Q_{nr}(d\mu; f)$ to f.

Lemma 4.5.1. *Let $-a = b = 1$ and let the assumptions of Theorem 4.1.2 prevail. Then the relation*

$$\lim_{n \to \infty} \left[\sum_{k=1}^{n} \sum_{j=0}^{m_{kn}-2} \lambda_{kjn} f^{(j)}(x_{kn}) - \int_{-1}^{1} f(x) S_n(x) d\mu(x) \right] = 0 \tag{4.5.6}$$

holds for all $f \in C^{m-2}[-1, 1]$.

Proof. Since (4.5.1) is exact for every polynomial $f \in \mathbf{P}_{N_n}$, by the well known Banach-Steinhaus theorem it suffices to show

$$\sum_{k=1}^{n} \sum_{j=0}^{m_k-2} |\lambda_{kj}| \le c < \infty. \tag{4.5.7}$$

This is indeed the case, because using (4.4.9) we have

$$\begin{aligned} \sum_{k=1}^{n} \sum_{j=0}^{m_k-2} |\lambda_{kj}| &\le 22 \sum_{k=1}^{n} \sum_{j=0}^{m_k-2} \frac{1}{j!} d_k^j \int_{x_{k-1}}^{x_{k+1}} d\mu(x) \le 22 \sum_{k=1}^{n} \sum_{j=0}^{\infty} \frac{2^j}{j!} \int_{x_{k-1}}^{x_{k+1}} d\mu(x) \\ &= 22 \sum_{j=0}^{\infty} \frac{2^j}{j!} \sum_{k=1}^{n} \int_{x_{k-1}}^{x_{k+1}} d\mu(x) \le 44 \int_{-1}^{1} d\mu(x) \sum_{j=0}^{\infty} \frac{2^j}{j!} \end{aligned}$$

and applying the Stering formula

$$j! = (2\pi j)^{1/2} (j/e)^j (1 + \varepsilon_j), \quad 0 < \varepsilon_j < e^{1/(4j)} - 1,$$

we obtain

$$\sum_{j=0}^{\infty} \frac{2^j}{j!} \le 1 + \sum_{j=1}^{\infty} (2\pi j)^{-1/2} (2e/j)^j \le c < \infty. \tag{4.5.8}$$

This prove (4.5.7) and the lemma. □

Lemma 4.5.2. *Under the assumptions of Lemma 4.5.1 if* $\int_c^d d\mu(x) > 0$ $([c,d] \subset [-1,1])$, *then for sufficiently large n the interval* $[c,d]$ *contains at least one zero* x_{kn}.

Proof. Suppose to the contrary that there would exist a subsequence $\{n_i\}_{i=1}^{\infty}$ with $n_1 < n_2 < \cdots$, such that the interval $[c,d]$ contains no zeros x_{k,n_i}. Then $S_{n_i}(x)$ does not change sign in $[c,d]$. Choose $f \in C^{m-2}[-1,1]$ so that

$$f(x)\begin{cases} > 0, & x \in (c,d), \\ = 0, & x \notin (c,d). \end{cases}$$

Hence

$$\begin{aligned} 0 &< \int_{-1}^{1} f(x)d\mu(x) = S_{n_i}(c+0)\int_{-1}^{1} f(x)S_{n_i}(x)d\mu(x) \\ &= S_{n_i}(c+0)\left[\int_{-1}^{1} f(x)S_{n_i}(x)d\mu(x) - \sum_{k=1}^{n}\sum_{j=0}^{m_{k,n_i}-2} \lambda_{k,j,n_i} f^{(j)}(x_{k,n_i})\right]. \end{aligned}$$

By Lemma 4.5.1 the last term tends to zero as $i \to \infty$, a contradiction. □

Lemma 4.5.3. *Let the assumptions of Lemma 4.5.1 prevail. If*

$$\mu \in C[-1,1], \tag{4.5.9}$$

then

$$\lim_{n\to\infty} \max_{0\le k\le n} \int_{x_{kn}}^{x_{k+1,n}} d\mu(x) = 0 \tag{4.5.10}$$

and

$$\lim_{n\to\infty} \max_{1\le k\le n} |\lambda_{k0n}| = 0. \tag{4.5.11}$$

Conversely, if all the m_{kn} *are even, then relation (4.5.11) implies (4.5.9).*

Proof. To prove the first part of the lemma by (4.4.9) it is enough to establish relation (4.5.10). To this end let

$$\int_{x_{r_n,n}}^{x_{r_n+1,n}} d\mu(x) = \max_{0\le k\le n} \int_{x_{kn}}^{x_{k+1,n}} d\mu(x), \quad 0 \le r_n \le n.$$

We may assume, passing to a subsequence if necessary, that as $n \to \infty$

$$x_{r_n,n} \to y_1, \quad x_{r_n+1,n} \to y_2.$$

It suffices to show $\int_{y_1}^{y_2} d\mu(x) = 0$. Suppose not and let $\int_{y_1}^{y_2} d\mu(x) > 0$. Then

$$\int_{y_1+\varepsilon}^{y_2-\varepsilon} d\mu(x) > 0$$

holds for some $\varepsilon > 0$. Meanwhile by definition each interval $(x_{r_n,n}, x_{r_n+1,n})$ contains no zeros x_{kn}. So for n large enough the interval $[y_1+\varepsilon, y_2-\varepsilon]$ contains no zeros x_{kn}, contradicting Lemma 4.5.2.

Conversely, assume that (4.5.11) is true. Let a point $y \in (-1,1)$ be arbitrary. Assume that $y \in (x_{r_n-1,n}, x_{r_n+1,n})$, $1 \le r_n \le n$. Then by (4.4.5) with $\lambda_{0,0,n} = \lambda_{n+1,0,n} = 0$

$$\int_{x_{r_n-1,n}}^{x_{r_n+1,n}} d\mu(x) \le \lambda_{r_n-1,0,n} + \lambda_{r_n,0,n} + \lambda_{r_n+1,0,n} \le 3 \max_{1\le k\le n} |\lambda_{k0n}|.$$

Thus

$$\mu(y+0) - \mu(y-0) \le \mu(x_{r_n+1,n}) - \mu(x_{r_n-1,n}) \le 3 \max_{1\le k\le n} |\lambda_{k0n}| \longrightarrow 0, \quad n \longrightarrow \infty.$$

This proves continuity of $\mu(x)$ at $x = y$. Similarly we can prove continuity of $\mu(x)$ at $x = -1$ and $x = 1$. □

Lemma 4.5.4. *Let the assumptions of Lemma 4.5.1 prevail and let $\mu \in C[-1,1]$. Then*

$$\lim_{n\to\infty} \sum_{k=1}^{n} \sum_{j=1}^{m_{kn}-2} |\lambda_{kjn}| = 0. \tag{4.5.12}$$

Proof. By virtue of (4.4.9), (4.5.8), and (4.5.10)

$$\begin{aligned}
\sum_{k=1}^{n} \sum_{j=1}^{m_k-2} |\lambda_{kj}| &\le 22 \sum_{k=1}^{n} \sum_{j=1}^{m_k-2} \frac{1}{j!} d_k^j \int_{x_{k-1,n}}^{x_{k+1,n}} d\mu(x) \\
&\le 22 \left[\max_{1\le k\le n} \int_{x_{k-1,n}}^{x_{k+1,n}} d\mu(x) \right] \sum_{k=1}^{n} \sum_{j=1}^{\infty} \frac{1}{j!} d_k^j \\
&\le 22 \left[\max_{1\le k\le n} \int_{x_{k-1,n}}^{x_{k+1,n}} d\mu(x) \right] \sum_{j=1}^{\infty} \frac{2^{j-1}}{j!} \sum_{k=1}^{n} d_k \\
&\le 44 \left[\max_{1\le k\le n} \int_{x_{k-1,n}}^{x_{k+1,n}} d\mu(x) \right] \sum_{j=1}^{\infty} \frac{2^j}{j!} \longrightarrow 0,
\end{aligned}$$

as $n \longrightarrow \infty$. □

Theorem 4.5.1. *Let the assumptions of Lemma 4.5.1 prevail and let $\mu \in C[-1,1]$. Then*

$$\lim_{n\to\infty} \left[Q_{nr}(d\mu; f) - \int_{-1}^{1} f(x) S_n(x) d\mu(x) \right] = 0, \quad f \in C^r[-1,1]. \tag{4.5.13}$$

In particular,

$$\lim_{n\to\infty} \left[\sum_{k=1}^{n} \lambda_{k0n} f(x_{kn}) - \int_{-1}^{1} f(x) S_n(x) d\mu(x) \right] = 0, \quad f \in C[-1,1] \tag{4.5.14}$$

and

$$\lim_{n\to\infty} \sum_{k=1}^{n} \sum_{j=1}^{r_{kn}} \lambda_{kjn} f^{(j)}(x_{kn}) = 0, \quad f \in C^r[-1,1]. \tag{4.5.15}$$

Proof. Clearly, the last two relations follow from the first. Now let us prove (4.5.13).

Let $N \in \mathbf{N}$ be fixed and let $P_N \in \mathbf{P}_N$ satisfy (1.5.2) and (1.5.3) with $n = N$ and $p = r$. Clearly for $N_n \geq N$ we have

$$\begin{aligned}\int_{-1}^{1} P_N(x)S_n(x)d\mu(x) &= \sum_{k=1}^{n}\sum_{j=0}^{m_k-2} \lambda_{kj}P_N^{(j)}(x_k)\\ &= Q_{nr}(d\mu;P_N) + \sum_{k=1}^{n}\sum_{j=r+1}^{m_k-2} \lambda_{kj}P_N^{(j)}(x_k).\end{aligned}$$

Hence

$$\begin{aligned}&\left|Q_{nr}(d\mu;f) - \int_{-1}^{1} f(x)S_n(x)d\mu(x)\right|\\ =& \left|\int_{-1}^{1}[P_N(x)-f(x)]S_n(x)d\mu(x) + Q_{nr}(d\mu;f-P_N) - \sum_{k=1}^{n}\sum_{j=r+1}^{m_k-2} \lambda_{kj}P_N^{(j)}(x_k)\right|\\ \leq& \left|\int_{-1}^{1}[f(x)-P_N(x)]S_n(x)d\mu(x)\right| + |Q_{nr}(d\mu;f-P_N)|\\ &+\left|\sum_{k=1}^{n}\sum_{j=r+1}^{m_k-2} \lambda_{kj}P_N^{(j)}(x_k)\right|\\ :=& S_1+S_2+S_3.\end{aligned} \tag{4.5.16}$$

By virtue of (1.5.2) we see

$$S_1 \leq c\int_{-1}^{1} \Delta_N(x)^r\omega(f^{(r)};\Delta_N(x))d\mu(x) \leq cN^{-r}\omega(f^{(r)};1/N)$$

and

$$\begin{aligned}S_2 &\leq \sum_{k=1}^{n}\sum_{j\leq r} |\lambda_{kj}|\cdot|f^{(j)}(x_k) - P_N^{(j)}(x_k)|\\ &\leq cN^{-r}\omega(f^{(r)};1/N)\sum_{k=1}^{n}\sum_{j\leq r}|\lambda_{kj}|\Delta_N(x_k)^{-j}.\end{aligned}$$

By means of (1.5.3)

$$S_3 \leq cN^{-r}\omega(f^{(r)};1/N)\sum_{k=1}^{n}\sum_{j=r+1}^{\min\{m_k-2,N\}}|\lambda_{kj}|\Delta_N(x_k)^{-j}.$$

Substituting S_1, S_2, and S_3 into (4.5.16), we get

$$\begin{aligned}&\left|Q_{nr}(d\mu;f) - \int_{-1}^{1} f(x)S_n(x)d\mu(x)\right|\\ \leq& cN^{-r}\omega(f^{(r)};1/N)\sum_{k=1}^{n}\sum_{j=0}^{\min\{m_k-2,N\}}|\lambda_{kj}|\Delta_N(x_k)^{-j}.\end{aligned} \tag{4.5.17}$$

But the sum of (4.5.17) is equal to

$$\sum_{k=1}^{n}\lambda_{k0}+\sum_{k=1}^{n}\sum_{j=1}^{\min\{m_k-2,N\}}|\lambda_{kj}|\Delta_N(x_k)^{-j}\le\int_{-1}^{1}d\mu(x)+N^{2N}\sum_{k=1}^{n}\sum_{j=1}^{\min\{m_k-2,N\}}|\lambda_{kj}|.$$

By (4.5.17) and (4.5.12) we have

$$\lim_{n\to\infty}\left|Q_{nr}(d\mu;f)-\int_{-1}^{1}f(x)S_n(x)d\mu(x)\right|\le\left[c\int_{-1}^{1}d\mu(x)\right]N^{-r}\omega(f^{(r)};1/N).$$

Then as $N\to\infty$ we get (4.5.13). □

The following general results concern the rate of convergence for $Q_{nr}(d\mu;f)$ to f.

Theorem 4.5.2. *Let the assumptions of Lemma 4.5.1 prevail and let* $\mu\in C[-1,1]$. *If* $f\in C^r[-1,1]$ *and*

$$r_{kn}=\min\{r,m_{kn}-2\},\quad 1\le k\le n,\quad n\in\mathbf{N},\tag{4.5.18}$$

then

$$\left|Q_{nr}(d\mu;f)-\int_{-1}^{1}f(x)S_n(x)d\mu(x)\right|$$
$$\le\ cn^{-r}\omega(f^{(r)};1/n)\sum_{k=1}^{n}\left[\int_{x_{k-1,n}}^{x_{k+1,n}}d\mu(x)\sum_{j=0}^{m_{kn}-2}d_{kn}^{j}\Delta_n(x_{kn})^{-j}\right].\tag{4.5.19}$$

Proof. Inequality (4.5.19) follows from (4.5.17) with $N=n$ and (4.4.9). □

Let

$$x_{kn}=\cos\theta_{kn},\ 0\le\theta_{kn}\le\pi,\ k=0,1,\ldots,n+1.\tag{4.5.20}$$

Lemma 4.5.5. *The following statements are equivalent.*

(a) We have

$$\theta_{kn}-\theta_{k+1,n}\le\frac{c_1}{n},\quad k=0,1,\ldots,n.\tag{4.5.21}$$

(b) We have

$$x_{k+1,n}-x_{kn}\le c_2\Delta_n(x_{kn}),\quad k=0,1,\ldots,n.\tag{4.5.22}$$

In addition, the following statements are equivalent.

(c) We have

$$\theta_{kn}-\theta_{k+1,n}\ge\frac{c_1}{n},\quad k=0,1,\ldots,n.\tag{4.5.23}$$

(d) We have

$$x_{k+1,n}-x_{kn}\ge c_2\Delta_n(x_{kn}),\quad k=0,1,\ldots,n.\tag{4.5.24}$$

Moreover, if inequality (4.5.21) is true then

$$d_{kn}\le c\Delta_n(x_{kn}),\quad k=1,2,\ldots,n;\tag{4.5.25}$$

if inequality (4.5.23) is true then

$$d_{kn}\ge c\Delta_n(x_{kn}),\quad k=1,2,\ldots,n.\tag{4.5.26}$$

Proof. Write

$$\begin{aligned} x_{k+1}-x_k &= \cos\theta_{k+1}-\cos\theta_k \\ &= 2\sin\frac{\theta_k-\theta_{k+1}}{2}\sin\frac{\theta_k+\theta_{k+1}}{2}. \end{aligned} \tag{4.5.27}$$

We have the estimations from above

$$\sin\frac{\theta_k-\theta_{k+1}}{2}\le\frac{\theta_k-\theta_{k+1}}{2}$$

and

$$\begin{aligned} \sin\frac{\theta_k+\theta_{k+1}}{2} &= \sin\left(\theta_k-\frac{\theta_k-\theta_{k+1}}{2}\right) \\ &= \sin\theta_k\cos\frac{\theta_k-\theta_{k+1}}{2}-\cos\theta_k\sin\frac{\theta_k-\theta_{k+1}}{2} \\ &\le \sin\theta_k+\theta_k-\theta_{k+1}. \end{aligned}$$

Inserting these estimations into (4.5.27), we obtain

$$x_{k+1}-x_k\le(\theta_k-\theta_{k+1})(\sin\theta_k+\theta_k-\theta_{k+1}). \tag{4.5.28}$$

On the other hand, we have the estimations from below

$$\sin\frac{\theta_k-\theta_{k+1}}{2}\ge\frac{1}{\pi}\left(\theta_k-\theta_{k+1}\right),$$

$$\begin{aligned} \sin\frac{\theta_k+\theta_{k+1}}{2} &\ge \frac{1}{\pi}\left(\theta_k+\theta_{k+1}\right)\ge\frac{1}{2\pi}\left(\theta_k+\theta_k-\theta_{k+1}\right) \\ &\ge \frac{1}{2\pi}\left(\sin\theta_k+\theta_k-\theta_{k+1}\right), \quad \theta_k+\theta_{k+1}\le\pi, \end{aligned}$$

and

$$\begin{aligned} \sin\frac{\theta_k+\theta_{k+1}}{2} &\ge \frac{1}{\pi}\left[(\pi-\theta_k)+(\pi-\theta_{k+1})\right]\ge\frac{1}{\pi}\left[\sin\left(\pi-\theta_k\right)+\left(\theta_k-\theta_{k+1}\right)\right] \\ &= \frac{1}{\pi}\left(\sin\theta_k+\theta_k-\theta_{k+1}\right), \quad \theta_k+\theta_{k+1}>\pi. \end{aligned}$$

Inserting these estimations into (4.5.27), we obtain

$$x_{k+1}-x_k\ge\frac{1}{\pi^2}\left(\theta_k-\theta_{k+1}\right)\left(\sin\theta_k+\theta_k-\theta_{k+1}\right), \tag{4.5.29}$$

Then the equivalence of Statements (a) and (b) as well as of Statements (c) and (d) follows from (4.5.28) and (4.5.29).

To prove the last part of the lemma by the same arguments as above we can obtain alternative estimations

$$x_k-x_{k-1}\le(\theta_{k-1}-\theta_k)(\sin\theta_k+\theta_{k-1}-\theta_k)$$

and

$$x_k - x_{k-1} \geq \frac{1}{\pi^2}(\theta_{k-1} - \theta_k)(\sin\theta_k + \theta_{k-1} - \theta_k),$$

from which by (4.5.21), (4.5.23), (4.5.28), and (4.5.29) we obtain (4.5.25) and (4.5.26), respectively. □

Lemma 4.5.6. *The following statements are equivalent.*

(a) We have

$$\frac{c_1}{n} \leq \theta_{kn} - \theta_{k+1,n} \leq \frac{c_2}{n}, \quad k = 0, 1, \ldots, n. \tag{4.5.30}$$

(b) We have

$$c_3\Delta_n(x_{kn}) \leq x_{k+1,n} - x_{kn} \leq c_4\Delta_n(x_{kn}), \quad k = 0, 1, \ldots, n. \tag{4.5.31}$$

(c) We have

$$\begin{cases} \frac{c_5(1-x_{1n}^2)^{1/2}}{n} \leq x_{1n} - x_{0n} \leq \frac{c_6(1-x_{1n}^2)^{1/2}}{n}, \\ \frac{c_5(1-x_{kn}^2)^{1/2}}{n} \leq x_{k+1,n} - x_{kn} \leq \frac{c_6(1-x_{kn}^2)^{1/2}}{n}, \quad k = 1, 2, \ldots, n. \end{cases} \tag{4.5.32}$$

Proof. (a)$\Longleftrightarrow$(b) Apply Lemma 4.5.5.

(b)$\Longleftrightarrow$(c) We observe that the relation

$$x_1 - x_0 = 1 + x_1 \sim \frac{(1-x_1^2)^{1/2}}{n} \quad \left(\text{or } x_{n+1} - x_n = 1 - x_n \sim \frac{(1-x_n^2)^{1/2}}{n}\right)$$

means that $x_1 - x_0 \sim 1/n^2$ (or $x_{n+1} - x_n \sim 1/n^2$). This shows that relation (4.5.31) with $k = 0$ (or $k = n$) is equivalent to relation (4.5.32) with $k = 0$ (or $k = n$). Meanwhile for each k, $1 \leq k \leq n-1$,

$$\Delta_n(x_k) \sim \frac{(1-x_k^2)^{1/2}}{n},$$

because $(1-x_k^2)^{1/2} \geq \frac{c}{n}$. This proves the equivalence of Statements (b) and (c). □

Theorem 4.5.3. *Let the assumptions of Lemma 4.5.1 prevail. Let $\mu \in C[-1,1]$ and $f \in C^r[-1,1]$. If (4.5.18) and (4.5.21) hold, then*

$$\left| Q_{nr}(d\mu; f) - \int_{-1}^{1} f(x)S_n(x)d\mu(x) \right| \leq cn^{-r}\omega(f^{(r)}; 1/n). \tag{4.5.33}$$

Proof. Recalling

$$\sum_{k=1}^{n} \int_{x_{k-1}}^{x_{k+1}} d\mu(x) \leq 2\int_{-1}^{1} d\mu(x),$$

relation (4.5.33) follows from (4.5.19) and (4.5.22). □

4.6 Convergence of Gaussian Quadrature Formulas on Infinite Intervals

In this section we deal with convergence of Gaussian quadrature formulas on **R**.

Given a number $T > 0$ let

$$D_{n,T} = \max_{|x_{kn}|\le T} \overline{d}_{kn}, \tag{4.6.1}$$

$$R_{n,T} = \max_{|x_{kn}|>T} \frac{\overline{d}_{kn}}{|x_{kn}|}, \tag{4.6.2}$$

and

$$D_n = \max_{1\le k\le n} \overline{d}_{kn}. \tag{4.6.3}$$

Theorem 4.6.1. *Let $d\mu$ be a measure supported on $\mathbf{R}$ and let $d\mu \in \mathbf{E}$. Let the assumptions of Theorem 4.1.2 prevail and let all the m_{kn} be even. Assume that for some fixed number $T \ge 1$*

$$\lim_{n\to\infty} D_{n,T} = 0 \tag{4.6.4}$$

and

$$\lim_{n\to\infty} R_{n,T} = 0. \tag{4.6.5}$$

If $f \in \mathbf{F}(d\mu)$ satisfies (4.2.9), then

$$\lim_{n\to\infty} Q_n(d\mu; f) = \int_{\mathbf{R}} f(x)d\mu(x), \tag{4.6.6}$$

where

$$Q_n(d\mu; f) = Q_{n0}(d\mu; f) = \sum_{k=1}^{n} \lambda_{k0n} f(x_{kn}).$$

Proof. First we observe that by Theorem 4.1.2 $\lambda_{k0} > 0,\ k = 1,2,\ldots,n$. If we can prove that relation (4.6.6) holds for every polynomial f then we conclude that the positive quadrature procedure $\{Q_n(d\mu;\cdot)\}$ belongs to the measure $d\mu$ and by Theorem 4.2.2 prove the theorem.

Let us prove that relation (4.6.6) holds for every monomial $f(x) = x^p,\ p \in \mathbf{N}_0$. Clearly, this relation holds for $f(x) = 1$. Assume $p \ge 1$. Choose n so large that $N_n \ge p$. Applying (4.1.3) and using the symbol (1.1.6)

$$\sum_{k=1}^{n}\sum_{j=0}^{m_k-2} \lambda_{kj}(p-j+1)_j x_k^{p-j} = \int_{\mathbf{R}} x^p d\mu(x), \tag{4.6.7}$$

which gives

$$\begin{aligned} \left|\sum_{k=1}^{n} \lambda_{k0} x_k^p - \int_{\mathbf{R}} x^p d\mu(x)\right| &= \left|\sum_{k=1}^{n}\sum_{j=1}^{m_k-2} \lambda_{kj}(p-j+1)_j x_k^{p-j}\right| \\ &\le \sum_{|x_k|\le T}\sum_{1\le j\le p} |\lambda_{kj}|(p-j+1)_j|x_k|^{p-j} + \sum_{|x_k|>T}\sum_{1\le j\le p} |\lambda_{kj}|(p-j+1)_j|x_k|^{p-j} \\ &= S_{1n} + S_{2n}. \end{aligned} \tag{4.6.8}$$

We must estimate S_{1n} and S_{2n}. It is particularly simple to estimate S_{1n}: by means of (4.1.10) and (4.6.1)

$$S_{1n} \le p!T^{p-1} \sum_{1\le j\le p} D^j_{n,T} \sum_{|x_k|\le T} \lambda_{k0} \le p!T^{p-1} \sum_{1\le j\le p} D^j_{n,T} \int_{\mathbf{R}} d\mu(x).$$

To estimate S_{2n} we use (4.1.10) and (4.6.2) to get

$$S_{2n} \le p! \sum_{|x_k|>T} \sum_{1\le j\le p} \left(\frac{\overline{d}_k}{|x_k|}\right)^j \lambda_{k0}|x_k|^p \le \left[p! \sum_{1\le j\le p} R^j_{n,T}\right] \sum_{k=1}^{n} \lambda_{k0}|x_k|^p.$$

Then relation (4.6.8) yields

$$\left|\sum_{k=1}^{n} \lambda_{k0}x_k^p - \int_{\mathbf{R}} x^p d\mu(x)\right| \le S_{1n} + \left[p! \sum_{1\le j\le p} R^j_{n,T}\right] \sum_{k=1}^{n} \lambda_{k0}|x_k|^p. \tag{4.6.9}$$

By (4.6.4)

$$\lim_{n\to\infty} S_{1n} = 0. \tag{4.6.10}$$

We have to prove

$$\lim_{n\to\infty} S_{2n} = 0.$$

By (4.6.5) it is sufficient to show

$$\sum_{k=1}^{n} \lambda_{k0}|x_k|^p \le c_p < \infty. \tag{4.6.11}$$

To this end we separate two cases when $p \in \mathbf{N}_2$ and $p \in \mathbf{N}_1$.

Case 1. $p \in \mathbf{N}_2$. In this case with the help of (4.6.5) we can choose n_0 so large that

$$p! \sum_{1\le j\le p} R^j_{n,T} \le \frac{1}{2}, \quad n \ge n_0.$$

Hence by (4.6.9)

$$\sum_{k=1}^{n} \lambda_{k0}x_k^p \le 2S_{1n} + 2\int_{\mathbf{R}} x^p d\mu(x), \quad n \ge n_0,$$

which according to (4.6.10) gives (4.6.11).

Case 2. $p \in \mathbf{N}_1$. In this case $p+1 \in \mathbf{N}_2$ and

$$\begin{aligned}\sum_{k=1}^{n} \lambda_{k0}|x_k|^p &\le \sum_{|x_k|\le 1} \lambda_{k0}|x_k|^p + \sum_{|x_k|>1} \lambda_{k0}|x_k|^p \\ &\le \sum_{|x_k|\le 1} \lambda_{k0} + \sum_{|x_k|>1} \lambda_{k0}x_k^{p+1} \le \sum_{k=1}^{n} \lambda_{k0} + \sum_{k=1}^{n} \lambda_{k0}x_k^{p+1} \le \int_{\mathbf{R}} x^p d\mu(x) + c_{p+1},\end{aligned}$$

which according to the conclusion of Case 1 again yields (4.6.11). □

Theorem 4.6.2. *Let $d\mu$ be a measure supported on* $\mathbf{R}$ *and let $d\mu \in \mathbf{E}$ and $f \in \mathbf{F}(d\mu)$. Let the assumptions of Theorem 4.1.2 prevail and let all the m_{kn} be even. Assume that g has all derivatives on* $\mathbf{R}$ *and satisfies that for some constant $C > 0$*

$$g^{(2j)}(x) \geq 0, \quad x \in \mathbf{R}, \quad j = 0, 1, \ldots, \tag{4.6.12}$$

$$|g^{(j)}(x)| \leq Cg(x), \quad x \in \mathbf{R}, \quad j = 1, 2, \ldots, m-2, \tag{4.6.13}$$

$$\lim_{|x| \to \infty} \frac{f(x)}{g(x)} = 0, \tag{4.6.14}$$

and $\int_{\mathbf{R}} g(x)d\mu(x)$ exists. If

$$\lim_{n \to \infty} D_n = 0 \tag{4.6.15}$$

then the relations

$$\lim_{n \to \infty} \sum_{k=1}^{n} \sum_{j=1}^{m_{kn}-2} \lambda_{kjn} f(x_{kn}) = 0 \tag{4.6.16}$$

and (4.6.6) hold.

Proof. Let us prove (4.6.16). By Theorem 4.4.3 relation (4.4.19) with (4.1.5) is true. According to (4.1.10) it follows from (4.6.13) that

$$\begin{aligned}
\left| \sum_{k=1}^{n} \sum_{j=1}^{m_k-2} \lambda_{kj} g^{(j)}(x_k) \right| &\leq \sum_{k=1}^{n} \sum_{j=1}^{m_k-2} \left| \lambda_{kj} g^{(j)}(x_k) \right| \\
\leq \quad C \sum_{k=1}^{n} \sum_{j=1}^{m_k-2} D_n^j \lambda_{k0} g(x_k) &\leq C \left(\sum_{j=1}^{m-2} D_n^j \right) \sum_{k=1}^{n} \lambda_{k0} g(x_k) \\
= \quad \frac{CD_n(1 - D_n^{m-2})}{1 - D_n} \sum_{k=1}^{n} \lambda_{k0} g(x_k),
\end{aligned}$$

which, coupled with (4.4.19) and (4.1.5), yields

$$\left[1 - \frac{CD_n(1 - D_n^{m-2})}{1 - D_n} \right] \sum_{k=1}^{n} \lambda_{k0} g(x_k) \leq \int_{\mathbf{R}} g(x) d\mu(x).$$

Using (4.6.15) we obtain

$$\limsup_{n \to \infty} \sum_{k=1}^{n} \lambda_{k0} g(x_k) \leq \int_{\mathbf{R}} g(x) d\mu(x). \tag{4.6.17}$$

Given an arbitrary number $\varepsilon > 0$, by (4.6.14) there exists a positive number T such that

$$|f(x)| \leq \varepsilon g(x), \quad |x| > T. \tag{4.6.18}$$

Again applying (4.1.10) and (4.4.5) we see that

$$\begin{aligned}
\left| \sum_{k=1}^{n} \lambda_{kj} f(x_k) \right| &\leq \sum_{k=1}^{n} |\lambda_{kj} f(x_k)| \leq CD_n^j \sum_{k=1}^{n} \lambda_{k0} |f(x_k)| \\
\leq \quad CD_n^j &\left[\sum_{|x_k| \leq T} \int_{x_{k-1}}^{x_{k+1}} |f(x_k)| \, d\mu(x) + \sum_{|x_k| > T} \lambda_{k0} |f(x_k)| \right].
\end{aligned} \tag{4.6.19}$$

Since $f \in \mathbf{F}(d\mu)$, by Theorem 4.2.1

$$\begin{aligned}
&\lim_{n\to\infty} \sum_{|x_k|\le T} \int_{x_{k-1}}^{x_{k+1}} |f(x_k)|\,d\mu(x) \\
= &\lim_{n\to\infty} \left[\sum_{|x_k|\le T} \int_{x_{k-1}}^{x_k} |f(x_k)|\,d\mu(x) + \sum_{|x_k|\le T} \int_{x_k}^{x_{k+1}} |f(x_k)|\,d\mu(x) \right] \\
= &\ 2\int_{|x|\le T} |f(x)|\,d\mu(x) \le 2\int_{\mathbf{R}} |f(x)|\,d\mu(x).
\end{aligned}$$

Meanwhile by (4.6.17) and (4.6.18) we see that

$$\begin{aligned}
&\limsup_{n\to\infty} \sum_{|x_k|>T} \lambda_{k0}|f(x_k)| \le \varepsilon \limsup_{n\to\infty} \sum_{|x_k|>T} \lambda_{k0}g(x_k) \\
\le &\ \varepsilon \limsup_{n\to\infty} \sum_{k=1}^{n} \lambda_{k0}g(x_k) \le \varepsilon \int_{\mathbf{R}} g(x)d\mu(x). \qquad (4.6.20)
\end{aligned}$$

The relation (4.6.16) follows from the above two relations, (4.6.19), and (4.6.15).

Let us prove (4.6.6). To this end we observe that for each polynomial P we can choose a number $N \in \mathbf{N}_2$, $N > \partial P$, so that the polynomial $g(x) = 1 + x^N$ satisfies (4.6.12), (4.6.13) with $C = N^N$, and (4.6.14) with $f = P$. In fact, relations (4.6.12) and (4.6.14) are obvious; relation (4.6.13) follows from the estimation

$$|g^{(j)}(x)| \le C|x|^{N-j} \le C(1+x^N) = Cg(x).$$

This shows that (4.6.16) holds for every polynomial $f = P$. Meanwhile, for a polynomial $P \in \mathbf{P}_{N_n}$ by (4.1.3)

$$\int_{\mathbf{R}} P(x)d\mu(x) = \sum_{k=1}^{n} \sum_{j=0}^{m_k-2} P^{(j)}(x_k) = Q_n(d\mu;P) + \sum_{k=1}^{n} \sum_{j=1}^{m_k-2} P^{(j)}(x_k). \qquad (4.6.21)$$

This also shows that the positive quadrature procedure $\{Q_n(d\mu;\cdot)\}$ belongs to the measure $d\mu$. So we can apply Theorem 4.2.2 to the function

$$f_T(x) = \begin{cases} f(x), & |x| \le T, \\ 0, & |x| > T \end{cases} \qquad (4.6.22)$$

to obtain

$$\lim_{n\to\infty} Q_n(d\mu;f_T) = \int_{\mathbf{R}} f_T(x)d\mu(x) = \int_{|x|\le T} f(x)d\mu(x). \qquad (4.6.23)$$

Thus by (4.6.18)

$$\left| \int_{|x|>T} f(x)d\mu(x) \right| \le \int_{|x|>T} |f(x)|d\mu(x) \le \varepsilon \int_{|x|>T} |g(x)|d\mu(x) \le \varepsilon \int_{\mathbf{R}} |g(x)|d\mu(x)$$

and hence

$$
\begin{aligned}
&\left|Q_n(d\mu;f) - \int_{\mathbf{R}} f(x)d\mu(x)\right| \\
\le\ &\left|Q_n(d\mu;f_T) - \int_{|x|\le T} f(x)d\mu(x)\right| + \left|\sum_{|x_k|>T} \lambda_{k0} f(x_k)\right| + \left|\int_{|x|>T} f(x)d\mu(x)\right| \\
\le\ &\left|Q_n(d\mu;f_T) - \int_{|x|\le T} f(x)d\mu(x)\right| + \left|\sum_{|x_k|>T} \lambda_{k0} f(x_k)\right| + \varepsilon\int_{\mathbf{R}} g(x)d\mu(x).
\end{aligned}
$$

By means of (4.6.20) and (4.6.23)

$$\limsup_{n\to\infty}\left|Q_n(d\mu;f) - \int_{\mathbf{R}} f(x)d\mu(x)\right| \le 2\varepsilon\int_{\mathbf{R}} g(x)d\mu(x).$$

Since ε is arbitrary, the relation (4.6.6) follows. □

Theorem 4.6.3. *Let $d\mu$ be a measure supported on $\mathbf{R}$ and let $d\mu \in \mathbf{E}$ and $f \in \mathbf{F}(d\mu)$. Let the assumptions of Theorem 4.1.2 prevail and let all the m_{kn} be even. Assume that relations (4.6.4), (4.6.12), and (4.6.13) hold. Furthermore, assume that*

$$\sup_n D_n = D < \infty, \tag{4.6.24}$$

$$\lim_{|x|\to\infty} \frac{f(x)}{g(c_0x)} = 0, \quad c_0 = 1/[D(1+C)], \tag{4.6.25}$$

and $\int_{\mathbf{R}} g(c_0x)d\mu(x)$ exists. Then relations (4.6.16) and (4.6.6) hold.

Proof. First we prove (4.6.16). Write $H(x) = g(c_0x)$. We see that

$$H^{(2j)}(x) = c_0^{2j} g^{(2j)}(c_0x) \ge 0, \quad x\in\mathbf{R}, \quad j = 0,1,\ldots, \tag{4.6.26}$$

and

$$|H^{(j)}(x)| = c_0^j|g^{(j)}(c_0x)| \le Cc_0^j g(c_0x) = Cc_0^j H(x), \quad j = 0,1,\ldots,m-2. \tag{4.6.27}$$

Hence by (4.1.10) and (4.6.27)

$$
\begin{aligned}
&\left|\sum_{k=1}^{n}\sum_{j=1}^{m_k-2} \lambda_{kj} H^{(j)}(x_k)\right| \le C\sum_{k=1}^{n}\sum_{j=1}^{m_k-2} (c_0D)^j \lambda_{k0} H(x_k) \\
\le\ & C\sum_{j=1}^{m-2} (c_0D)^j \sum_{k=1}^{n} \lambda_{k0} H(x_k) = C\frac{c_0D[1-(c_0D)^{m-2}]}{1-c_0D}\sum_{k=1}^{n} \lambda_{k0} H(x_k) \\
=\ & \left[1-(1+C)^{2-m}\right]\sum_{k=1}^{n} \lambda_{k0} H(x_k).
\end{aligned} \tag{4.6.28}
$$

Recalling (4.6.26) and applying Theorem 4.4.3 to the function H, we obtain

$$\sum_{k=1}^{n}\sum_{j=0}^{m_k-2} \lambda_{kj} H^{(j)}(x_k) \le \int_{\mathbf{R}} H(x)d\mu(x),$$

which, together with (4.6.28), gives

$$\sum_{k=1}^{n} \lambda_{k0} H(x_k) \le (1+C)^{m-2} \int_{\mathbf{R}} H(x)d\mu(x).$$

Thus

$$\limsup_{n\to\infty} \sum_{k=1}^{n} \lambda_{k0} H(x_k) \le (1+C)^{m-2} \int_{\mathbf{R}} H(x)d\mu(x). \tag{4.6.29}$$

Given an arbitrary number $\varepsilon > 0$, by (4.6.25) there exists a positive number T such that

$$|f(x)| \le \varepsilon H(x), \quad |x| > T. \tag{4.6.30}$$

Applying (4.1.10) we see that for $j \ge 1$

$$\begin{aligned} \left|\sum_{k=1}^{n} \lambda_{kj} f(x_k)\right| &\le \sum_{|x_k|\le T} \lambda_{k0} \overline{d}_k^j |f(x_k)| + \sum_{|x_k|>T} \lambda_{k0} \overline{d}_k^j |f(x_k)| \\ &\le D_{n,T}^j \sum_{|x_k|\le T} \lambda_{k0} |f(x_k)| + D^j \sum_{|x_k|>T} \lambda_{k0} |f(x_k)| \\ &= S_{1n} + D^j \sum_{|x_k|>T} \lambda_{k0} |f(x_k)|. \end{aligned} \tag{4.6.31}$$

Since $f \in \mathbf{F}(d\mu)$, by Corollary 4.4.1 and Theorem 4.2.1

$$\begin{aligned} \lim_{n\to\infty} S_{1n} &= \left(\lim_{n\to\infty} D_{n,T}^j\right) \lim_{n\to\infty} \sum_{|x_k|\le T} \lambda_{k0} |f(x_k)| \\ &\le \left(\lim_{n\to\infty} D_{n,T}^j\right) \lim_{n\to\infty} \sum_{|x_k|\le T} \int_{x_{k-1}}^{x_{k+1}} |f(x_k)| d\mu(x) \\ &= 2\left(\lim_{n\to\infty} (D_{n,T}^j\right) \int_{\mathbf{R}} |f_T(x)| d\mu(x) = 0, \end{aligned} \tag{4.6.32}$$

where f_T is defined in (4.6.22). Meanwhile by (4.6.30) and (4.6.29) we see that

$$\begin{aligned} &\limsup_{n\to\infty} \sum_{|x_k|>T} \lambda_{k0} |f(x_k)| \le \varepsilon \limsup_{n\to\infty} \sum_{|x_k|>T} \lambda_{k0} H(x_k) \\ \le\ &\varepsilon \limsup_{n\to\infty} \sum_{k=1}^{n} \lambda_{k0} H(x_k) \le \varepsilon (1+C)^{m-2} \int_{\mathbf{R}} H(x)d\mu(x). \end{aligned} \tag{4.6.33}$$

The relation (4.6.16) follows from (4.6.31)–(4.6.33).

Next let us prove (4.6.6). For $P \in \mathbf{P}_{N_n}$ we have (4.6.21) and in the same way as that in Theorem 4.6.2 we can choose a polynomial g satisfying (4.6.12), (4.6.13), and (4.6.25) with $f = P$. By (4.6.16)

$$\lim_{n\to\infty} Q_n(d\mu; P) = \int_{\mathbf{R}} P(x)d\mu(x).$$

Hence by using Theorem 4.2.2 relation (4.6.23) holds. Thus

$$\begin{aligned}
&\left|Q_n(d\mu;f)-\int_{\mathbf{R}}f(x)d\mu(x)\right|\\
\leq\ &\left|Q_n(d\mu;f_T)-\int_{|x|\leq T}f(x)d\mu(x)\right|+\left|\sum_{|x_k|>T}\lambda_{k0}f(x_k)\right|+\left|\int_{|x|>T}f(x)d\mu(x)\right|\\
\leq\ &\left|Q_n(d\mu;f_T)-\int_{|x|\leq T}f(x)d\mu(x)\right|+\sum_{|x_k|>T}\lambda_{k0}|f(x_k)|+\varepsilon\int_{\mathbf{R}}H(x)d\mu(x).
\end{aligned}$$

By means of (4.6.33) and (4.6.23)

$$\limsup_{n\to\infty}\left|Q_n(d\mu;f)-\int_{\mathbf{R}}f(x)d\mu(x)\right|\leq\varepsilon[1+(1+C)^{m-2}]\int_{\mathbf{R}}H(x)d\mu(x).$$

Since ε is arbitrary, relation (4.6.6) follows. □

4.7 Notes

4.7.1. A Gaussian quadrature formula is closely related to a power orthogonal polynomial; for its existence and uniqueness we refer to Section 3.5.

4.7.2. The classical Erdős-Turán inequality (4.3.26) was discovered by Erdős and Turán [22, Lemma IV]. This inequality has many applications, say, it is used to estimate lower bounds for the Lebesgue function of Lagrange interpolation [135, Chapter III, pp. 71–123]. Its weighted form is given by Lubinsky [51]. Theorem 4.3.1 and its corollaries are due to Shi [113].

4.7.3. The generalized Markov-Stieltjes inequality for a Gaussian quadrature formula has many applications in the convergence of product integration rules and an estimation of the rate of convergence of a Gaussian quadrature formula for singular integrands. For full information on this subject see the introduction and the references in [35]. The Markov-Stieltjes inequality for a Gauss quadrature formula with simple nodes was conjectured by Chebyshev [13] and independently proved by A. A. Markov [53] and Stieltjes [131]. The Markov-Stieltjes inequality was extended to a Gaussian quadrature formula with nodes of odd multiplicity by Gori [35]. The Markov-Stieltjes inequality in the present form was proved by Shi [93].

4.7.4. Theorem 4.4.3 was discovered by A. A. Markov [53, p. 81] for the case when $m_k\equiv 2$ and extended by Zhou [161, Lemma 3.3] to the case when all the m_k are even.

4.7.5. The results in Section 4.5 are due to Shi [110, 124].

4.7.6. Theorem 4.6.2 is due to Zhou [161, Theorem 3.1].

Chapter 5

Christoffel Type Functions

5.1 Introduction and Definitions

Throughout of this chapter let $m_k \equiv m \in \mathbf{N}$ $(m \geq 2)$, $\mathbf{M}_1 = \{j \in \mathbf{N}_0 : m - j \in \mathbf{N}_1\}$, and $\mathbf{M}_2 = \{j \in \mathbf{N}_0 : m - j \in \mathbf{N}_2\}$. Put $\mathbf{P}_N^* = \{P(x) = c(x - y_1)\cdots(x - y_r) : \ c, y_1, \ldots, y_r \in \mathbf{R}, \ r \leq N\}$ and $\mathbf{P}_N^*(x) = \{P \in \mathbf{P}_N^* : P(x) = 1\}$ for $x \in \mathbf{R}$. We agree $\mathbf{P}_0^* = \mathbf{P}_0$. We also use the simple symbol $x_k = x_{kn} = x_{kn}(d\mu, m)$, etc, unless otherwise indicated.

As we know, orthogonal polynomials $(m = 2)$ have a long history of study and a classical theory. One of the important contents of this theory is the *Christoffel function*

$$\lambda_n(d\mu; x) = \min_{P \in \mathbf{P}_{n-1}, P(x)=1} \int_{\mathbf{R}} P(t)^2 d\mu(t), \tag{5.1.1}$$

which is closely related to the *Christoffel numbers*

$$\lambda_{kn}(d\mu) = \lambda_n(d\mu; x_{kn}(d\mu)), \quad k = 1, 2, \ldots, n. \tag{5.1.2}$$

The study and applications of the Christoffel functions can be found in [67]. In this chapter we shall extend the Christoffel functions to the case of the m orthogonal polynomials. We shall investigate their properties in Section 5.2 and determine the *Christoffel type functions* for the *generalized Jacobi weight* in Section 5.3.

Given a fixed point $x \in \mathbf{R}$, an index j, $0 \leq j \leq m - 2$, and $n \in \mathbf{N}$, for $P \in \mathbf{P}_{n-1}$ with $P(x) = 1$ let the polynomial

$$A_j(P, x; t) = A_{jnm}(P, x; t) = \frac{1}{j!}(t - x)^j B_j(P, x; t) P(t)^m \tag{5.1.3}$$

with $B_j(P, x; \cdot) \in \mathbf{P}_{m-j-2}$ satisfying the condition

$$A_j^{(i)}(P, x; x) = \delta_{ij}, \quad i = 0, 1, \ldots, m - 2. \tag{5.1.4}$$

It is easy to see that $A_j(P, x; t)$ must exist and be unique.

Definition 5.1.1. *The Christoffel type function $\lambda_{jn}(d\mu, m; x)$ with respect to $d\mu$ is defined by*

$$\lambda_{jn}(d\mu, m; x) = \inf_{P \in \mathbf{P}_{n-1}^*(x)} \int_{\mathbf{R}} A_j(P, x; t) sgn\left[(t - x)P(t)\right]^m d\mu(t) \tag{5.1.5}$$

for $j \in \mathbf{M}_2$ *and by*

$$\lambda_{jn}(d\mu,m;x) = \int_{\mathbf{R}} A_j(P,x;t) sgn\,[(t-x)P(t)]^m d\mu(t) \tag{5.1.6}$$

for $j \in \mathbf{M}_1$, *where the polynomial P in (5.1.6) is the solution of (5.1.5) in the case when* $j \in \mathbf{M}_2$.

Remark 5.1.1. We shall see that there is a unique polynomial $P \in \mathbf{P}^*_{n-1}(x)$ such that the equation (5.1.5) holds for *every* $j \in \mathbf{M}_2$ (Theorem 5.2.1). So the definition of $\lambda_{jn}(d\mu,m;x)$ for $j \in \mathbf{M}_1$ is reasonable.

Remark 5.1.2. It is particularly simple to determine $\lambda_{j1}(d\mu,m;x)$: in this case we have that $\mathbf{P}^*_0(x) = \{1\}$, $A_j(P,x;t) = (t-x)^j/j!$, and

$$\lambda_{j1}(d\mu,m;x) = \frac{1}{j!}\int_{\mathbf{R}} (t-x)^j \text{sgn}\,(t-x)^m d\mu(t).$$

So in what follows we always assume $n \geq 2$.

We also need the *generalized Christoffel function* introduced by Nevai.

Definition 5.1.2. [65, p. 106] *The generalized Christoffel function* $\lambda_n(d\mu,m;x)$, $0 < m < \infty$, *with respect to* $d\mu$ *is defined by*

$$\lambda_n(d\mu,m;x) = \inf_{P\in\mathbf{P}_{n-1}, P(x)=1} \int_{\mathbf{R}} |P(t)|^m d\mu(t). \tag{5.1.7}$$

5.2 Properties

The expression and the main properties of the polynomial $B_j(P,x;t)$ are as follows.

Lemma 5.2.1. *We have*

$$B_j(P,x;t) = \sum_{i=0}^{m-j-2} b_i(t-x)^i, \tag{5.2.1}$$

where

$$b_i = b_i(P,x) = \frac{1}{i!}\left[P(t)^{-m}\right]^{(i)}_{t=x}, \quad i = 0,1,\ldots. \tag{5.2.2}$$

Moreover, for $P \in \mathbf{P}^*_{n-1}(x)$ *and* $j \in \mathbf{M}_2$

$$b_{m-j-2} > 0, \quad B_j(P,x;t) > 0, \quad t \in \mathbf{R}. \tag{5.2.3}$$

Proof. Apply (2.2.5), (2.2.6), and (2.2.20). □

Remark 5.2.1. By (5.2.3) we see that the integrand in (5.1.5) is nonnegative for $j \in \mathbf{M}_2$:

$$A_j(P,x;t)\text{sgn}\,[(t-x)P(t)]^m = |A_j(P,x;t)|.$$

Remark 5.2.2. For $m \geq 3$ the restriction $P \in \mathbf{P}^*_{n-1}(x)$ can not be replaced by the condition $P \in \mathbf{P}_{n-1}$ with $P(x) = 1$ used in the case when $m = 2$, for otherwise the function $\lambda_{jn}(d\mu,m;x) = -\infty$ may occur. For example choose $n = 3$, $m = 4$, $x = 0$, and $P(t) = 1 + dt^2$ $(d > 0)$. By a simple calculation we see

$$[P(t)^{-4}]' = -8dt(1+dt^2)^{-5},$$

$$[P(t)^{-4}]'' = -8d(1+dt^2)^{-5} + 80d^2t^2(1+dt^2)^{-6},$$

$$b_0 = 1,\ b_1 = 0,\ b_2 = -4d,$$

and

$$A_0(P,0;t) = (1-4dt^2)(1+dt^2)^4.$$

It is easy to see that for any measure $d\mu$

$$\lim_{d\to\infty}\int_{\mathbf{R}} A_0(P,0;t)d\mu(t) = -\infty.$$

Remark 5.2.3. The definition (5.1.5) is not suitable for $j \in \mathbf{M}_1$. For example for $m \in \mathbf{N}_2$, $j = m-3$, and

$$-2 < x_1 < x_2 < \cdots < x_{n-1} < x = -1,$$

we have

$$A_{m-3}(P,x;t) = [1+b_1(t-x)]P(t)^m, \quad P(t) = \prod_{i=1}^{n-1}\frac{t-x_i}{x-x_i}.$$

Since by Theorem 2.2.2

$$b_1 = a_{1n} = \sum_{i=1}^{n-1}\frac{m}{x_i - x} \le \frac{m}{x_{n-1}-x},$$

we have

$$A_{m-3}(P,x;t) \le \left[1+\frac{m(t-x)}{x_{n-1}-x}\right]P(t)^m = \frac{m(t_0-t)}{x-x_{n-1}}P(t)^m, \quad t\in[-1,1],$$

where

$$t_0 = \frac{x-x_{n-1}}{m} - 1 \in (-1,1).$$

Then

$$\begin{aligned}
&\int_{-1}^{1} A_{m-3}(P,x;t)d\mu(t) \\
\le\ & \frac{m}{x-x_{n-1}}\left[\int_{-1}^{t_0}(t_0-t)P(t)^m d\mu(t) - \int_{t_0}^{1}(t-t_0)P(t)^m d\mu(t)\right] \\
\le\ & \frac{m}{x-x_{n-1}}\left[P(t_0)^m\int_{-1}^{t_0}(t_0-t)d\mu(t) - P(t_0)^m\int_{t_0}^{1}(t-t_0)d\mu(t)\right].
\end{aligned}$$

Thus as $x_{n-1}\to x$ we have $t_0 \to -1$ and hence

$$\lim_{x_{n-1}\to x}\int_{-1}^{1} A_{m-3}(P,x;t)d\mu(t) = -\infty.$$

This shows that the definition (5.1.5) is not suitable for this case.

Let

$$P_\lambda(t) = P(t) + \lambda(t-x)Q(t), \quad P\in\mathbf{P}_{n-1}^*(x), \quad Q\in\mathbf{P}_{n-2}, \tag{5.2.4}$$

and put $f(\lambda;t) = A_j(P_\lambda,x;t)$ and $g(\lambda;t) = B_j(P_\lambda,x;t)$.

Lemma 5.2.2. *Let $d\mu$ be a measure on $\mathbf{R}$. For a fixed point $x \in \mathbf{R}$ and a fixed index $j \in \mathbf{M}_2$ let a polynomial $P \in \mathbf{P}^*_{n-1}(x)$ satisfy the equation*

$$\int_{\mathbf{R}} A_j(P,x;t) sgn\,[(t-x)P(t)]^m d\mu(t) = \lambda_{jn}(d\mu,m;x). \tag{5.2.5}$$

If the polynomial P_λ in (5.2.4) satisfies Condition A: there is a number $\delta > 0$ such that $P_\lambda \in \mathbf{P}^*_{n-1}(x)$ holds for every $\lambda \in [0,\delta]$, then

$$\int_{\mathbf{R}} [(t-x)P(t)]^{m-1} q(t) sgn\,[(t-x)P(t)]^m d\mu(t) \geq 0, \tag{5.2.6}$$

where

$$q(t) = (t-x)^{j-m+1}[g'_\lambda(0;t)P(t) + m(t-x)g(0;t)Q(t)] \tag{5.2.7}$$

and

$$q \in \mathbf{P}_{\max\{\partial P-1, \partial Q\}}. \tag{5.2.8}$$

Proof. By (5.2.5) and Condition A we conclude

$$\liminf_{\lambda \to +0} \int_{\mathbf{R}} \frac{f(\lambda;t) - f(0;t)}{\lambda} \text{sgn}\,[(t-x)P(t)]^m d\mu(t) \geq 0.$$

Thus to prove (5.2.6) it suffices to establish

$$\begin{aligned} &\liminf_{\lambda \to +0} \int_{\mathbf{R}} \frac{f(\lambda;t) - f(0;t)}{\lambda} \text{sgn}\,[(t-x)P(t)]^m d\mu(t) \\ = &\int_{\mathbf{R}} f'_\lambda(0;t) \text{sgn}\,[(t-x)P(t)]^m d\mu(t) \end{aligned} \tag{5.2.9}$$

and

$$f'_\lambda(0;t) = \frac{1}{j!}[(t-x)P(t)]^{m-1} q(t). \tag{5.2.10}$$

To this end we observe that the function

$$d_i(P_\lambda,x) = \frac{1}{i!}\,[P_\lambda(t)^m]^{(i)}_{t=x} = \frac{1}{i!}\sum_{\nu=0}^{m} \binom{m}{\nu} \left[P(t)^{m-\nu}(t-x)^\nu Q(t)^\nu\right]^{(i)}_{t=x} \lambda^\nu$$

is a polynomial in λ of degree at most m. From the identity

$$[P_\lambda(t)^m P_\lambda(t)^{-m}]^{(i)} = 0, \quad i \geq 1,$$

applying Newton-Leibniz rule and using (5.2.2), we obtain

$$\sum_{\nu=0}^{i} d_\nu(P_\lambda,x) b_{i-\nu}(P_\lambda,x) = 0, \quad i \geq 1.$$

Hence

$$b_0(P_\lambda,x) = 1, \quad b_i(P_\lambda,x) = -\sum_{\nu=1}^{i} d_\nu(P_\lambda,x) b_{i-\nu}(P_\lambda,x), \quad i \geq 1.$$

Thus $b_i(P_\lambda, x)$ is also a polynomial in λ of degree at most im. By (5.1.3), (5.2.1), and (5.2.4) both $g'_\lambda(\lambda;t)$ and $f'_\lambda(\lambda;t)$ exist, and are polynomials in λ of degree at most $m(m-j-1)$. Meanwhile

$$f'_\lambda(\lambda;t) = \frac{1}{j!}(t-x)^j[g'_\lambda(\lambda;t)P_\lambda(t)^m + m(t-x)Q(t)P_\lambda(t)^{m-1}g(\lambda;t)]. \tag{5.2.11}$$

Then the integral

$$\int_{\mathbf{R}} f'_\lambda(\lambda;t)\mathrm{sgn}\,[(t-x)P(t)]^m d\mu(t)$$

converges uniformly for $\lambda \in [0,\delta]$. Hence relation (5.2.9) is true. On the other hand, by (5.2.11)

$$f'_\lambda(0;t) = \frac{1}{j!}(t-x)^j P(t)^{m-1}[g'_\lambda(0;t)P(t) + m(t-x)Q(t)g(0;t)]$$

is a polynomial in t. According to (5.1.4)

$$A_j^{(i)}(P_\lambda, x; x) = A_j^{(i)}(P, x; x) = \delta_{ij}, \quad i = 0, 1, \ldots, m-2,$$

which shows that the polynomial $A_j(P_\lambda, x;t) - A_j(P, x;t) = f(\lambda;t) - f(0;t)$ in t contains the factor $(t-x)^{m-1}$, so does the polynomial $f'_\lambda(0;t)$. Thus relation (5.2.10) with (5.2.7) is true. Recalling that $g'_\lambda(0;\cdot), g(0;\cdot) \in \mathbf{P}_{m-j-2}$, relation (5.2.7) implies (5.2.8). □

Lemma 5.2.3. *Let $d\mu$ be a measure on $\mathbf{R}$. Let a point $x \in \mathbf{R}$ be fixed and $j \in \mathbf{M}_2$. Then there exists a polynomial $P \in \mathbf{P}^*_{n-1}(x)$ such that relation (5.2.5) holds.*

Moreover, if a polynomial $P \in \mathbf{P}^*_{n-1}(x)$ is a solution of (5.2.5), then

(a) the zeros of the polynomial P are distinct;

(b) $\partial P \geq n-2$;

(c) the polynomial P satisfies the orthogonality relation:

$$\int_{\mathbf{R}} [(t-x)P(t)]^{m-1} q(t) sgn\,[(t-x)P(t)]^m d\mu(t) = 0, \quad \forall q \in \mathbf{P}_{n-2}. \tag{5.2.12}$$

Proof. To prove the first part of the lemma assume that the polynomials $P_N \in \mathbf{P}^*_{n-1}(x)$, $N = 1, 2, \ldots$, satisfy

$$\lim_{N\to\infty} \int_{\mathbf{R}} A_j(P_N, x;t)\mathrm{sgn}\,[(t-x)P_N(t)]^m d\mu(t) = \lambda_{jn}(d\mu, m; x).$$

Then (see Remark 5.2.1)

$$\int_{\mathbf{R}} |A_j(P_N, x;t)| d\mu(t) \leq c < +\infty, \quad \forall N \in \mathbf{N}.$$

Write

$$A_j(P_N, x;t) = \sum_{k=0}^{mn-2} a_{kN} t^k.$$

By Theorem of equivalent norms of finite dimensional spaces the previous inequalities imply that

$$|a_{kN}| \leq c_1 < +\infty, \quad k = 0, 1, \ldots, mn-2, \quad \forall N \in \mathbf{N}.$$

According to Bolzano-Weierstrass theorem, by passing to a subsequence if necessary, we may suppose that $P_N \to P$ $(N \to \infty)$. Then $P \in \mathbf{P}^*_{n-1}(x)$ and relation (5.2.5) holds.

Let us prove the second part of the lemma. Assume

$$P(t) = \prod_{k=1}^{r} \left(\frac{t-t_k}{x-t_k}\right)^{p_k},$$

where $-\infty < t_1 < t_2 < \cdots < t_r < +\infty$, $p_1, p_2, \ldots, p_r \in \mathbf{N}$. We claim that

$$\int_{\mathbf{R}} \frac{(t-x)^{m-1}P(t)^m}{t-t_k} \text{sgn}\,[(t-x)P(t)]^m d\mu(t) = 0, \quad k = 1,2,\ldots,r; \tag{5.2.13}$$

further if $\partial P \le n-2$ then the additional equation

$$\int_{\mathbf{R}} (t-x)^{m-1}P(t)^m \text{sgn}\,[(t-x)P(t)]^m d\mu(t) = 0 \tag{5.2.14}$$

holds.

To prove (5.2.13) choose $Q(t) = \pm P(t)/(t-t_k)$, $1 \le k \le r$, which obviously satisfies Condition A. In this case relation (5.2.7) is of the form

$$q(t) = C(t)Q(t), \tag{5.2.15}$$

where the function

$$C(t) = (t-x)^{j-m+1}[mg(0;t)(t-x) \pm g'_\lambda(0;t)(t-t_k)]$$

is a polynomial in t, because the functions q, Q, and the function in the bracket are polynomials in t, and $Q(x) \ne 0$. The relation (5.2.8) shows $\partial q \le \partial Q$ and hence $C(t) \equiv C$. This constant C is not zero, since by (5.2.3)

$$C = C(t_k) = mg(0;t_k)(t_k-x)^{j-m+2} = mB_j(P,x;t_k)(t_k-x)^{j-m+2} > 0.$$

Substituting q into (5.2.6), we obtain (5.2.13).

Similarly, to prove (5.2.14) choose $Q(t) \doteq \pm P(t)$, which obviously satisfies Condition A. In this case relation (5.2.15) holds, where the function

$$C(t) = (t-x)^{j-m+1}[mg(0;t)(t-x) \pm g'_\lambda(0;t)]$$

is also a constant by the same argument as above. To determine this constant we observe that

$$\partial g'_\lambda(0;\cdot) \le m-j-2 < m-j-1 = \partial[g(0;\cdot)(\cdot - x)]$$

and hence by (5.2.3)

$$C(t) \equiv C = \lim_{t\to\infty} C(t) = mb_{m-j-2} > 0.$$

Substituting this q into (5.2.6), we obtain (5.2.14).

(a) Suppose to the contrary that $p_k > 1$ holds for some k, $1 \le k \le r$. Choose $Q(t) = -(t-x)P(t)/(t-t_k)^2$, which obviously satisfies Condition A. In this case by (5.2.7) we can write q in the form

$$q(t) = C(t)\frac{P(t)}{(t-t_k)^2}, \tag{5.2.16}$$

where the function

$$C(t) = (t-x)^{j-m+1}[g'_\lambda(0;t)(t-t_k)^2 - mg(0;t)(t-x)^2]$$

is a polynomial in t by the same reason as above. The relation (5.2.8) shows $\partial q \leq \partial Q$ and hence $\partial C \leq 1$. Thus we may write the linear function C in the form $C(t) = C_1(t-x) + C_2(t-t_k)$. To determine the sign of the constant C_1 we use (5.2.3) to get

$$C_1 = \frac{C(t_k)}{t_k - x} = -mg(0;t_k)(t_k-x)^{j-m+2} = -mB_j(P,x;t_k)(t_k-x)^{j-m+2} < 0.$$

Then by (5.2.16) and (5.2.13)

$$\begin{aligned}
&\int_{\mathbf{R}} [(t-x)P(t)]^{m-1} q(t) \operatorname{sgn}\, [(t-x)P(t)]^m d\mu(t) \\
= \ & C_1 \int_{\mathbf{R}} \frac{[(t-x)P(t)]^m}{(t-t_k)^2} \operatorname{sgn}\, [(t-x)P(t)]^m d\mu(t) \\
&+ C_2 \int_{\mathbf{R}} \frac{(t-x)^{m-1}P(t)^m}{t-t_k} \operatorname{sgn}\, [(t-x)P(t)]^m d\mu(t) \\
= \ & C_1 \int_{\mathbf{R}} \frac{|(t-x)P(t)|^m}{(t-t_k)^2} d\mu(t) < 0,
\end{aligned}$$

contradicting (5.2.6). This proves Statement (a).

(b) Suppose not and let $\partial P < n-2$. Choose $Q(t) = -(t-x)P(t)$, which belongs to $\mathbf{P}_{n-2}$ and obviously satisfies Condition A. In the present case by (5.2.7) we get $q(t) = C(t)P(t)$, where

$$C(t) = (t-x)^{j-m+1}[g'_\lambda(0;t) - mg(0;t)(t-x)^2].$$

Again we conclude $\partial C \leq 1$ and we may write the linear function C in the form $C(t) = C_1(t-x) + C_2$. To determine the sign of the constant C_1 we observe that

$$\partial g'_\lambda(0;\cdot) \leq m-j-2 < m-j = \partial[(\cdot - x)^2 g(0;\cdot)]$$

and hence by (5.2.3)

$$C_1 = \lim_{t\to\infty} \frac{C(t)}{t-x} = -mb_{m-j-2} < 0.$$

By (5.2.14) this leads to a contradiction

$$\begin{aligned}
&\int_{\mathbf{R}} [(t-x)P(t)]^{m-1} q(t) \operatorname{sgn}\, [(t-x)P(t)]^m d\mu(t) \\
= \ & C_1 \int_{\mathbf{R}} |(t-x)P(t)|^m d\mu(t) \\
&+ C_2 \int_{\mathbf{R}} (t-x)^{m-1} P(t)^m \operatorname{sgn}\, [(t-x)P(t)]^m d\mu(t) \\
= \ & C_1 \int_{\mathbf{R}} |(t-x)P(t)|^m d\mu(t) < 0
\end{aligned}$$

and proves Statement (b).

(c) If $r = n-1$ then relation (5.2.13) means (5.2.12), since the set

$$\{P(t)/(t-t_1),\ldots,P(t)/(t-t_{n-1})\}$$

spans the space $\mathbf{P}_{n-2}$; if $r = n-2$ then relations (5.2.13) and (5.2.14) imply (5.2.12), since the set

$$\{P(t)/(t-t_1),\ldots,P(t)/(t-t_{n-2}),P(t)\}$$

again spans the space $\mathbf{P}_{n-2}$. □

Remark 5.2.4. The present proof of Lemma 5.2.3 fills a gap in the proof of the same lemma (Lemma 3) in [108].

The first main result in this section is the following

Theorem 5.2.1. *Let $d\mu$ be a measure on $\mathbf{R}$ and let $x \in \mathbf{R}$ be fixed.*

(a) There exists a unique polynomial $P \in \mathbf{P}^*_{n-1}(x)$ such that (5.2.5) holds for every $j \in \mathbf{M}_2$.

(b) $\partial P \geq n-2$ and the zeros of the polynomial P are distinct.

(c) Equation (5.2.5) is true if and only if orthogonality relation (5.2.12) holds.

(d) We have

$$\lambda_{m-2,n}(d\mu,m;x) = \min_{Q\in\mathbf{P}_{n-1},Q(x)=1} \frac{1}{(m-2)!}\int_{\mathbf{R}} |Q(t)|^m|t-x|^{m-2}d\mu(t). \tag{5.2.17}$$

Proof. We begin by showing Statement (d). To this end introduce the $(n-1)$-dimensional space

$$G_x = \{(t-x)Q(t) : Q \in \mathbf{P}_{n-2}\}. \tag{5.2.18}$$

Let us consider the extremal problem: to find $P \in \mathbf{P}_{n-1}$ such that $P(x) = 1$ and

$$\int_{\mathbf{R}} |P(t)|^m|t-x|^{m-2}d\mu(t) = \min_{Q\in\mathbf{P}_{n-1},Q(x)=1} \int_{\mathbf{R}} |Q(t)|^m|t-x|^{m-2}d\mu(t). \tag{5.2.19}$$

It is easy to see that relation (5.2.19) is true if and only if the polynomial $R = 1-P$ $(\in G_x)$ satisfies the equation

$$\int_{\mathbf{R}} |1-R(t)|^m|t-x|^{m-2}d\mu(t) = \min_{Q\in G_x} \int_{\mathbf{R}} |1-Q(t)|^m|t-x|^{m-2}d\mu(t). \tag{5.2.20}$$

But this is a problem of best L^m approximation to the function 1 with respect to the measure $|t-x|^{m-2}d\mu(t)$ from the $(n-1)$-dimensional subspace G_x. By [128, Corollary 2.2, p. 98, Corollary 3.5, p. 111, Theorem 1.11, p. 56] we conclude that there is a *unique* polynomial $R \in G_x$ satisfying (5.2.20) and further relation (5.2.20) holds if and only if

$$\int_{\mathbf{R}} [1-R(t)]^{m-1}q(t)|t-x|^{m-2}\text{sgn}\,[1-R(t)]^m d\mu(t) = 0, \quad \forall q \in G_x. \tag{5.2.21}$$

Recalling $R = 1-P$, relation (5.2.21) is equivalent to (5.2.12). This means by (5.2.18) that there is a *unique* polynomial $P \in \mathbf{P}_{n-1}$ with $P(x) = 1$ satisfying (5.2.19) and further relation (5.2.19) holds if and only if (5.2.12) is valid. The orthogonality relation (5.2.12) shows that the polynomial $(t-x)P(t)$ in t changes sign at least $n-1$ times and hence $P(t)$ changes sign at least $n-2$ times. But $P \in \mathbf{P}_{n-1}$. So P has distinct real zeros only and hence $P \in \mathbf{P}_{n-1}^*(x)$. By (5.1.3), (5.2.1), and (5.2.2) we see

$$A_{m-2}(P,x;t) = \frac{1}{(m-2)!}(t-x)^{m-2}P(t)^m. \tag{5.2.22}$$

This proves Statements (d).

Meanwhile, we have proved that the solution of the orthogonality relation (5.2.12) is *unique*. By Lemma 5.2.3 Statements (a)–(c) follow. □

As an immediate consequence of Theorem 5.2.1 we state

Corollary 5.2.1. *Let $d\mu$ be a measure on $\mathbf{R}$ and let $P \in \mathbf{P}_{n-1}$ with $P(x) = 1$ satisfy (5.2.12) or (5.2.19). Then the relation (5.1.6) holds for every $j = 0,1,\ldots,m-2$.*

Remark 5.2.5. Corollary 5.2.1 provides an alternative definition of $\lambda_{jn}(d\mu,m;x)$.

Corollary 5.2.2. *Let $d\mu$ be a measure on $\mathbf{R}$. We have*

$$\lambda_{0n}(d\mu,2;x) = \lambda_n(d\mu;x). \tag{5.2.23}$$

Corollary 5.2.3. *Let $d\mu$ be a measure on $\mathbf{R}$. If $P \in \mathbf{P}_{n-1}^*(x)$ satisfies (5.2.5) then the interval $\Delta(d\mu)$ contains at least $n-2$ zeros of P.*

Proof. Suppose to the contrary that the interval $\Delta(d\mu)$ contains $r(\le n-3)$ zeros of P, say, $t_1,\ldots,t_r$. For $q(t) = (t-x)(t-t_1)\cdots(t-t_r)$ we see that the polynomial $[(t-x)P(t)]^{m-1}q(t)$ does not change sign in $\Delta(d\mu)$, which implies that its integral over $\mathbf{R}$ is not zero, contradicting (5.2.12). □

The second main result in this section is the following

Theorem 5.2.2. *Let $d\mu$ be a measure on $\mathbf{R}$ and let a polynomial $P \in \mathbf{P}_{n-1}^*(x)$ have the form*

$$P(t) = c\prod_{i=1}^{\partial P}(t-t_i), \quad t_1 < t_2 < \cdots < t_{\partial P}. \tag{5.2.24}$$

Then the following statements are equivalent:

(a) The polynomial P satisfies (5.2.5) for $j \in \mathbf{M}_2$.

(b) The polynomial P satisfies the orthogonality relation (5.2.12).

(c) We have the Gaussian quadrature formula

$$\int_{\mathbf{R}} f(t)sgn\,[(t-x)P(t)]^m d\mu(t) = \sum_{k=0}^{\partial P}\sum_{j=0}^{m-2} c_{kj} f^{(j)}(t_k) \tag{5.2.25}$$

exact for all $f \in \mathbf{P}_{(m-1)(\partial P+1)+n-2}$, where $t_0 = x$ and

$$c_{0j} = \lambda_{jn}(d\mu,m;x), \quad j = 0,1,\ldots,m-2. \tag{5.2.26}$$

(d) The polynomial P satisfies (5.2.19).

Proof. (a)$\Longleftrightarrow$(b) Use Theorem 5.2.1(c).

(b)$\Longleftrightarrow$(c) We separate the cases when $\partial P = n-1$ and $\partial P = n-2$.

Case 1. $\partial P = n-1$. In this case let $A_{kj} \in \mathbf{P}_{mn-2}$ be the fundamental polynomials of Hermite interpolation at the nodes $t_0, t_1, \ldots, t_{n-1}$ with the corresponding multiplicity $m_0 = m-1$, $m_1 = \cdots = m_{n-1} = m$. Then for $f \in \mathbf{P}_{mn-2}$

$$f(t) = \sum_{k=0}^{\partial P} \sum_{j=0}^{m_k-1} f^{(j)}(t_k) A_{kj}(t). \tag{5.2.27}$$

Multiplying this with sgn $[(t-x)P(t)]^m$ and then integrating the obtained formula, we have

$$\int_{\mathbf{R}} f(t) \operatorname{sgn}\, [(t-x)P(t)]^m d\mu(t) = \sum_{k=0}^{\partial P} \sum_{j=0}^{m_k-1} c_{kj} f^{(j)}(t_k), \tag{5.2.28}$$

where

$$c_{kj} = \int_{\mathbf{R}} A_{kj}(t) \operatorname{sgn}\, [(t-x)P(t)]^m d\mu(t). \tag{5.2.29}$$

The formula (5.2.28) becomes (5.2.25) if and only if

$$c_{k,m_k-1} = c_{k,m-1} = 0, \quad k = 0, 1, \ldots, \partial P. \tag{5.2.30}$$

Since

$$A_{k,m-1}(t) = \frac{1}{(m-1)!(t_k - x)^{m-1} P'(t_k)^m} [(t-x)P(t)]^{m-1} \frac{P(t)}{t-t_k}, \quad k = 1, 2, \ldots, n-1,$$

relation (5.2.30) is equivalent to (5.2.12).

Case 2. $\partial P = n-2$. In this case we have to consider the Gaussian quadrature formula at the nodes $t_0, t_1, \ldots, t_{n-2}$ with the corresponding multiplicity $m_0 = m_1 = \cdots = m_{n-2} = m$. By Theorem 4.1.1 relation (5.2.25) holds for all $f \in \mathbf{P}_{m(n-1)-1}$ if and only if the orthogonality relation (5.2.12) is valid.

(b)$\Longleftrightarrow$(d) Apply Theorem 5.2.1(c) and (d). □

The relationship between the Christoffel type functions and the Christoffel numbers is as follows.

Theorem 5.2.3. *Let $d\mu$ be a measure on $\mathbf{R}$ and for any point $x \in \mathbf{R}$ let $P(x;\cdot) \in \mathbf{P}^*_{n-1}(x)$ satisfy (5.2.12). Let $x_{kn} = x_{kn}(d\mu, m)$.*

(a) We have

$$\lambda_{kjn}(d\mu, m) = [sgn\, P_n'(d\mu, m; x_{kn})^m] \lambda_{jn}(d\mu, m; x_{kn}),$$
$$k = 1, 2, \ldots, n, \quad j = 0, 1, \ldots, m-2. \tag{5.2.31}$$

(b) We have

$$P(x_{kn}; t) = \ell_{kn}(d\mu, m; t) = \frac{P_n(d\mu, m; t)}{P_n'(d\mu, m; x_{kn})(x - x_{kn})}, \quad k = 1, 2, \ldots, n. \tag{5.2.32}$$

(c) We have

$$\begin{aligned} P(x_{k,n-1};t) &= \ell_{k,n-1}(d\mu,m;t) \\ &= \frac{P_{n-1}(d\mu,m;t)}{P'_{n-1}(d\mu,m;x_{k,n-1})(x-x_{k,n-1})}, \quad k=1,2,\ldots,n-1. \end{aligned} \tag{5.2.33}$$

(d) The equality $\partial P(x;\cdot) = n-2$ is true if and only if equality $x = x_{k,n-1}$ holds for some index k, $1 \le k \le n-1$.

Proof. (a) Let an index k, $1 \le k \le n$, be fixed. We notice that in the present case the orthogonality relation (3.2.6) becomes

$$\int_{\mathbf{R}} P_n(d\mu,m;t)^{m-1} q(t) \mathrm{sgn}\, P_n(d\mu,m;t)^m d\mu(t) = 0, \quad \forall q \in \mathbf{P}_{n-1},$$

or equivalently

$$\int_{\mathbf{R}} [(t-x_k)P(t)]^{m-1} q(t) \mathrm{sgn}\, [(t-x_k)P(t)]^m d\mu(t) = 0, \quad \forall q \in \mathbf{P}_{n-1}, \tag{5.2.34}$$

where $P(t) = \ell_{kn}(d\mu,m;t)$. According to Theorem 5.2.1 equation (5.2.34) means

$$\lambda_{jn}(d\mu,m;x_k) = \int_{\mathbf{R}} A_j(P,x_k;t) \mathrm{sgn}\, [(t-x_k)P(t)]^m d\mu(t).$$

Inserting $f(t) = A_j(P,x_k;t)$ into (4.1.3) the above relation immediately gives

$$\begin{aligned} \lambda_{kjn}(d\mu,m) &= \int_{\mathbf{R}} A_j(P,x_k;t) \mathrm{sgn}\, P_n(d\mu,m;t)^m d\mu(t) \\ &= [\mathrm{sgn}\, P'_n(d\mu,m;x_{kn})^m] \int_{\mathbf{R}} A_j(P,x_k;t) \mathrm{sgn}\, [(t-x_k)P(t)]^m d\mu(t) \\ &= [\mathrm{sgn}\, P'_n(d\mu,m;x_{kn})^m] \lambda_{jn}(d\mu,m;x_k). \end{aligned}$$

This proves (5.2.31).

(b) The relation (5.2.32) is a direct consequence of the above conclusion.

(c) To prove (5.2.33) we use (5.2.34) with replacing n by $n-1$

$$\int_{\mathbf{R}} [(t-x_k)P(t)]^{m-1} q(t) \mathrm{sgn}\, [(t-x_k)P(t)]^m d\mu(t) = 0, \quad \forall q \in \mathbf{P}_{n-2},$$

where $P(t) = \ell_{k,n-1}(d\mu,m;t)$. This by Theorem 5.2.1 means (5.2.33).

(d) We observe that by (5.2.33) equality $\partial P(x;\cdot) = n-2$ is true if equality $x = x_{k,n-1}$ holds for some index k, $1 \le k \le n-1$. Conversely, if equality $\partial P(x;\cdot) = n-2$ is true, then by (5.2.12) the $(n-1)$th polynomial $(t-x)P(x;t)$ is the m orthogonal polynomial with respect to $d\mu$ and hence equality $x = x_{k,n-1}$ holds for some index k, $1 \le k \le n-1$. □

Theorem 5.2.4. *Let $d\mu$ be a measure on $\mathbf{R}$ and for any point $x \in \mathbf{R}$ let $P(x;\cdot) \in \mathbf{P}^*_{n-1}(x)$ satisfy (5.2.5) for $j \in \mathbf{M}_2$. If $x \ne y$ then the zeros of the polynomials $(t-x)P(x;t)$ and $(t-y)P(y;t)$ are separate each other; more precisely, between two consecutive zeros of the polynomial $(t-x)P(x;t)$ (or $(t-y)P(y;t)$) there is precisely one zero of $(t-y)P(y;t)$ (or $(t-x)P(x;t)$).*

In particular, if the polynomial $P(x;\cdot)$ has the form (5.2.24) and if $x \notin \{x_{1n}(d\mu,m),\ldots,x_{nn}(d\mu,m)\}$, then each open interval

$$(x_{kn}(d\mu,m),x_{k+1,n}(d\mu,m)), \quad k=1,2,\ldots,n-1,$$

contains precisely one point of $\{x,t_1,\ldots,t_{\partial P}\}$.

Proof. By Theorem 5.2.1

$$\int_{\mathbf{R}}[(t-x)P(x;t)]^{m-1}q(t)\text{sgn }[(t-x)P(x;t)]^m d\mu(t)=0, \quad \forall q\in\mathbf{P}_{n-2},$$

and

$$\int_{\mathbf{R}}[(t-y)P(y;t)]^{m-1}q(t)\text{sgn }[(t-y)P(y;t)]^m d\mu(t)=0, \quad \forall q\in\mathbf{P}_{n-2}.$$

Then for arbitrary numbers c and d

$$\int_{\mathbf{R}}\{c[(t-x)P(x;t)]^{m-1}\text{sgn }[(t-x)P(x;t)]^m$$
$$-d[(t-y)P(y;t)]^{m-1}\text{sgn }[(t-y)P(y;t)]^m\}q(t)d\mu(t)=0, \quad \forall q\in\mathbf{P}_{n-2},$$

or equivalently

$$\int_{\mathbf{R}}\{c|(t-x)P(x;t)|^{m-2}(t-x)P(x;t)$$
$$-d|(t-y)P(y;t)|^{m-2}(t-y)P(y;t)\}q(t)d\mu(t)=0, \quad \forall q\in\mathbf{P}_{n-2}.$$

This shows that the term in the braces has at least $n-1$ sign changes for every nonzero pair $\{c,d\}$. Applying Lemma 3.3.1 we obtain the first part of the theorem. The second follows from the first and Theorem 5.2.3. □

Theorem 5.2.5. *Let $d\mu$ be a measure on $\mathbf{R}$ and for any point $x\in\mathbf{R}$ let $P(x;\cdot)\in\mathbf{P}^*_{n-1}(x)$ satisfy (5.2.5) for $j\in\mathbf{M}_2$. Then both $P(x;\cdot)$ and $\lambda_{jn}(d\mu,m;x)$ are continuous with respect to x.*

Proof. To prove our theorem it is convenient to apply Theorem 5.2.2(d): the inequality

$$\int_{\mathbf{R}}|P(x;t)|^m|t-x|^{m-2}d\mu(t)\le\int_{\mathbf{R}}|Q(t)|^m|t-x|^{m-2}d\mu(t) \tag{5.2.35}$$

holds for all $Q\in\mathbf{P}_{n-1}$ with $Q(x)=1$. Let $x\to\xi$ for some point ξ and let $Q_0\in\mathbf{P}_{n-1}$ with $Q_0(\xi)=1$ be arbitrary. Put $Q(t)=Q_0(t-x+\xi)$. Clearly $Q(x)=1$ and inequality (5.2.35) holds. Taking $x\to\xi$ in this inequality, suppose, passing to a subsequence if necessary, that $P(x;\cdot)\to P_0$. Then

$$\int_{\mathbf{R}}|P_0(t)|^m|t-\xi|^{m-2}d\mu(t)\le\int_{\mathbf{R}}|Q_0(t)|^m|t-\xi|^{m-2}d\mu(t).$$

Since $P_0(\xi)=1$ and $Q_0\in P_{n-1}$ with $Q_0(\xi)=1$ is arbitrary, by uniqueness $P_0=P(\xi;\cdot)$. This prove that $P(x;\cdot)$ is continuous with respect to x. Hence $\lambda_{jn}(d\mu,m;x)$ is also continuous with respect to x. □

Theorem 5.2.6. *Let $d\mu$ be a measure on* **R**. *Then for $j \in \mathbf{M}_2$ and $i > j$ we have the estimation*

$$|\lambda_{in}(d\mu, m; x)| \le c d^{i-j} \lambda_{jn}(d\mu, m; x), \tag{5.2.36}$$

where

$$d = \begin{cases} \max\{|x - x_{k-1,n}|, |x - x_{k+2,n}|\}, & x \in [x_{kn}, x_{k+1,n}], \quad 2 \le k \le n-2, \\ |x - x_{2n}|, & x \le x_{1n}, \\ |x - x_{n-2,n}|, & x \ge x_{nn}. \end{cases} \tag{5.2.37}$$

Further, if $d\mu$ is supported in $[-1,1]$ and condition (4.5.21) is valid, then

$$|\lambda_{in}(d\mu, m; x)| \le c\Delta_n(x)^{i-j} \lambda_{jn}(d\mu, m; x), \quad x \in [-1,1] \setminus [(x_{1n}, x_{2n}) \cup (x_{n-1,n}, x_{nn})]. \tag{5.2.38}$$

Proof. Let $P \in \mathbf{P}^*_{n-1}(x)$ satisfy (5.2.5) for $j \in \mathbf{M}_2$, where $x \in [x_{kn}, x_{k+1,n}]$, $2 \le k \le n-2$. By Theorem 5.2.4 the interval $[x_{k-1}, x_k)$ must contain a zero of P, say, y, and the interval $(x_{k+1}, x_{k+2}]$ must contain a zero of P, say, z. Then by (5.1.5), (5.1.6), and (2.4.20)

$$|\lambda_{in}(d\mu, m; x)| \le c(\max\{|x-y|, |x-z|\})^{i-j} \lambda_{jn}(d\mu, m; x) \le c d^{i-j} \lambda_{jn}(d\mu, m; x).$$

If $x \le x_1$ then by Theorem 5.2.4 the interval $(-\infty, x)$ contains no any zero of P and the interval $(x_1, x_2]$ must contain a zero of P, say, y. Then by (5.1.5), (5.1.6), and (2.4.37)

$$|\lambda_{in}(d\mu, m; x)| \le c|x-y|^{i-j} \lambda_{jn}(d\mu, m; x) \le c d^{i-j} \lambda_{jn}(d\mu, m; x).$$

Similarly, if $x \ge x_n$ then we can obtain (5.2.36).

Relation (5.2.38) follows from (5.2.37) and (4.5.22). □

Remark 5.2.6. It is difficult to give the estimation of $\lambda_{in}(d\mu, m; x)$ for $x \in (x_{1n}, x_{2n}) \cup (x_{n-1,n}, x_{nn})$, because in this case the number d may be large enough.

Lemma 5.2.4. [65, Lemma 6.3.8, p. 108] *Let*

$$K_n(v; x, t) = \frac{T_n(x)T_{n-1}(t) - T_{n-1}(x)T_n(t)}{\pi(x-t)}, \quad n \ge 2. \tag{5.2.39}$$

Then

$$|K_n(v; x, t)| \le c \min\left\{ n, \frac{(1-x^2)^{1/2} + (1-t^2)^{1/2}}{|x-t|} \right\}, \quad x, t \in [-1,1], \tag{5.2.40}$$

where c is an absolute constant.

Lemma 5.2.5. [24, (3.7), p. 102; p. 104] *Let*

$$\pi_{n-1}(x,t) = \frac{K_n(v; x, t)}{K_n(v; x, x)}.$$

Then

$$|\pi_{n-1}(x,t)| \le 4, \quad x, t \in [-1,1], \quad n \ge 3, \tag{5.2.41}$$

and

$$K_n(v; x, x) \sim n, \quad |x| \le 1. \tag{5.2.42}$$

By [24, Theorem 3.1, p. 19] the polynomial $\pi_{n-1}(x,t)$ in t has simple real zeros only and hence

$$\pi_{n-1}(x,\cdot) \in \mathbf{P}^*_{n-1}(x). \tag{5.2.43}$$

Lemma 5.2.6. *We have*

$$|b_i(\pi_{n-1}(x,\cdot);x)| \le c\Delta_n(x)^{-i}, \quad |x| \le 1, \quad i = 0,1,\dots. \tag{5.2.44}$$

Proof. The estimation (5.2.41) implies

$$\left| \left[\frac{d^i}{dt^i} \pi_{n-1}(x,t)^m \right]_{t=x} \right| \le c\Delta_n(x)^{-i}, \quad |x| \le 1, \quad i = 0,1,\dots$$

Then by the recurrence relation

$$\begin{aligned} & b_i(\pi_{n-1}(x,\cdot);x) \\ = & -\sum_{\nu=0}^{i-1} b_\nu(\pi_{n-1}(x,\cdot),x) \left\{ \frac{1}{(i-\nu)!} \left[\frac{d^{i-\nu}}{dt^{i-\nu}} \pi_{n-1}(x,t)^m \right]_{t=x} \right\}, \\ & i = 1,2,\dots, \end{aligned} \tag{5.2.45}$$

we can get (5.2.44), where (5.2.45) may be derived by Newton-Leibniz rule from the identity

$$[\pi_{n-1}(x,t)^{-m} \pi_{n-1}(x,t)^m]^{(i)}_{t=x} = 0, \quad i \ge 1. \quad \square$$

Theorem 5.2.7. *Let $d\mu$ be a measure supported in $[-1,1]$ and satisfy*

$$\mu \in Lip_M \gamma, \quad 0 < \gamma \le 1. \tag{5.2.46}$$

Then for $j \in \mathbf{M}_2$

$$\lambda_{jn}(d\mu,m;x) \le cn^{2-m-\gamma} \Delta_n(x)^{2+j-m}. \tag{5.2.47}$$

Proof. Choose

$$P(x;t) = \pi_{n-1}(x,t).$$

By (5.2.40), (5.2.42), and (5.2.44)

$$\begin{aligned} & \lambda_{jn}(d\mu,m;x) \\ \le & \int_{-1}^{1} |A_j(\pi_{n-1},x;t)| d\mu(t) \\ \le & c \sum_{i=0}^{m-j-2} \Delta_n(x)^{-i} \int_{-1}^{1} |(t-x)^{j+i} \pi_{n-1}(x,t)^m| d\mu(t) \\ \le & c \sum_{i=0}^{m-j-2} n^{-j-i} \Delta_n(x)^{-i} \int_{-1}^{1} |\pi_{n-1}(x,t)|^{m-j-i} d\mu(t). \end{aligned} \tag{5.2.48}$$

We have to estimate the integral

$$\sigma_q = \int_{-1}^{1} |\pi_{n-1}(x,t)|^q d\mu(t), \quad q \ge 2.$$

We divide the integral σ_q into the three parts:

$$\begin{aligned}\sigma_q &= \int_{x-1/n}^{x+1/n} |\pi_{n-1}(x,t)|^q d\mu(t) + \int_{-1}^{x-1/n} |\pi_{n-1}(x,t)|^q d\mu(t) \\ &\quad + \int_{x+1/n}^{1} |\pi_{n-1}(x,t)|^q d\mu(t) \\ &= S_1 + S_2 + S_3.\end{aligned}$$

It is simple to estimate S_1, since by (5.2.41) and (5.2.46)

$$S_1 \le M2^{2q+1}n^{-\gamma}.$$

If $1+x \le 1/n$ then $S_2 = 0$; otherwise applying (5.2.40), (5.2.42), and (5.2.46), and using partial integration, we obtain

$$\begin{aligned}S_2 &\le cn^{-q}\int_{-1}^{x-1/n} (x-t)^{-q} d\mu(t) \\ &= cn^{-q}\int_{-1}^{x-1/n} (x-t)^{-q} d[\mu(t)-\mu(x)] \\ &\le cn^{-q}\left\{\frac{\mu(x)-\mu(-1)}{(1+x)^q} + qM\int_{-1}^{x-1/n} (x-t)^{\gamma-q-1}dt\right\} \\ &\le cMn^{-q}\{(1+x)^{\gamma-q} + q(q-\gamma)^{-1}n^{q-\gamma}\} \\ &\le cMn^{-\gamma}.\end{aligned}$$

Similarly, if $1-x \le 1/n$ then $S_3 = 0$; otherwise

$$S_3 \le cMn^{-\gamma}.$$

Thus $\sigma_q \le cn^{-\gamma}$ and by (5.2.48) we obtain (5.2.47). □

Theorem 5.2.8. *Let $d\mu$ and $d\nu$ be measures on* **R**.
(a) If $d\mu \le d\nu$ then

$$\lambda_{jn}(d\mu,m;x) \le \lambda_{jn}(d\nu,m;x), \quad x \in \mathbf{R}, \quad j \in \mathbf{M}_2. \tag{5.2.49}$$

(b) If $m \in \mathbf{N}_2$ then

$$\lambda_{0n}(d\mu,m;x) \ge \lambda_{\frac{mn}{2}}(d\mu;x). \tag{5.2.50}$$

Proof. (a) Inequality (5.2.49) follows directly from (5.1.5).
(b) By (5.1.1)

$$\lambda_{\frac{mn}{2}}(d\mu;x) = \min_{Q\in\mathbf{P}_{\frac{mn}{2}-1}} \frac{1}{Q(x)^2}\int_{\mathbf{R}} Q(t)^2 d\mu(t).$$

Then

$$Q(x)^2 \le \lambda_{\frac{mn}{2}}(d\mu;x)^{-1}\int_{\mathbf{R}} Q(t)^2 d\mu(t), \quad Q \in \mathbf{P}_{\frac{mn}{2}-1}. \tag{5.2.51}$$

Let $P \in \mathbf{P}_{n-1}^*(x)$ satisfy (5.2.5) with $j=0$. Since $A_0(P,x;t) \geq 0$ on $\mathbf{R}$, by Theorem 1.5.2 it may be written as

$$A_0(P,x;t) = R(t)^2 + Q(t)^2, \quad R, Q \in \mathbf{P}_{\frac{mn}{2}-1}.$$

Thus by (5.2.51)

$$\begin{aligned} A_0(P,x;t) &= R(t)^2 + Q(t)^2 \leq \lambda_{\frac{mn}{2}}(d\mu;t)^{-1} \int_{\mathbf{R}} [R(s)^2 + Q(s)^2] d\mu(s) \\ &= \lambda_{\frac{mn}{2}}(d\mu;t)^{-1} \int_{\mathbf{R}} A_0(P,x;s) d\mu(s) \\ &= \lambda_{\frac{mn}{2}}(d\mu;t)^{-1} \lambda_{0n}(d\mu,m;x). \end{aligned} \tag{5.2.52}$$

Putting $t = x$ we get

$$1 \leq \lambda_{\frac{mn}{2}}(d\mu;x)^{-1} \lambda_{0n}(d\mu,m;x),$$

which is equivalent to (5.2.50). □

5.3 Christoffel Type Functions for Generalized Jacobi Weights

In this section we shall give an estimation of $\lambda_{jn}(u,m;x)$ for a weight $u \sim W$, a.e., where W is a *generalized Jacobi weight*:

$$\begin{aligned} W(x) &= \prod_{i=1}^{r} |x - t_i|^{p_i}, \quad |x| < 1, \quad W(x) = 0, \quad |x| \geq 1, \\ -1 &= t_1 < t_2 < \cdots < t_r = 1 \ (r \geq 2), \quad p_i > -1, \ i = 1, 2, \ldots, r. \end{aligned} \tag{5.3.1}$$

Lemma 5.3.1. [65, Theorem 6.3.28, p. 120 and Theorem 9.22, p. 166] *If* $0 < m < \infty$ *and*

$$u \sim W, \quad a.e. \tag{5.3.2}$$

then

$$\lambda_n(u,m;x) \sim \lambda_n(u;x) \sim \frac{1}{n} \widehat{W}_n(x), \quad |x| \leq 1 \tag{5.3.3}$$

and

$$\theta_{kn}(u) - \theta_{k+1,n}(u) \sim \frac{1}{n}, \quad k = 0, 1, \ldots, n, \tag{5.3.4}$$

where

$$\widehat{W}_n(x) = \left[(1+x)^{1/2} + \frac{1}{n}\right]^{2p_1+1} \left[(1-x)^{1/2} + \frac{1}{n}\right]^{2p_r+1} \prod_{i=2}^{r-1} \left[|x - t_i| + \frac{1}{n}\right]^{p_i} \tag{5.3.5}$$

and

$$x_{kn}(u) = x_{kn}(u,2) = \cos\theta_{kn}(u), \quad 0 \leq \theta_{kn}(u) \leq \pi, \quad k = 0, 1, \ldots, n+1.$$

Lemma 5.3.2. *Let* $A, B, p \geq 0$ *and* $AB + p > 0$. *Then*

$$(A+B)^p \leq c(A^p + B^p), \tag{5.3.6}$$

where

$$c = \begin{cases} 1, & 0 \leq p \leq 1, \\ 2^{p-1}, & p > 1. \end{cases}$$

Proof. Suppose without loss of generality that $A \geq B$. If $B = 0$ or $p = 0$ then inequality (5.3.6) is trivial. Let $B, p > 0$. In this case we consider the equivalent inequality with $x = A/B$

$$f(x) = c(x^p + 1) - (x+1)^p \geq 0, \quad x \geq 1.$$

Since

$$f'(x) = p[cx^{p-1} - (x+1)^{p-1}] = px^{p-1}\left[c - \left(\frac{x+1}{x}\right)^{p-1}\right] \geq 0, \quad x \geq 1,$$

we have

$$f(x) \geq f(1) = 2c - 2^p \geq 0, \quad x \geq 1,$$

as stated. □

The following result will play an important role in this section.

Theorem 5.3.1. *If* $u(t) \leq CW(t)$ *and*

$$N \geq N_0 = 8 + 2\left[p_1 + p_r + \frac{1}{2}\sum_{i=2}^{r-1}|p_i|\right], \tag{5.3.7}$$

where the symbol $[y]$ *denotes the integral part of the real number y, then*

$$S = \int_{-1}^{1} |\pi_{n-1}(x,t)|^N u(t)dt \leq c(W,C,N)n^{-1}\widehat{W}_n(x) \tag{5.3.8}$$

holds for all $x \in [-1,1]$ *and* $n = 1, 2, \ldots$.

Proof. The proof follows and properly modifies the ideas of Nevai in [65, pp. 109–111]. By (5.2.41) it is enough to show inequality (5.3.8) for $N = N_0$. Now let $n \geq 3$ and write $M = N_0 n$. Assume that a point $x \in [-1,1]$ is fixed and $|x - x_j| = \min_{1 \leq k \leq M}|x - x_k|$, where $x_k = x_{kM}(W)$ and $1 \leq j \leq M$. Suppose without loss of generality that $1 \leq j \leq M/2$. (The case when $j > M/2$ may be discussed analogously and symmetrically.) Using the Gaussian quadrature formula (4.1.3) with $m_k \equiv 2$ we obtain

$$S \leq C\int_{-1}^{1} \pi_{n-1}(x,t)^{N_0} W(t)dt = C\sum_{k=1}^{M} \lambda_k \pi_{n-1}(x,x_k)^{N_0},$$

where $\lambda_k = \lambda_M(W; x_k)$, $k = 1, 2, \ldots, M$. Using (5.2.40)–(5.2.42) and (5.3.6), we get

$$\begin{aligned} S &\leq c\left[\lambda_j \pi_{n-1}(x,x_j)^{N_0} + \sum_{k \neq j} \lambda_k n^{-N_0} K_n(v; x, x_k)^{N_0}\right] \\ &\leq c\left\{\lambda_j + M^{-N_0}\sum_{k \neq j}\lambda_k\left[\frac{(1-x^2)^{1/2} + (1-x_k^2)^{1/2}}{x - x_k}\right]^{N_0}\right\} \\ &\leq c\left\{\lambda_j + M^{-N_0}\left[(1-x^2)^{N_0/2}S_1 + S_2\right]\right\}, \end{aligned} \tag{5.3.9}$$

here

$$S_1 = \sum_{k\neq j} \lambda_k (x-x_k)^{-N_0}, \quad S_2 = \sum_{k\neq j} \lambda_k (1-x_k^2)^{N_0/2}(x-x_k)^{-N_0}.$$

We break the estimations of S_1 and S_2 into a series of inequalities. According to Lemma 5.3.1 we have

$$\begin{aligned} c_1 \frac{|k-j|\min\{k+j, 2M+2-k-j\}}{M^2} &\leq |x_k - x_j| \\ \leq \quad c_0 \frac{|k-j|\min\{k+j, 2M+2-k-j\}}{M^2}&, \quad k = 0, 1, \ldots, M+1, \end{aligned} \tag{5.3.10}$$

where $c_0 = c_0(W)$ and $c_1 = c_1(W)$. For simplicity write

$$v_{ki} = (|x_k - t_i| + M^{-1})^{p_i}, \quad i = 2, \ldots, r-1, \quad V_k = \prod_{i=2}^{r-1} v_{ki}.$$

Claim 1. We have

$$v_{ki} \leq [(1+2c_0)|k-j|]^{|p_i|} v_{ji}, \quad 1 \leq k \leq M, \quad k \neq j, \quad 2 \leq i \leq r-1, \tag{5.3.11}$$

and

$$V_k \leq [(1+2c_0)|k-j|]^{|p_2|+\ldots+|p_{r-1}|} V_j, \quad 1 \leq k \leq M, \quad k \neq j, \tag{5.3.12}$$

where c_0 is given in (5.3.10).

In fact, it follows from (5.3.10) that

$$\begin{aligned} |x_k - t_i| + M^{-1} &\leq |x_k - x_j| + |x_j - t_i| + M^{-1} \\ \leq \quad 2c_0|k-j|M^{-1} &+ |x_j - t_i| + M^{-1} \leq (1+2c_0)|k-j|(|x_j - t_i| + M^{-1}). \end{aligned}$$

Hence

$$(|x_k - t_i| + M^{-1})^{|p_i|} \leq [(1+2c_0)|k-j|]^{|p_i|}(|x_j - t_i| + M^{-1})^{|p_i|},$$

that is,

$$v_{ki}^{\operatorname{sgn} p_i} \leq [(1+2c_0)|k-j|]^{|p_i|} v_{ji}^{\operatorname{sgn} p_i}. \tag{5.3.13}$$

If $p_i \geq 0$ then (5.3.13) becomes (5.3.11); if $p_i < 0$ then inequality (5.3.13) by interchanging the indices k and j in it again becomes (5.3.11). The inequality (5.3.12) follows directly from (5.3.11). This proves Claim 1.

Claim 2. If $\delta = N_0 - \sum_{i=2}^{r-1} |p_i|$ and

$$-1 < \gamma_1, \gamma_2 < \delta + N_0 - 1, \tag{5.3.14}$$

then

$$\sigma(\gamma_1, \gamma_2) = \sum_{k\neq j} h_k \leq c(\gamma_1, \gamma_2, \delta, N_0) M^{\gamma_2} j^{\gamma_1 - N_0}, \tag{5.3.15}$$

where $h_k = k^{\gamma_1}(M+1-k)^{\gamma_2}|k-j|^{-\delta}(\min\{k+j, 2M+2-k-j\})^{-N_0}$.

In fact, since $j \le M/2$, we may conclude

$$\min\{k+j, 2M+2-k-j\} \ge \frac{3}{5}(k+j), \quad k \le 3M/4. \tag{5.3.16}$$

In order to prove (5.3.16) it suffices to show $2M+2-k-j \ge 3(k+j)/5$. This is the case, because

$$\begin{aligned} 2M+2-k-j &= \frac{8}{5}\left(\frac{3M}{4}+\frac{M}{2}\right)+2-k-j \\ &\ge \frac{8}{5}(k+j)-(k+j) = \frac{3}{5}(k+j). \end{aligned}$$

This proves (5.3.16). Thus by (5.3.16)

$$h_k \le ck^{\gamma_1}M^{\gamma_2}|k-j|^{-\delta}(k+j)^{-N_0}, \quad k \le 3M/4. \tag{5.3.17}$$

Since $\gamma_1 > -1$ and $\delta > 1$, we have that by (5.3.17)

$$\sum_{k\le j/2} h_k \le cM^{\gamma_2} j^{-\delta-N_0} \sum_{k\le j/2} k^{\gamma_1} \le cM^{\gamma_2} j^{\gamma_1+1-\delta-N_0} \le cM^{\gamma_2} j^{\gamma_1-N_0}$$

and

$$\sum_{j/2<k\le 3j/2} h_k \le cM^{\gamma_2} j^{\gamma_1-N_0} \sum_{j/2<k\le 3j/2} |k-j|^{-\delta} \le cM^{\gamma_2} j^{\gamma_1-N_0}.$$

We see that the inequality $k > 3j/2$ implies $k-j \ge k/3$ and that inequalities (5.3.14) imply $\gamma_1 - N_0 - \delta < -1$. Thus using (5.3.14) and (5.3.17) we obtain

$$\sum_{3j/2<k\le 3M/4} h_k \le cM^{\gamma_2} \sum_{3j/2<k\le 3M/4} k^{\gamma_1-N_0-\delta}$$

$$\le \begin{cases} cM^{\gamma_2} \le cM^{\gamma_2} j^{\gamma_1-N_0}, & \gamma_1 - N_0 \ge 0, \\ cM^{\gamma_2} j^{\gamma_1-N_0} \sum_{3j/2<k\le 3M/4} k^{-\delta} \le cM^{\gamma_2} j^{\gamma_1-N_0}, & \gamma_1 - N_0 < 0. \end{cases}$$

At last, it follows from the three inequalities $j \le M/2$, $N_0 > \gamma_1 - \delta + 1$, and $\delta > 1$ that

$$\begin{aligned} \sum_{k>3M/4} h_k &\le cM^{\gamma_1-\delta-N_0} \sum_{k>3M/4} (M+1-k)^{\gamma_2} \le cM^{\gamma_1+\gamma_2+1-\delta-N_0} \\ &\le cM^{\gamma_2} j^{\gamma_1+1-\delta-N_0} \le cM^{\gamma_2} j^{\gamma_1-N_0}. \end{aligned}$$

This proves Claim 2.

Now let us estimate S_1 and S_2. To apply Claim 2 corresponding to S_1 and S_2 choose two sets of values of γ_1 and γ_2

$$\gamma_1 = 2p_1+1, \quad \gamma_2 = 2p_r+1,$$

and

$$\gamma_1 = N_0+2p_1+1, \quad \gamma_2 = N_0+2p_r+1,$$

respectively. It is easy to check that each of them satisfies (5.3.14). It follows from (5.3.3) and (5.3.10) that

$$\begin{aligned}
S_1 &\le c\sum_{k\neq j}\left\{M^{-1}V_k\left(\frac{k}{M}\right)^{2p_1+1}\left(\frac{M+1-k}{M}\right)^{2p_r+1}\right.\\
&\quad\left.\times\left[\frac{|k-j|\min\{k+j,2M+2-k-j\}}{M^2}\right]^{-N_0}\right\}\\
&= cM^{2N_0-2p_1-2p_r-3}\sum_{k\neq j}\{V_k k^{2p_1+1}(M+1-k)^{2p_r+1}\\
&\quad\times[|k-j|\min\{k+j,2M+2-k-j\}]^{-N_0}\}.
\end{aligned}$$

The kth term in the sum by (5.3.12) does not exceed

$$\begin{aligned}
&cV_j|k-j|^{|p_2|+\cdots+|p_{r-1}|}k^{2p_1+1}(M+1-k)^{2p_r+1}\\
&\times[|k-j|\min\{k+j,2M+2-k-j\}]^{-N_0}.
\end{aligned}$$

Substituting this estimation into the above inequality and using (5.3.15) with $\gamma_1=2p_1+1$ and $\gamma_2=2p_r+1$ gives

$$\begin{aligned}
S_1 &\le cM^{2N_0-2p_1-2p_r-3}V_j\sum_{k\neq j}k^{2p_1+1}(M+1-k)^{2p_r+1}|k-j|^{-\delta}\\
&\quad\times[\min\{k+j,2M+2-k-j\}]^{-N_0}\\
&= cM^{2N_0-2p_1-2p_r-3}V_j\sigma(2p_1+1,2p_r+1)\\
&\le cM^{2N_0-2p_1-2}j^{2p_1+1-N_0}V_j.
\end{aligned}$$

Similarly, it follows from (5.3.3), (5.3.10), (5.3.12), and (5.3.15) with $\gamma_1=N_0+2p_1+1$ and $\gamma_2=N_0+2p_r+1$ that

$$\begin{aligned}
S_2 &\le c\sum_{k\neq j}\left\{M^{-1}V_k\left(\frac{k}{M}\right)^{2p_1+1}\left(\frac{k}{M}\right)^{N_0}\left(\frac{M+1-k}{M}\right)^{2p_r+1}\left(\frac{M+1-k}{M}\right)^{N_0}\right.\\
&\quad\left.\times\left[\frac{|k-j|\min\{k+j,2M+2-k-j\}}{M^2}\right]^{-N_0}\right\}\\
&\le cM^{-2p_1-2p_r-3}V_j\sigma(N_0+2p_1+1,N_0+2p_r+1)\\
&\le cM^{N_0-2p_1-2}j^{2p_1+1}V_j.
\end{aligned}$$

We must remember that $x\sim x_j$ and $j\le M/2$, which imply $1+x\sim 1+x_j\sim (j/M)^2$ and $1-x\sim 1-x_j\sim 1$. So substituting S_1 and S_2 into (5.3.9) and applying (5.3.10) and (5.3.3) gives

$$\begin{aligned}
S &\le c\left[\lambda_j+M^{N_0-2p_1-2}j^{2p_1+1-N_0}V_j(1-x^2)^{N_0/2}+M^{-2p_1-2}j^{2p_1+1}V_j\right]\\
&= c\left[\lambda_j+M^{-1}(j/M)^{2p_1+1-N_0}(1-x^2)^{N_0/2}V_j+M^{-1}(j/M)^{2p_1+1}V_j\right]\\
&\le c\left[\lambda_j+M^{-1}(1+x_j)^{(2p_1+1-N_0)/2}(1+x)^{N_0/2}V_j+M^{-1}(1+x_j)^{(2p_1+1)/2}V_j\right]\\
&\le c\left[\lambda_j+M^{-1}(1+x_j)^{(2p_1+1)/2}V_j\right]\\
&\le c\lambda_j=c\lambda_M(x_j)\le c\lambda_M(x)\le cM^{-1}\widehat{W}_M(x)\le cn^{-1}\widehat{W}_n(x). \quad\square
\end{aligned}$$

Theorem 5.3.2. *Let*

$$C_1 W \le u \le C_2 W. \tag{5.3.18}$$

Then for a fixed number $N > 0$

$$\int_{-1}^{1} |\pi_{n-1}(x,t)|^N u(t)dt \ge C_1 C_2^{-1} c(W,N)\lambda_n(W;x), \quad |x| \le 1, \quad n \in \mathbf{N}. \tag{5.3.19}$$

Proof. We distinguish two cases when $N < 2$ and $N \ge 2$.

Case 1. $N < 2$. In this case by the definition of $\lambda_n(W;x)$ it follows from (5.3.18) and (5.2.41) that

$$\begin{aligned} \lambda_n(W;x) &\le \int_{-1}^{1} |\pi_{n-1}(x,t)|^2 W(t)dt \le C_1^{-1} \int_{-1}^{1} |\pi_{n-1}(x,t)|^2 u(t)dt \\ &\le 4^{2-N} C_1^{-1} \int_{-1}^{1} |\pi_{n-1}(x,t)|^N u(t)dt, \end{aligned}$$

which gives (5.3.19).

Case 2. $N \ge 2$. In this case let $p = 1 + [N/2]$. Then $p \ge N/2$ and hence

$$\begin{aligned} &\int_{-1}^{1} |\pi_{n-1}(x,t)|^N u(t)dt = \int_{-1}^{1} |\pi_{n-1}(x,t)|^{N-2p+2p} u(t)dt \\ \ge\ & 4^{N-2p} \int_{-1}^{1} |\pi_{n-1}(x,t)|^{2p} u(t)dt \ge 4^{N-2p} \lambda_{p(n-1)+1}(u;x) \\ \ge\ & 4^{N-2p} \lambda_{pn}(u;x) \ge 4^{N-2p} C_1 \lambda_{pn}(W;x) \\ \ge\ & 4^{N-2p} C_1 c(pn)^{-1} \widehat{W}_{pn}(x) \ge 4^{N-2p} C_1 c(pn)^{-1} p^{-s} \widehat{W}_n(x) \\ \ge\ & C_1 c(W,N) \lambda_n(W;x) \ge C_1 C_2^{-1} c(W,N) \lambda_n(u;x), \end{aligned}$$

where

$$s = \max\{0, 2p_1 + 1\} + \max\{0, 2p_r + 1\} + \sum_{i=2}^{r-1} \max\{0, p_i\},$$

because

$$\left(|y| + \frac{1}{pn}\right)^q \ge \begin{cases} \left(|y| + \frac{1}{n}\right)^q, & q < 0, \\ p^{-q}\left(|y| + \frac{1}{n}\right)^q, & q \ge 0. \end{cases} \quad \square$$

As a direct consequence of Theorems 5.3.1 and 5.3.2 we state the following

Corollary 5.3.1. *If (5.3.7) and (5.3.18) are true then*

$$C_1 C_2^{-1} c_1(W,N)\lambda_n(W;x) \le \int_{-1}^{1} |\pi_{n-1}(x,t)|^N u(t)dt \le C_2 c_2(W,N)\lambda_n(W;x) \tag{5.3.20}$$

holds for all $x \in [-1,1]$ and $n \in \mathbf{N}$.

Lemma 5.3.3. *Let $P \in \mathbf{P}_n$. Then*

$$\max_{|x| \le 1} \left| P'(x) \widehat{W}_n(x) \delta_n(x) \right| \le cn \max_{|x| \le 1} \left| P(x) \widehat{W}_n(x) \right|. \tag{5.3.21}$$

Moreover, if (5.3.2) is true then

$$\max_{|x|\le 1}\left|P(x)\widehat{W}_n(x)\right| \le cn\int_{-1}^{1}|P(x)|u(x)dx \tag{5.3.22}$$

and

$$\int_{-1}^{1}|P'(x)|(1-x^2)^{1/2}u(x)dx \le cn\int_{-1}^{1}|P(x)|u(x)dx. \tag{5.3.23}$$

Here $\delta_n(x) = [(1-x)^{1/2}+n^{-1}][(1+x)^{1/2}+n^{-1}]$ *and c is a constant independent of n and P.*

Proof. The inequalities (5.3.21) and (5.3.23) follow directly from [65, Theorem 9.19, p. 164] and [65, Theorem 9.16, p. 163], respectively. To prove (5.3.22) by (5.3.3) we have

$$|P(x)| \le \lambda_n(W,1;x)^{-1}\int_{-1}^{1}|P(t)|W(t)dt \le c[n^{-1}\widehat{W}_n(x)]^{-1}\int_{-1}^{1}|P(t)|u(t)dt,$$

which is equivalent to (5.3.22). □

Lemma 5.3.4. *Let* $y_n = \cos\alpha_n$ *and* $z_n = \cos\beta_n$. *If* $|\alpha_n-\beta_n| \le C/n$ *then with the constants associated with the symbol* $\sim$ *depending on W and C only*

$$\widehat{W}_n(y_n) \sim \widehat{W}_n(z_n). \tag{5.3.24}$$

Proof. Clearly

$$\begin{aligned}\left[(1-y_n)^{1/2}+n^{-1}\right]-\left[(1-z_n)^{1/2}+n^{-1}\right] &= 2^{1/2}[\sin(\alpha_n/2)-\sin(\beta_n/2)]\\ \le\ |\alpha_n-\beta_n| \le C/n &\le C[(1-z_n)^{1/2}+n^{-1}]\end{aligned}$$

and hence

$$(1-y_n)^{1/2}+n^{-1} \le (1+C)[(1-z_n)^{1/2}+n^{-1}]. \tag{5.3.25}$$

Interchanging y_n and z_n in (5.3.25) we get

$$(1-y_n)^{1/2}+n^{-1} \ge (1+C)^{-1}[(1-z_n)^{1/2}+n^{-1}].$$

Hence

$$[(1-y_n)^{1/2}+n^{-1}]^{p_r} \sim [(1-z_n)^{1/2}+n^{-1}]^{p_r}.$$

Similarly

$$[(1+y_n)^{1/2}+n^{-1}]^{p_1} \sim [(1+z_n)^{1/2}+n^{-1}]^{p_1}.$$

Meanwhile for $2 \le i \le r-1$

$$[|y_n-t_i|+n^{-1}]-[|z_n-t_i|+n^{-1}] \le |y_n-z_n| \le |\alpha_n-\beta_n| \le C/n \le C[|z_n-t_i|+n^{-1}],$$

which by the same argument as above again yields

$$[|y_n-t_i|+n^{-1}]^{p_i} \sim [|z_n-t_i|+n^{-1}]^{p_i}.$$

This proves (5.3.24). □

After these preparations we can estimate the Christoffel type functions $\lambda_{jn}(u,m)$ with $u \sim W$, a.e.. The main result is as follows.

Theorem 5.3.3. *Let relation (5.3.2) prevail. Then with the constants associated with the symbol $\sim$ depending on u and m,*

$$\lambda_{jn}(u,m;x) \sim \lambda_n(u;x)\Delta_n(x)^j \sim \frac{1}{n}\widehat{W}_n(x)\Delta_n(x)^j, \quad x \in [-1,1], \quad j \in \mathbf{M}_2. \qquad (5.3.26)$$

Proof. Let $P \in \mathbf{P}^*_{n-1}(x)$ satisfy (5.2.5) with $d\mu(t) = u(t)dt$. Since $A_j(P,x;\cdot) \in \mathbf{P}_{mn}$, using (5.3.22) we obtain

$$\begin{aligned} |A_j(P,x;t)\widehat{W}_{mn}(t)| &\leq cmn\int_{-1}^{1} |A_j(P,x;s)|u(s)ds \\ &= cmn\int_{-1}^{1} A_j(P,x;s)\text{sgn}\,[(t-x)P(t)]^m u(s)ds \\ &= cmn\lambda_{jn}(u,m;x), \quad |t| \leq 1. \end{aligned}$$

Applying (5.3.21) j times and using the above inequality, we obtain

$$\begin{aligned} &\max_{|t|\leq 1}\left|A_j^{(j)}(P,x;t)\widehat{W}_{mn}(t)\delta_{mn}(t)^j\right| \\ \leq\ & c(mn)^j \max_{|t|\leq 1}\left|A_j(P,x;t)\widehat{W}_{mn}(t)\right| \leq c(mn)^{j+1}\lambda_{jn}(u,m;x). \end{aligned}$$

By (5.1.4) the above relation gives

$$\widehat{W}_{mn}(x)\delta_{mn}(x)^j \leq c(mn)^{j+1}\lambda_{jn}(u,m;x).$$

It is easy to see that $\delta_n(x) \sim n\Delta_n(x)$, $\Delta_{mn}(x) \sim \Delta_n(x)$, and $\widehat{W}_{mn}(x) \sim \widehat{W}_n(x)$ for a fixed m. By (5.3.3) we obtain

$$\lambda_{jn}(u,m;x) \geq c(mn)^{-j-1}\widehat{W}_{mn}(x)\delta_{mn}(x)^j \geq cn^{-1}\widehat{W}_n(x)\Delta_n(x)^j \geq c\lambda_n(u;x)\Delta_n(x)^j.$$

In order to prove the second half of (5.3.26) choose

$$N = 4+2m+2\left[p_1+p_r+\frac{1}{2}\sum_{i=2}^{r-1}|p_i|\right]. \qquad (5.3.27)$$

Then for each i, $0 \leq i \leq m-2$, by Theorem 5.3.1

$$\int_{-1}^{1} |\pi_{s-1}(x,t)|^N(1-t^2)^{i/2}u(t)dt \leq \frac{c}{s}\delta_s(x)^i\widehat{W}_s(x), \quad x \in [-1,1], \quad s \in \mathbf{N}. \qquad (5.3.28)$$

Put

$$s = 1+\left[\frac{2(n-1)}{N}\right]$$

and

$$P(t) = \pi_{s-1}(x,t)^{N/2}. \qquad (5.3.29)$$

Then $\partial P = N(s-1)/2 \leq n-1$ and hence by (5.3.28)

$$\begin{aligned} &\int_{-1}^{1} P(t)^2(1-t^2)^{i/2}u(t)dt = \int_{-1}^{1} \pi_{s-1}(x,t)^N(1-t^2)^{i/2}u(t)dt \\ \leq\ & \frac{c}{s}\delta_s(x)^i\widehat{W}_s(x) \leq \frac{c}{n}\delta_n(x)^i\widehat{W}_n(x), \quad i = 0,1,\ldots,m-2, \qquad (5.3.30) \end{aligned}$$

where $c = c(u,m,N) = c(u,m)$.

By the definition of $\lambda_{jn}(u,m;x)$, using relation (5.2.43) and noticing $\partial P \le n-1$, it follows from (5.1.3), (5.2.1), and (5.3.29) that

$$\begin{aligned} \lambda_{jn}(u,m;x) &\le \frac{1}{j!}\int_{-1}^{1} |(t-x)^j B_j(P,x;t)P(t)^m|u(t)dt \\ &\le \frac{1}{j!}\sum_{i=0}^{m-2-j} |b_i(P;x)| \int_{-1}^{1} |t-x|^{i+j}|\pi_{s-1}(x,t)|^{Nm/2}u(t)dt. \end{aligned} \tag{5.3.31}$$

Let us estimate

$$S_i = \int_{-1}^{1} |t-x|^{i+j}|\pi_{s-1}(x,t)|^{Nm/2}u(t)dt, \quad 0 \le i \le m-2-j.$$

Using (5.2.40)–(5.2.42) we get

$$\begin{aligned} S_i &= \int_{-1}^{1} |\pi_{s-1}(x,t)|^{Nm/2-i-j}[|t-x|K_s(v;x,t)]^{i+j}K_s(v;x,x)^{-i-j}u(t)dt \\ &\le cs^{-i-j}\int_{-1}^{1} \pi_{s-1}(x,t)^N[(1-x^2)^{1/2}+(1-t^2)^{1/2}]^{i+j}u(t)dt \\ &= cs^{-i-j}\sum_{\nu=0}^{i+j}\binom{i+j}{\nu}(1-x^2)^{(i+j-\nu)/2}\int_{-1}^{1} \pi_{s-1}(x,t)^N(1-t^2)^{\nu/2}u(t)dt. \end{aligned}$$

By (5.3.28) we obtain

$$\begin{aligned} S_i &\le cs^{-i-j-1}\widehat{W}_s(x)\sum_{\nu=0}^{i+j}\binom{i+j}{\nu}[(1-x^2)^{1/2}]^{i+j-\nu}\delta_s(x)^\nu \\ &= cs^{-i-j-1}\widehat{W}_s(x)\left[(1-x^2)^{1/2}+\delta_s(x)\right]^{i+j} \\ &\le cs^{-1}\widehat{W}_s(x)\Delta_s(x)^{i+j} \\ &\le cn^{-1}\widehat{W}_n(x)\Delta_n(x)^{i+j}. \end{aligned}$$

By (5.3.3)

$$S_i \le c\lambda_n(W;x)\Delta_n(x)^{i+j} \le c\lambda_n(u;x)\Delta_n(x)^{i+j},$$

which, coupled with (5.2.44) and (5.3.31), yields

$$\lambda_{jn}(u,m;x) \le c\lambda_n(u;x)\Delta_n(x)^j, \quad x\in[-1,1], \quad j\in \mathbf{M}_2.$$

This proves (5.3.26). □

The estimations of the functions $\lambda_{jn}(u,m;x)$ for $j \in \mathbf{M}_1$ will be given in Corollary 7.2.4, since they use Theorem 7.2.11.

5.4 Notes

5.4.1. The main results in Sections 5.1 and 5.2 are due to Shi [108].

5.4.2. The main result (Theorem 5.3.3) in Section 5.3 with $m = 2$ was proved by Nevai [65, Theorems 9.20–9.22, pp. 164–167]. Using his ideas Shi [112, 121, 125] gave the general result.

5.4.3. It would be interesting to extend the inequalities (5.2.36), (5.2.38), and (5.3.27) to the whole interval $[-1,1]$.

5.4.4. It is also interesting to give estimations of the functions $\lambda_{jn}(\ell^{(\alpha)}, m; x)$ and $\lambda_{jn}(h, m; x)$, where $\ell^{(\alpha)}$ and h are defined in (1.4.1) and (1.4.7), respectively.

Chapter 6

Explicit Expressions

6.1 Introduction

As Turán pointed out in [142, p. 46] that, little is known about $P_n(w,m;x)$ for $m \geq 4$, apart from the well known fact (see Corollary 6.3.1) $P_n(v,m;x) = 2^{1-n}T_n(x)$, where v is the Chebyshev weight. So he raised a problem connecting to this direction [142, p. 73].

Problem 74. Give the m orthogonal polynomials ($m = 4$) in an explicit form for weights other than $v(x)$.

He also raised a problem of an explicit formula for $\lambda_{kjn}(v,m)$ [142, p. 47].

Problem 26. Give an explicit formula for $\lambda_{kjn}(v,m)$ and determine its asymptotic behavior as $n \longrightarrow \infty$.

In Section 3 we investigate explicit expressions of power orthogonal polynomials.

In Sections 4, 5, and 6 we study three kinds of explicit forms of Gaussian quadrature formulas : a Gauss-Turán form , a form in terms of divided differences , and a trigonometric analogue , respectively. We also discuss how to transform the last two forms into the first.

In Section 7 we discuss the weights depending on the index n introduced by Gori and Micchelli which admit explicit Gaussian quadrature formulas.

In this chapter let

$$\omega_n(x) = \prod_{k=1}^{n}(x-x_k), \quad \Omega_{n-1}(x) = \prod_{k=1}^{n-1}(x-y_k), \tag{6.1.1}$$

where

$$x_1 \leq x_2 \leq \cdots \leq x_n, \quad y_1 \leq y_2 \leq \cdots \leq y_{n-1}. \tag{6.1.2}$$

6.2 Properties of the Chebyshev Polynomials

We collect here, partly without proof, some properties of the Chebyshev polynomials which will be needed later. We use the notations (1.3.8)–(1.3.11).

Theorem 6.2.1. [37, Formula 1.320, pp. 25–26] *We have*

$$\cos^m n\theta = \begin{cases} \frac{1}{2^{m-1}}\sum_{i=0}^{(m-1)/2}\binom{m}{(m-1)/2-i}\cos(2i+1)n\theta, & m \in \mathbf{N}_1, \\ \frac{1}{2^{m-1}}\sum_{i=0}^{\prime m/2}\binom{m}{m/2-i}\cos 2in\theta, & m \in \mathbf{N}_2, \end{cases} \tag{6.2.1}$$

and

$$T_n(x)^m = \begin{cases} \frac{1}{2^{m-1}} \sum_{i=0}^{(m-1)/2} \binom{m}{(m-1)/2-i} T_{(2i+1)n}(x), & m \in \mathbf{N}_1, \\ \frac{1}{2^{m-1}} \sum_{i=0}^{\prime m/2} \binom{m}{m/2-i} T_{2in}(x), & m \in \mathbf{N}_2. \end{cases} \tag{6.2.2}$$

Here the prime indicates that the first term in the sum is to be halved.

Theorem 6.2.2. *We have*

$$\int_{-1}^{1} x^k T_n(x)^m \frac{dx}{\sqrt{1-x^2}}$$
$$= \begin{cases} \frac{\pi}{2^{m+k-1}} \sum_{i=0}^{(m-1)/2} \binom{m}{\frac{m-1}{2}-i} \binom{k}{\frac{k-(2i+1)n}{2}}, & m \in \mathbf{N}_1,\ k-n \in \mathbf{N}_{02}, \\ \frac{\pi}{2^{m+k-1}} \sum_{i=0} m/2' \binom{m}{\frac{m}{2}-i} \binom{k}{\frac{k-2in}{2}}, & m \in \mathbf{N}_2,\ k \in \mathbf{N}_{02}, \\ 0, & \textit{otherwise}. \end{cases} \tag{6.2.3}$$

In particular

$$\int_{-1}^{1} x^k T_n(x) \frac{dx}{\sqrt{1-x^2}} = \begin{cases} \frac{\pi}{2^k} \binom{k}{(k-n)/2}, & k-n \in \mathbf{N}_{02}, \\ 0, & \textit{otherwise}. \end{cases} \tag{6.2.4}$$

Proof. The formula (6.2.4) may be found in [37, Formula 3.631-17, p. 374]. To prove (6.2.3) we separate the cases when $m \in \mathbf{N}_1$ and $m \in \mathbf{N}_2$.

Case 1. $m \in \mathbf{N}_1$. We denote by S the left-hand side of (6.2.3). By (6.2.2) and (6.2.4)

$$\begin{aligned} S &= \frac{1}{2^{m-1}} \sum_{i=0}^{(m-1)/2} \binom{m}{(m-1)/2-i} \int_{-1}^{1} x^k T_{(2i+1)n}(x) \frac{dx}{\sqrt{1-x^2}} \\ &= \frac{1}{2^{m-1}} \sum_{i=0}^{(m-1)/2} \binom{m}{(m-1)/2-i} \frac{\pi}{2^k} \binom{k}{(k-(2i+1)n)/2} \\ &= \frac{\pi}{2^{m+k-1}} \sum_{i=0}^{(m-1)/2} \binom{m}{(m-1)/2-i} \binom{k}{(k-(2i+1)n)/2} \end{aligned}$$

for $k-n \in \mathbf{N}_{02}$, and $S = 0$ otherwise.

Case 2. $m \in \mathbf{N}_2$. By (6.2.2) and (6.2.4)

$$\begin{aligned} S &= \frac{1}{2^{m-1}} \sum_{i=0} m/2' \binom{m}{m/2-i} \int_{-1}^{1} x^k T_{2in}(x) \frac{dx}{\sqrt{1-x^2}} \\ &= \frac{1}{2^{m-1}} \sum_{i=0} m/2' \binom{m}{m/2-i} \frac{\pi}{2^k} \binom{k}{(k-2in)/2} \\ &= \frac{\pi}{2^{m+k-1}} \sum_{i=0} m/2' \binom{m}{m/2-i} \binom{k}{(k-2in)/2} \end{aligned}$$

for $k \in \mathbf{N}_{02}$ and $S = 0$ otherwise. □

Remark 6.2.1. It should be noted that by (1.1.8) the term $\binom{k}{p} = 0$ for $-p \in \mathbf{N}$.

Theorem 6.2.3. *Let $j,k,m \in \mathbf{N}_0$. Then*

$$\int_{-1}^{1} T_n(x)^j T_{kn}(x)[sgn\ T_n(x)^m]\frac{dx}{\sqrt{1-x^2}} = \begin{cases} 0, & k+j+m \in \mathbf{N}_1 \text{ or } k-j \in \mathbf{N}_2, \\ \frac{\pi}{2^j}\binom{j}{(j+k)/2}, & \text{otherwise.} \end{cases} \tag{6.2.5}$$

Proof. Denote by S the left-hand side of (6.2.5). By making the change of variable $x = \cos(\theta/n)$ we have

$$\begin{aligned} S &= \frac{1}{n}\int_0^{n\pi} \cos^j\theta\cos k\theta[\text{sgn } \cos^m\theta]d\theta \\ &= \frac{1}{n}\left[\sum_{i=0}^{n-1}(-1)^{i(j+k+m)}\right]\left[1+(-1)^{j+k+m}\right]\int_0^{\pi/2}\cos^j\theta\cos k\theta d\theta. \end{aligned}$$

Using this and the formula in [37, Formula 3.621-9, p. 372]

$$\int_0^{\pi/2}\cos^j\theta\cos k\theta d\theta = \frac{\pi}{2^{j+1}}\binom{j}{(j+k)/2} \tag{6.2.6}$$

we obtain (6.2.5). □

Theorem 6.2.4. *Let $k,j,m \in \mathbf{N}_0$. Let ℓ_{in}'s denote the fundamental polynomials of Lagrange interpolation based on the zeros of T_n. Then for each i, $1 \le i \le n$,*

$$\int_{-1}^{1} T_n(x)^j T_{kn}(x)\ell_{in}(x)[sgn\ T_n(x)^m]\frac{dx}{\sqrt{1-x^2}} = \begin{cases} 0, & k+j+m \in \mathbf{N}_1 \text{ or } k-j \in \mathbf{N}_2, \\ \frac{\pi}{2^j n}\binom{j}{(j+k)/2}, & \text{otherwise.} \end{cases} \tag{6.2.7}$$

Proof. Expand T_{kn} and ℓ_{in} in the forms

$$T_{kn}(x) = T_k(T_n(x)) = \sum_{p=0}^{[k/2]} c_{k-2p}T_n(x)^{k-2p} \tag{6.2.8}$$

and

$$\ell_{in}(x) = C_0 + \sum_{q=1}^{n-1} C_q T_q(x), \tag{6.2.9}$$

respectively. To determine the constant C_0 we recall that

$$\lambda_{in}(v) = \int_{-1}^{1}\ell_{in}(x)\frac{dx}{\sqrt{1-x^2}} = \frac{\pi}{n}$$

(see, [14, Problem 13, p. 114]). Multiplying (6.2.9) with $dx/\sqrt{1-x^2}$, then integrating between -1 and 1, we obtain $C_0 = 1/n$. Thus formula (6.2.9) becomes

$$\ell_{in}(x) = \frac{1}{n} + \sum_{q=1}^{n-1} C_q T_q(x). \tag{6.2.10}$$

Denote by S the left-hand side of (6.2.7). Using (6.2.8) and (6.2.10), we see

$$\begin{aligned} S &= \frac{1}{n}\int_{-1}^{1} T_n(x)^j T_{kn}(x)[\operatorname{sgn} T_n(x)^m]\frac{dx}{\sqrt{1-x^2}} \\ &+\sum_{p=0}^{[k/2]}\sum_{q=1}^{n-1} c_pC_q\int_{-1}^{1} T_n(x)^{j+k-2p}T_q(x)[\operatorname{sgn} T_n(x)^m]\frac{dx}{\sqrt{1-x^2}}. \end{aligned} \tag{6.2.11}$$

It is particularly simple to compute S for the case when $k+j+m\in\mathbf{N}_1$; in this case by (6.2.5) it follows from orthogonality that $S=0$. To determine S for the case when $k+j+m\in\mathbf{N}_2$ we observe that in this case

$$T_n(x)^{j+k-2p}[\operatorname{sgn} T_n(x)^m]=|T_n(x)|^{j+k-2p}=\sum_{r=0}^{\infty} b_rT_{2rn}(x)$$

and hence the last term of (6.2.11) is zero. Then equation (6.2.11) reads

$$S=\frac{1}{n}\int_{-1}^{1} T_n(x)^j T_{kn}(x)[\operatorname{sgn} T_n(x)^m]\frac{dx}{\sqrt{1-x^2}}.$$

It remains to apply Theorem 6.2.3. □

Theorem 6.2.5. *Let $f\in\mathbf{P}_{n-1}$ and let x_k's denote the zeros of T_n. Then*

$$\begin{aligned} &\int_{-1}^{1} f(x)T_n(x)^j\frac{dx}{\sqrt{1-x^2}} \\ &= \begin{cases} \frac{j!}{(j!!)^2}\int_{-1}^{1} f(x)\frac{dx}{\sqrt{1-x^2}}=\frac{j!\pi}{(j!!)^2n}\sum_{k=1}^{n} f(x_k), & j\in\mathbf{N}_{02}, \\ 0, & j\in\mathbf{N}_1. \end{cases} \end{aligned} \tag{6.2.12}$$

Proof. Apply (6.2.7) with $k=m=0$. □

Theorem 6.2.6. *Let $k,j\in\mathbf{N}$. Then*

$$\int_{-1}^{1}|T_n(x)|^{k-1}|U_{n-1}(x)|^{j-1}(1-x^2)^{(j-2)/2}dx=B(k/2,j/2). \tag{6.2.13}$$

Proof. Clearly, relation (6.2.13) is equivalent to

$$\int_0^{\pi}|\cos n\theta|^{k-1}|\sin n\theta|^{j-1}d\theta=B(k/2,j/2).$$

By making the change of variable $\tau=\theta/n$ the above relation becomes

$$\int_0^{\pi}|\cos\tau|^{k-1}|\sin\tau|^{j-1}d\tau=B(k/2,j/2),$$

which can be found in [37, Formula 3.621-5, p. 369]. □

Let V_n and W_n be defined in (1.3.10) and (1.3.11), respectively. Denote

$$\begin{cases} v(x) &= w^{(-1/2,-1/2)}(x), \\ u_m(x) &= w^{((m-1)/2,(m-1)/2)}(x), \\ v_m(x) &= w^{(-1/2,(m-1)/2)}(x), \\ w_m(x) &= w^{((m-1)/2,-1/2)}(x). \end{cases} \tag{6.2.14}$$

Theorem 6.2.7. *We have*

$$V_n(2x^2-1) = \frac{T_{2n+1}(x)}{x}, \tag{6.2.15}$$

$$W_n(2x^2-1) = U_{2n}(x), \tag{6.2.16}$$

$$W_n(-x) = (-1)^n V_n(x). \tag{6.2.17}$$

Proof. Formulas (6.2.15) and (6.2.16) may be obtained by setting $\cos\theta = 2x^2-1$ in (1.3.10) and (1.3.11), respectively. Formula (6.2.17) directly follows from the identity

$$P_n(w(-\cdot),m;-x) = (-1)^n P_n(w,m;x), \tag{6.2.18}$$

which may derived from (3.1.1). □

6.3 Explicit Expressions of Power Orthogonal Polynomials

The following theorem provides explicit expressions of a kind of power orthogonal polynomials.

Theorem 6.3.1. *Let $m,p\in\{0\}\cup[1,\infty)$ and $m+p\geq 1$. Then for $n\in\mathbf{N}$*

$$\begin{aligned}&\int_{-1}^{1}\left|2^{1-n}T_n(x)\right|^m\left|2^{1-n}U_{n-1}(x)\right|^p(1-x^2)^{(p-1)/2}dx\\ =\ &\min_{P\in\widehat{\mathbf{P}}_n,Q\in\widehat{\mathbf{P}}_{n-1}}\int_{-1}^{1}|P(x)|^m|Q(x)|^p(1-x^2)^{(p-1)/2}dx\end{aligned} \tag{6.3.1}$$

Proof. We treat the case when $m,p\geq 1$; all other cases are similar and simpler. By (3.2.6) it is enough to show that the equation

$$\int_{-1}^{1}|T_n(x)|^{m-1}|U_{n-1}(x)|^{p-1}[\text{sgn } T_n(x)U_{n-1}(x)]q(x)(1-x^2)^{(p-1)/2}dx = 0 \tag{6.3.2}$$

holds for every $q\in\mathbf{P}_{2n-2}$. Since $T_n(x)U_{n-1}(x) = U_{2n-1}(x)/2$, it suffices to show that

$$\begin{aligned}&\int_{-1}^{1}|T_n(x)|^{m-1}|U_{n-1}(x)|^{p-1}[\text{sgn } U_{2n-1}(x)]U_{k-1}(x)(1-x^2)^{(p-1)/2}dx = 0,\\ &k=1,2,\ldots,2n-1.\end{aligned}$$

By making the change of variable $x=\cos\theta$ and integrating over the interval twice, the above equations become

$$\int_{-\pi}^{\pi}|\cos n\theta|^{m-1}|\sin n\theta|^{p-1}[\text{sgn } \sin 2n\theta]\sin k\theta d\theta = 0,\quad k=1,2,\ldots,2n-1.$$

Since $\sin k\theta$ is a linear combination of the functions $e^{\pm ik\theta}$ ($i=\sqrt{-1}$), it will be enough to establish

$$\begin{aligned}I=\int_{-\pi}^{\pi}|\cos n\theta|^{m-1}|\sin n\theta|^{p-1}[\text{sgn } \sin 2n\theta]e^{ik\theta}d\theta = 0,\\ k=\pm1,\pm2,\ldots,\pm(2n-1).\end{aligned}$$

Remembering the periodicity of the functions, by making the change of variable $\theta = \tau + \pi/n$ we see

$$\begin{aligned} I &= \int_{-\pi}^{\pi} |\cos(n\tau+\pi)|^{m-1}|\sin(n\tau+\pi)|^{p-1}[\operatorname{sgn}\,\sin(2n\tau+2\pi)]e^{ik(\tau+\pi/n)}d\tau \\ &= e^{ik\pi/n}I. \end{aligned}$$

Clearly, $e^{ik\pi/n} \neq 1$, which means $I = 0$. □

As the two important special cases when $p = 0$ or $m = 0$ we state the following results.

Corollary 6.3.1. *We have*

$$\omega_n(v,m;x) = 2^{1-n}T_n(x). \tag{6.3.3}$$

Corollary 6.3.2. *We have*

$$\omega_n(u_m,m;x) = 2^{-n}U_n(x). \tag{6.3.4}$$

Let us consider the extremal problem

$$\int_{-1}^{1} |\omega_n(x)|^m w(x)dx = \min_{P\in\widehat{\mathbf{P}}_n} \int_{-1}^{1} |P(x)|^m w(x)dx. \tag{6.3.5}$$

According to Corollary 6.3.1 the polynomial $2^{1-n}T_n(x)$ is the solution of (6.3.5) with the Chebyshev weight

$$w(x) = \frac{1}{\sqrt{1-x^2}} \tag{6.3.6}$$

for all $m \in \mathbf{N}$. This elegant property is very useful, say, this makes it possible to give an explicit formula for the Christoffel numbers $\lambda_{kjn}(v,m)$ (see Section 6.4). It is natural to ask whether or not there are other weights having this property? Clearly, a linear transformation of the weight (6.3.6)

$$w(x) = \begin{cases} \frac{1}{\sqrt{(x-a)(b-x)}}, & x \in (a,b), \\ 0, & x \notin (a,b) \end{cases} \tag{6.3.7}$$

also admits this property. It turns out that the Chebyshev weight (6.3.6) is the only weight (up to a linear transformation). In fact, we shall prove slightly more:

Theorem 6.3.2. *Let a weight w satisfy $w(x) > 0$ for almost every $x \in [-1,1]$ and $\int_{-1}^{1} w(x)dx = \pi$. If the equation (6.3.5) holds for the following pairs (m,n):*

$$m = \begin{cases} m_1, m_2, \ldots, & n \in \{1,2,4\}, \\ 2,4, & n \notin \{1,2,4\}, \end{cases} \tag{6.3.8}$$

where $\{m_k\}_{k=1}^{\infty}$ is a strictly increasing sequence of even numbers such that

$$\sum_{k=1}^{\infty} \frac{1}{m_k} = \infty, \tag{6.3.9}$$

then relation (6.3.6) is valid.

To prove this theorem we need some auxiliary lemmas.

The following result is due to Pólya [76].

Lemma 6.3.1. *Let a weight w satisfy $w(x) > 0$ for almost every $x \in [-1,1]$ and let $u \in C[-1,1]$. Let $n \in \mathbf{N}$ be fixed. If*

$$\int_{-1}^{1} |\omega_n(x)u(x)|^{p_j} w(x)dx = \min_{P \in \widehat{\mathbf{P}}_n} \int_{-1}^{1} |P(x)u(x)|^{p_j} w(x)dx \tag{6.3.10}$$

holds for every p_j and

$$\lim_{j\to\infty} p_j = +\infty \tag{6.3.11}$$

then

$$\|\omega_n u\| = \min_{P \in \widehat{\mathbf{P}}_n} \|Pu\|. \tag{6.3.12}$$

Lemma 6.3.2. *Let a weight w satisfy $w(x) > 0$ for almost every $x \in [-1,1]$. Let $\{m_k\}_{k=1}^{\infty}$ be a strictly increasing sequence of even numbers satisfying (6.3.9). If for a point $c \in (-1,1)$ the equation*

$$\int_{-1}^{1} (x-c)^m w(x)dx = \min_t \int_{-1}^{1} (x-t)^m w(x)dx \tag{6.3.13}$$

holds for every $m = m_k$, then relation (6.3.13) holds for every $m \in \mathbf{N}_2$. Moreover,

$$c = 0 \tag{6.3.14}$$

and

$$w(-x) = w(x). \tag{6.3.15}$$

Proof. First, applying Lemma 6.3.1 with $u = 1$ and $n = 1$ we obtain $x - c = T_1(x) = x$, which yields (6.3.14).

Next, by Theorem 3.2.2 equation (6.3.13) with $c = 0$ is equivalent to the orthogonality relation

$$\int_{-1}^{1} x^{m-1} w(x)dx = 0. \tag{6.3.16}$$

Clearly, relation (6.3.15) implies that equation (6.3.13) with $c = 0$ holds for every $m \in \mathbf{N}_2$. So it is enough to prove (6.3.15). Let ξ and δ satisfy $0 < \xi < \xi + \delta < 1$. Put

$$f_\delta(x) = \begin{cases} 0, & x \in [0,\xi], \\ \frac{x-\xi}{\delta}, & x \in [\xi, \xi+\delta], \\ 1, & x \in [\xi+\delta, 1] \end{cases}$$

and $f_\delta(x) = -f_\delta(-x)$, $x \in [-1,0]$. Obviously, $f_\delta \in C[-1,1]$ and f_δ is an odd function. By Műntz theorem [14, p. 197] it follows from (6.3.9) that given an arbitrary number $\varepsilon > 0$ there is a polynomial of the form $P_\varepsilon(x) = \sum a_k x^{m_k - 1}$ such that

$$|f_\delta(x) - P_\varepsilon(x)| \le \varepsilon, \quad x \in [-1,1]. \tag{6.3.17}$$

Since (6.3.13) or (6.3.16) holds for every $m = m_k$, we obtain

$$\int_{-1}^{1} P_\varepsilon(x)w(x)dx = 0,$$

which, coupled with (6.3.17), yields

$$\left|\int_{-1}^{1} f_\delta(x)w(x)dx\right| \le \varepsilon.$$

Noting that $f_\delta(x)$ is independent of ε and ε is arbitrary, we have

$$\int_{-1}^{1} f_\delta(x)w(x)dx = 0.$$

Furthermore, as $\delta \to 0$ we get

$$\int_{-1}^{-\xi} w(x)dx = \int_{\xi}^{1} w(x)dx.$$

Differentiating this relation with respect to ξ gives $w(-\xi) = w(\xi)$. This is equivalent to (6.3.15). □

Lemma 6.3.3. *Let a weight w satisfy $w(x) > 0$ for almost every $x \in [-1,1]$. Let $\{m_k\}_{k=1}^{\infty}$ be a strictly increasing sequence of even numbers satisfying (6.3.9). Further, assume that relation (6.3.15) is valid. If for a number d the equation*

$$\int_{-1}^{1} (x^2-d)^m w(x)dx = \min_t \int_{-1}^{1} (x^2-t)^m w(x)dx \tag{6.3.18}$$

holds for every $m = m_k$, then equation (6.3.18) holds for every $m \in \mathbf{N}_2$. Moreover,

$$d = \frac{1}{2} \tag{6.3.19}$$

and

$$\sqrt{1-x^2}w(x) = |x|w\left(\sqrt{1-x^2}\right). \tag{6.3.20}$$

Furthermore, for every even function f

$$\int_{-1}^{1} f(x)w(x)dx = \int_{-1}^{1} f\left(\sqrt{1-x^2}\right)w(x)dx. \tag{6.3.21}$$

Proof. First, applying Lemma 6.3.1 with $u = 1$ and $n = 2$ we obtain $x^2 - d = 2^{-1}T_2(x) = x^2 - \frac{1}{2}$, which gives (6.3.19).

Next, by (6.3.15) and (6.3.19) equation (6.3.18) becomes

$$\int_{0}^{1} \left(x^2 - \frac{1}{2}\right)^m w(x)dx = \min_t \int_{0}^{1} (x^2-t)^m w(x)dx$$

or by making the change of variable $x = [(1+y)/2]^{1/2}$

$$\int_{-1}^{1} y^m w^*(y)dy = \min_t \int_{-1}^{1} (y-t)^m w^*(y)dy, \tag{6.3.22}$$

where

$$w^*(y) = \left(\frac{1+y}{2}\right)^{-1/2} w\left(\left(\frac{1+y}{2}\right)^{1/2}\right). \tag{6.3.23}$$

Since equation (6.3.22) holds for every $m = m_k$, applying Lemma 6.3.2 to (6.3.22) we conclude that equations (6.3.22) and (6.3.18) hold for every $m \in \mathbf{N}_2$. Moreover, for the weight $w^*(y)$ we have

$$w^*(-y) = w^*(y). \tag{6.3.24}$$

The above equation, using the substitution $y = 2x^2 - 1$, gives (6.3.20). In order to obtain (6.3.21) we apply (6.3.20) and use the substitution $x = \sqrt{1-y^2}$:

$$\begin{aligned}\int_{-1}^{1} f(x)w(x)dx &= 2\int_{0}^{1} f(x)w(x)dx = 2\int_{0}^{1} f(x)\frac{xw\left(\sqrt{1-x^2}\right)}{\sqrt{1-x^2}}dx \\ &= 2\int_{0}^{1} f\left(\sqrt{1-y^2}\right)w(y)dy = \int_{-1}^{1} f\left(\sqrt{1-x^2}\right)w(x)dx. \quad \square\end{aligned}$$

Proof of Theorem 6.3.2. Let

$$\mu_j = \int_{-1}^{1} x^j w(x)dx, \quad j = 0, 1, \ldots.$$

According to Theorem 2.2 in [24, p. 64] it is sufficient to show that the equality

$$\mu_j = \int_{-1}^{1} x^j \frac{dx}{\sqrt{1-x^2}} \tag{6.3.25}$$

holds for every $j \in \mathbf{N}_0$.

First, the equation (6.3.25) with $j = 0$ is satisfied; using the assumptions of the theorem for $n = 1$ and applying Lemma 6.3.2 we have (6.3.15) and hence

$$\mu_j = \int_{-1}^{1} x^j \frac{dx}{\sqrt{1-x^2}} = 0, \quad j \in \mathbf{N}_1.$$

Next, for $j \in \mathbf{N}_2$ we separate the cases when $j \leq 8$ and $j > 8$.

Case 1. $j \leq 8$. In this case using the assumptions of the theorem for $n = 2$ and applying Lemma 6.3.3 we obtain equations (6.3.19)–(6.3.21) and conclude that equation (6.3.18) holds for every $m \in \mathbf{N}_2$. By (3.2.6), (6.3.18), and (6.3.19) equation

$$\int_{-1}^{1} \left(x^2 - \frac{1}{2}\right)^{m-1} w(x)dx = 0$$

holds for every $m \in \mathbf{N}_2$. This equation with $m = 2, 4, 6$ gives that

$$\mu_2 = \frac{\pi}{2}, \tag{6.3.26}$$

$$\mu_6 = \frac{3}{2}\mu_4 - \frac{3}{4}\mu_2 + \frac{\pi}{8} = \frac{3}{2}\mu_4 - \frac{\pi}{4}, \tag{6.3.27}$$

$$\mu_{10} = \frac{5}{2}\mu_8 - \frac{5}{2}\mu_6 + \frac{5}{4}\mu_4 - \frac{5}{16}\mu_2 + \frac{\pi}{32} = \frac{1}{2}(5\mu_8 - 5\mu_4 + \pi). \tag{6.3.28}$$

On the other hand, the solution ω_3 of equation (6.3.5) by (6.3.15) takes the forms

$$\omega_3(x) = x(x^2 - p)$$

with certain constant p and by (3.2.6) satisfies the orthogonality relations

$$\int_{-1}^{1} \omega_3(x)xw(x)dx = \int_{-1}^{1} \omega_3(x)^3 xw(x)dx = 0.$$

By a calculation we obtain, using (6.3.26) and (6.3.27),

$$\mu_4 = p\mu_2 = \frac{\pi p}{2}, \tag{6.3.29}$$

$$\mu_{10} = 3p\mu_8 - 3p^2\mu_6 + p^3\mu_4 = 3p\mu_8 + \frac{\pi p^2}{4}(2p^2 - 9p + 3). \tag{6.3.30}$$

Using the assumptions of the theorem for $n = 4$ and applying Lemma 6.3.1 the solution of equation (6.3.5) is

$$\omega_4(x) = 2^{-3}T_4(x) = x^4 - x^2 + \frac{1}{8}.$$

By (6.3.15) equation (6.3.5) reads

$$\int_0^1 \left(x^4 - x^2 + \frac{1}{8}\right)^m w(x)dx = \min_{s,t} \int_0^1 (x^4 + sx^2 + t)^m w(x)dx.$$

Again making the change of variable $x = [(1+y)/2]^{1/2}$, and using (6.3.23) and (6.3.24), the above equation becomes

$$\int_{-1}^{1} \left(y^2 - \frac{1}{2}\right)^m w^*(y)dy = \min_t \int_{-1}^{1} (y^2 - t)^m w^*(y)dy.$$

Now using the assumptions of the theorem for $n = 4$ and applying Lemma 6.3.3 the above equation holds for every $m \in \mathbf{N}_2$. Hence equation (6.3.18) holds for every $m \in \mathbf{N}_2$. Then by (3.2.6) we obtain the orthogonality relation

$$\int_{-1}^{1} \omega_4(x)w(x)dx = 0,$$

which by (6.3.26) gives

$$\mu_4 = \frac{3\pi}{8}. \tag{6.3.31}$$

Substituting (6.3.31) into (6.3.27) and (6.3.29) gives $\mu_6 = 5\pi/16$ and $p = 3/4$. Then equations (6.3.28) and (6.3.30) become

$$\mu_{10} = \frac{5}{2}\mu_8 - \frac{7\pi}{16}$$

and

$$\mu_{10} = \frac{9}{4}\mu_8 - \frac{189\pi}{512},$$

respectively. These equations give $\mu_8 = \frac{35\pi}{128}$. Comparing μ_j with (6.2.4) we prove (6.3.25) for $j \le 8$.

Case 2. $j > 8$. In this case using the assumptions of the theorem for $m = 2, 4$ and (3.2.6) we obtain

$$\int_{-1}^{1} x^i \omega_n(x) w(x) dx = 0, \quad i = 0, 1, \ldots, n-1, \tag{6.3.32}$$

and

$$\int_{-1}^{1} x^i \omega_n(x)^3 w(x) dx = 0, \quad i = 0, 1, \ldots, n-1, \tag{6.3.33}$$

respectively. We claim that given $3n$ values $\mu_0, \mu_1, \ldots, \mu_{3n-1}$, one can uniquely determine the following n values $\mu_{3n}, \mu_{3n+1}, \ldots, \mu_{4n-1}$. In fact, let $\omega_n(x) = \sum_{j=0}^{n} c_j x^j$ with $c_n = 1$. Equations (6.3.32) and (6.3.33) become

$$\sum_{j=0}^{n-1} c_j \mu_{i+j} = -\mu_{i+n}, \quad i = 0, 1, \ldots, n-1, \tag{6.3.34}$$

and

$$\sum_{j=0}^{n} \sum_{p=0}^{n} \sum_{q=0}^{n} c_j c_p c_q \mu_{i+j+p+q} = 0, \quad i = 0, 1, \ldots, n-1, \tag{6.3.35}$$

respectively. Since the determinant corresponding to the coefficient matrix of equations (6.3.34) $\det\{\mu_{i+j}\}_{i,j=0}^{n-1} > 0$, we can uniquely determine the solution $c_0, \ldots, c_{n-1}$, from $2n$ values $\mu_0, \mu_1, \ldots, \mu_{2n-1}$. Then from $3n$ values $\mu_0, \mu_1, \ldots, \mu_{3n-1}$ we can successively calculate, using (6.3.35), the following n values $\mu_{3n}, \mu_{3n+1}, \ldots, \mu_{4n-1}$. This proves our claim. According to this claim using the initial 9 values $\mu_0, \ldots, \mu_8$, we can uniquely determine all moments $\mu_0, \mu_1, \ldots$ by induction, because $4n \ge 3(n+1)$ when $n \ge 3$. Since the initial 9 values of the moments and the equations (6.3.32)–(6.3.35) to determine their moments successively are the same for the weight w and the Chebyshev weight, we can obtain (6.3.25) and hence (6.3.6). □

Let $m, p \in \{0\} \cup [1, \infty)$, $m + p \ge 1$, and $q \ge 0$. We shall characterize the conditions such that the solutions ω_n and Ω_{n-1} of the extremal problem

$$\begin{aligned} &\int_{-1}^{1} |\omega_n(x)|^m |\Omega_{n-1}(x)|^p (1-x^2)^q w(x) dx \\ &= \min_{P \in \widehat{\mathbf{P}}_n, Q \in \widehat{\mathbf{P}}_{n-1}} \int_{-1}^{1} |P(x)|^m |Q(x)|^p (1-x^2)^q w(x) dx \end{aligned} \tag{6.3.36}$$

are independent of m, p, and q.

The following lemma plays a crucial role for this problem.

Lemma 6.3.4. *Let the function $g \in C[-1,1]$ be strictly monotone on $[a,b] \subset [-1,1]$ and satisfy*

$$\min\{|g(a)|, |g(b)|\} = 0, \quad \max\{|g(a)|, |g(b)|\} = \|g\| = \max_{-1 \le x \le 1} |g(x)|. \tag{6.3.37}$$

Let $w(x) > 0$ a.e. in $[a,b]$. Then the following statements are equivalent each to other.

(a) The relation

$$\int_{-1}^{1}[sgn\ g(x)]|g(x)|^m w(x)dx = 0 \qquad (6.3.38)$$

holds for every $m = dM + r$, $M = 0, 1, \ldots$, where $d > 0$ and $r \geq 0$ are fixed constants.

(b) The relation (6.3.38) holds for any $m \geq 0$.

(c) The relation

$$\int_{|g(x)|\geq c}[sgn\ g(x)]w(x)dx = 0 \qquad (6.3.39)$$

holds for any $c \geq 0$.

(d) The relation

$$\int_{|g(x)|\geq c}[sgn\ g(x)]|g(x)|^m w(x)dx = 0 \qquad (6.3.40)$$

holds for any $c, m \geq 0$.

Moreover, one of Statements (a)–(d) implies

$$\max_{-1\leq x\leq 1} g(x) + \min_{-1\leq x\leq 1} g(x) = 0. \qquad (6.3.41)$$

Proof. (a)$\Longrightarrow$(c) Let c be an arbitrary number satisfying $0 < c < \|g\|$ and choose δ so that $0 < \delta < c$. Put for $x \in [a,b]$

$$f_\delta(x) = \begin{cases} 0, & \text{if } |g(x)| \leq c - \delta, \\ \text{sgn } g(x), & \text{if } |g(x)| \geq c, \\ \text{linear }, & \text{if } c - \delta \leq |g(x)| \leq c \end{cases}$$

so that $f_\delta \in C[a,b]$ and for $x \in [-1,1] \setminus [a,b]$

$$f_\delta(x) = [\text{sgn } g(x)]|f_\delta(y)|, \quad \text{if } |g(x)| = |g(y)|, \quad y \in [a,b].$$

Clearly, $f_\delta \in C[-1,1]$. Let us consider the related function on $[a,b]$

$$F_\delta(x) = \begin{cases} 0, & \text{if } |g(x)| \leq c - \delta, \\ \frac{|f_\delta(x)|}{|g(x)|^r}, & \text{if } |g(x)| > c - \delta, \end{cases}$$

which is continuous on $[a,b]$. The span of the set of functions $\{|g|^{dM} : M \in \mathbf{N}_0\}$ forms an algebra, that is, a product of generalized polynomials $\sum a_M|g|^{dM}$ is another generalized polynomial [14, p. 190]. Since this span separates points in $[a,b]$ (i.e., there is a function, say, $|g|^d$ such that $|g(x)|^d \neq |g(y)|^d$ for $x \neq y$), by Stone theorem [14, p. 191] this span is dense in the space $C[a,b]$. So for a given number $\varepsilon > 0$ there is a generalized polynomial $\sum a_M|g(x)|^{dM}$ such that

$$|F_\delta(x) - \sum a_M|g(x)|^{dM}| \leq \varepsilon, \quad x \in [a,b].$$

Hence for $P_\varepsilon(x) = [\text{sgn } g(x)] \sum a_M|g(x)|^{dM+r}$ we have

$$|f_\delta(x) - P_\varepsilon(x)| \leq \varepsilon\|g\|^r, \quad x \in [a,b].$$

By the definition of f_δ, we even have

$$\|f_\delta - P_\varepsilon\| \le \varepsilon \|g\|^r. \tag{6.3.42}$$

It follows from Statement (a) that

$$\int_{-1}^{1} P_\varepsilon(x) w(x) dx = 0,$$

which, coupled with (6.3.42), yields

$$\left| \int_{-1}^{1} f_\delta(x) w(x) dx \right| \le \varepsilon \|g\|^r \int_{-1}^{1} w(x) dx.$$

Noting that f_δ is independent of ε and ε is arbitrary, we have

$$\int_{-1}^{1} f_\delta(x) w(x) dx = 0.$$

Furthermore, as $\delta \to 0$ we get (6.3.39). It is easy to see that (6.3.39) remains true for $c = 0$ and $c = \|g\|$.

(c)$\Longrightarrow$(b) Let $m > 0$ and $0 \le c < C$. Clearly, by (6.3.39)

$$\begin{aligned} & \int_{c \le |g(x)| < C} [\operatorname{sgn} g(x)] w(x) dx \\ = & \int_{|g(x)| \ge c} [\operatorname{sgn} g(x)] w(x) dx - \int_{|g(x)| \ge C} [\operatorname{sgn} g(x)] w(x) dx = 0. \end{aligned}$$

Hence

$$\begin{aligned} & \left| \int_{c \le |g(x)| < C} [\operatorname{sgn} g(x)] |g(x)|^m w(x) dx \right| \\ = & \left| \int_{c \le |g(x)| < C} [\operatorname{sgn} g(x)] [|g(x)|^m - c^m] w(x) dx + c^m \int_{c \le |g(x)| < C} [\operatorname{sgn} g(x)] w(x) dx \right| \\ = & \left| \int_{c \le |g(x)| < C} [\operatorname{sgn} g(x)] [|g(x)|^m - c^m] w(x) dx \right| \\ \le & (C^m - c^m) \int_{c \le |g(x)| < C} w(x) dx. \end{aligned}$$

Then for $n \in \mathbf{N}$

$$\begin{aligned} & \left| \int_{-1}^{1} [\operatorname{sgn} g(x)] |g(x)|^m w(x) dx \right| \\ = & \left| \sum_{k=0}^{n} \int_{k\|g\|/n \le |g(x)| < (k+1)\|g\|/n} [\operatorname{sgn} g(x)] |g(x)|^m w(x) dx \right| \\ \le & \frac{\|g\|^m}{n^m} \sum_{k=0}^{n} |(k+1)^m - k^m| \int_{k\|g\|/n \le |g(x)| < (k+1)\|g\|/n} w(x) dx \\ \le & \frac{(m+1)2^m \|g\|^m}{n^{\min\{1,m\}}} \sum_{k=0}^{n} \int_{k\|g\|/n \le |g(x)| < (k+1)\|g\|/n} w(x) dx \\ = & \frac{(m+1)2^m \|g\|^m}{n^{\min\{1,m\}}} \int_{-1}^{1} w(x) dx. \end{aligned}$$

Here the last inequality follows by the mean-value theorem for the derivative from the inequalities

$$(k+1)^m - k^m \le \begin{cases} (n+1)^m - n^m \le m(n+1)^{m-1}, & m \ge 1, \\ 1, & m < 1 \end{cases} \quad (0 \le k \le n).$$

As $n \to \infty$ we get (6.3.38).

(b)$\Longrightarrow$(d) If we replace m by $M+m$ in (6.3.38) and consider $|g|^m w$ as a weight, then applying the implication (a)$\Longrightarrow$(c) one can get Statement (d).

(d)$\Longrightarrow$(a) Equation (6.3.40) with $c = 0$ becomes (6.3.38).

To prove (6.3.41) we notice that if (6.3.41) does not hold then for the number

$$c = \min\{\max_{-1 \le x \le 1} g(x), -\min_{-1 \le x \le 1} g(x)\}$$

we have

$$\left| \int_{|g(x)| \ge c} [\text{sgn } g(x)] w(x) dx \right| \ge \int_{|g(x)| \ge c,\ x \in [a,b]} w(x) dx > 0,$$

a contradiction to (6.3.39). □

Lemma 6.3.5. *Let $w > 0$ a.e. in $[-1,1]$ and let $u \in C[-1,1]$. Let the function $g(x) = \omega_n(x)u(x)$ satisfy the assumptions of Lemma 6.3.4. Then the following statements are equivalent.*

(a) The equation

$$\int_{-1}^{1} |g(x)|^m w(x) dx = \min_{P \in \widehat{\mathbf{P}}_n} \int_{-1}^{1} |P(x)u(x)|^m w(x) dx \tag{6.3.43}$$

holds for every $m = dM + r + 1$, $M = 0, 1, \ldots$, where $d > 0$ and $r \ge 0$ are fixed constants.

(b) The equation (6.3.43) holds for any $m \ge 1$.

(c) The orthogonality relations

$$\int_{|g(x)| \ge c} \frac{|g(x)|^m}{x - x_k} w(x) dx = 0, \quad k = 1, 2, \ldots, n, \tag{6.3.44}$$

hold for any $m \ge 1$ and any $c \ge 0$.

Proof. (a)$\Longleftrightarrow$(b) If we rewrite (6.3.43) in the form

$$\int_{-1}^{1} |\omega_n(x)|^m |u(x)|^m w(x) dx = \min_{P \in \widehat{\mathbf{P}}_n} \int_{-1}^{1} |P(x)|^m |u(x)|^m w(x) dx,$$

we can see that by Lemma 3.2.2 equation (6.3.43) means

$$\int_{-1}^{1} [\text{sgn } g(x)] |g(x)|^{m-1} \frac{g(x)}{x - x_k} w(x) dx = 0, \quad k = 1, 2, \ldots, n,$$

or equivalently

$$\int_{-1}^{1} [\text{sgn } |g(x)|(x - x_k)] \, \big| \, |g(x)| \text{sgn } (x - x_k) \big|^{m-1} \left[\left| \frac{g(x)}{x - x_k} \right| w(x) \right] dx = 0,$$
$$k = 1, 2, \ldots, n. \tag{6.3.45}$$

By Lemma 6.3.4 with the weight $|g(x)/(x-x_k)|w(x)$ equations (6.3.45) hold for every $m = dM + r + 1$, $M = 0, 1, \ldots$, if and only if equations (6.3.45) hold for any $m \geq 1$, which is equivalent to Statement (b).

(b)$\Longleftrightarrow$(c) Again by Lemma 6.3.4 equations (6.3.45) hold for any $m \geq 1$ if and only if equations (6.3.44) hold for any $m \geq 1$ and any $c \geq 0$. □

Theorem 6.3.3. *Let $w(x) = (1-x^2)^{-1/2}$ and $p \geq 1$. If for $n \in \mathbf{N}$*

$$p - 2q = 0, \tag{6.3.46}$$

$$\omega_n(x) = 2^{1-n} T_n(x), \tag{6.3.47}$$

and

$$\Omega_{n-1}(x) = 2^{1-n} U_{n-1}(x), \tag{6.3.48}$$

then equation (6.3.36) holds for any $m \geq 1$.

Conversely, if equation (6.3.36) with $n = 2$ holds for every $m = dM + r + 1$, $M = 0, 1, \ldots$, where $d > 0$ and $r \geq 0$ are fixed constants, then equations (6.3.46)–(6.3.48) are valid.

Proof. The first part of the theorem is given by Theorem 6.3.1. To prove the second by Lemma 6.3.1 with $u = 1$ and $n = 2$ we get $\omega_2(x) = 2^{-1} T_2(x) = x^2 - \frac{1}{2}$. Then equation (6.3.36) with $n = 2$ reads

$$\int_{-1}^{1} \left| x^2 - \frac{1}{2} \right|^m |\Omega_1(x)|^p (1-x^2)^{q-1/2} dx = \min_{Q \in \widehat{\mathbf{P}}_1} \int_{-1}^{1} \left| x^2 - \frac{1}{2} \right|^m |Q(x)|^p (1-x^2)^{q-1/2} dx.$$

It is easy to see that $\Omega_1(x) = x$ and hence equation (6.3.36) gives

$$\int_{-1}^{1} \left| x^2 - \frac{1}{2} \right|^m |x|^p (1-x^2)^{q-1/2} dx = \min_{P \in \widehat{\mathbf{P}}_2} \int_{-1}^{1} |P(x)|^m |x|^p (1-x^2)^{q-1/2} dx,$$

which holds for every $m = dM + r + 1$, $M = 0, 1, \ldots$. Equation (6.3.20) for the weight $w(x) = |x|^p (1-x^2)^{q-1/2}$ becomes

$$(1-x^2)^{1/2} |x|^p (1-x^2)^{q-1/2} = |x| (1-x^2)^{p/2} |x|^{2q-1}.$$

Hence

$$(1-x^2)^{(2q-p)/2} = |x|^{2q-p},$$

which implies (6.3.46).

Using (6.3.46) and applying Theorem 6.3.1, we obtain (6.3.47) and (6.3.48). □

Moreover, we will prove that the weight $w(x) = (1-x^2)^{-1/2}$ is the only weight having this property (up to a linear transformation). More precisely, we have the following.

Theorem 6.3.4. *Let a weight w satisfy $w(x) > 0$ for almost every $x \in [-1, 1]$ and $\int_{-1}^{1} w(x) dx = \pi$. Let $p_j, q_j \geq 0$ $(j = 0, 1, \ldots)$ satisfy*

$$p_0 - 2q_0 = 0 \tag{6.3.49}$$

and (6.3.11). If for each $n \in \mathbf{N}$ equation (6.3.36) holds for every pair $(p, q) = (p_j, q_j)$ and for every $m = dM + r + 1$, $M = 0, 1, \ldots$, where $d > 0$ and $r \geq 0$ are fixed constants, then relations (6.3.6), (6.3.47), (6.3.48), and

$$p_j - 2q_j = 0, \quad j = 1, 2, \ldots, \tag{6.3.50}$$

are valid.

Proof. If equation (6.3.36) holds then it remains true provided q and $w(x)$ are replaced by $q-s(s\leq q)$ and $(1-x^2)^s w(x)$, respectively; so we need the restriction (6.3.49) to guarantee the uniqueness of the weight.

Again by Lemma 6.3.1 with $u=1$ we get (6.3.47).

Equality (6.3.15) follows from (6.3.36) and (6.3.47) with $n=1$ by Lemma 6.3.2. Further, it follows from (6.3.36) and (6.3.47) with $n=2$ by (6.3.15) that $\Omega_1(x)=x$. Thus equation (6.3.20) for the weight $|x|^p(1-x^2)^q w(x)$ reads

$$(1-x^2)^{1/2}|x|^p(1-x^2)^q w(x)=|x|(1-x^2)^{p/2}|x|^{2q}w((1-x^2)^{1/2}).$$

That is,

$$(1-x^2)^{(2q-p+1)/2}w(x)=|x|^{2q-p+1}w((1-x^2)^{1/2}),$$

which holds for every pair $(p,q)=(p_j,q_j)$. This by (6.3.49) gives (6.3.50) and hence equation (6.3.36) becomes

$$\int_{-1}^{1}|T_n(x)|^m|\Omega_{n-1}(x)(1-x^2)^{1/2}|^p w(x)dx$$

$$=\min_{Q\in\widehat{\mathbf{P}}_{n-1}}\int_{-1}^{1}|T_n(x)|^m|Q(x)(1-x^2)^{1/2}|^p w(x)dx,$$

which holds for every $p=p_j$. By (6.3.11) we apply Lemma 6.3.1 to obtain

$$\|\Omega_{n-1}(x)(1-x^2)^{1/2}\|=\min_{Q\in\widehat{\mathbf{P}}_{n-1}}\|Q(x)(1-x^2)^{1/2}\|$$

which gives (6.3.48).

To prove (6.3.6) using (6.3.47) and (6.3.48), and applying Lemma 6.3.5 we conclude that equations

$$\int_{|T_n(x)|\geq c}\frac{|T_n(x)|^m}{x-x_k}|U_{n-1}(x)(1-x^2)^{1/2}|^p w(x)dx=0,\quad k=1,2,\ldots,n,$$

hold for any $m\geq 1$ and any $c\geq 0$, here the x_k's are the zeros of $T_n(x)$. By making the change of variable $x=\cos\theta$ we get

$$\int_{|\cos n\theta|\geq c}f(\theta)d\theta=0,\quad k=1,2,\ldots,n,$$

where $\theta_k=(2n-2k+1)\pi/(2n)$, $k=1,2,\ldots,n$ and

$$f(\theta)=\frac{|\cos n\theta|^m}{\cos\theta-\cos\theta_k}|\sin n\theta|^p\sin\theta\, w(\cos\theta).$$

That is,

$$\int_{\theta_1+\tau}^{\pi}f(\theta)d\theta+\sum_{k=1}^{n-1}(-1)^k\int_{\theta_{k+1}+\tau}^{\theta_k-\tau}f(\theta)d\theta+(-1)^n\int_0^{\theta_n-\tau}f(\theta)d\theta=0,$$

where $0 \le \tau \le \pi/(2n)$. Differentiating this equation with respect to τ yields

$$-f(\theta_1+\tau)-\sum_{k=1}^{n-1}(-1)^k[f(\theta_k-\tau)+f(\theta_{k+1}+\tau)]-(-1)^n f(\theta_n-\tau)=0.$$

Since $|\sin n\theta|$ takes the same values at the points $\theta=\theta_k\pm\tau$, we obtain

$$-g(\theta_1+\tau)-\sum_{k=1}^{n-1}(-1)^k[g(\theta_k-\tau)+g(\theta_{k+1}+\tau)]-(-1)^n g(\theta_n-\tau)=0,$$

where

$$g(\theta)=\frac{|\cos n\theta|^m}{\cos\theta-\cos\theta_k}\sin\theta\, w(\cos\theta).$$

This means

$$\int_{|\cos n\theta|\ge c} g(\theta)d\theta=0.$$

By making the change of variable $\cos\theta=x$ we get

$$\int_{|T_n(x)|\ge c}\frac{|T_n(x)|^m}{x-x_k}w(x)dx=0,\quad k=1,2,\ldots,n,$$

which holds for any $m\ge 1$ and any $c\ge 0$. By Lemma 6.3.5 equation

$$\int_{-1}^{1}|2^{1-n}T_n(x)|^m w(x)dx=\min_{P\in\widehat{\mathbf{P}}_n}\int_{-1}^{1}|P(x)|^m w(x)dx$$

holds for every $m\ge 1$. Then we must have (6.3.6) by Theorem 6.3.2. □

Theorem 6.3.5. *Let*

$$u^{(\alpha,\beta)}(x)=|x|^{2\alpha+1}(1-x^2)^{\beta}. \tag{6.3.51}$$

Then

$$\begin{cases} P_{2n}(u^{(\alpha,\beta)},m;x) &= (-2)^{-n}P_n(w^{(\alpha,\beta)},m;1-2x^2),\\ P_{2n+1}(u^{(\alpha,\beta)},m;x) &= (-2)^{-n}xP_n(w^{(\alpha+m/2,\beta)},m;1-2x^2).\end{cases} \tag{6.3.52}$$

Proof. We observe that the polynomial $P_{2n}(u^{(\alpha,\beta)},m)$ is even and $P_{2n+1}(u^{(\alpha,\beta)},m)$ odd, since the weight $u^{(\alpha,\beta)}$ is even. By (3.2.6) it is enough to prove

$$\int_{-1}^{1}\left|P_n(w^{(\alpha,\beta)},m;1-2x^2)\right|^{m-1}\left[\operatorname{sgn} P_n(w^{(\alpha,\beta)},m;1-2x^2)\right]$$
$$\times q(1-2x^2)u^{(\alpha,\beta)}(x)dx=0,\quad q\in\mathbf{P}_{n-1} \tag{6.3.53}$$

and

$$\int_{-1}^{1}\left|xP_n(w^{(\alpha+m/2,\beta)},m;1-2x^2)\right|^{m-1}\left\{\operatorname{sgn}\left[xP_n(w^{(\alpha+m/2,\beta)},m;1-2x^2)\right]\right\}$$
$$\times xq(1-2x^2)u^{(\alpha,\beta)}(x)dx=0,\quad q\in\mathbf{P}_{n-1}. \tag{6.3.54}$$

By making the change of variable $t = 1-2x^2$, we have

$$x = \left(\frac{1-t}{2}\right)^{1/2}, \quad 1-x^2 = \frac{1+t}{2}, \quad dx = -2^{-3/2}(1-t)^{-1/2}dt,$$

and

$$u^{(\alpha,\beta)}(x) = 2^{-\alpha-\beta-1/2}(1-t)^{\alpha+1/2}(1+t)^{\beta} = 2^{-\alpha-\beta-1/2}w^{(\alpha+1/2,\beta)}(t).$$

Thus the left-hand side of (6.3.53) is equal to

$$\begin{aligned} & 2\int_0^1 \left|P_n(w^{(\alpha,\beta)},m;1-2x^2)\right|^{m-1}\left[\operatorname{sgn} P_n(w^{(\alpha,\beta)},m;1-2x^2)\right] \\ & \times q(1-2x^2)u^{(\alpha,\beta)}(x)dx \\ = \ & 2^{-\alpha-\beta-1}\int_{-1}^1 \left|P_n(w^{(\alpha,\beta)},m;t)\right|^{m-1}\left[\operatorname{sgn} P_n(w^{(\alpha,\beta)},m;t)\right] \\ & \times q(t)w^{(\alpha,\beta)}(t)dt = 0; \end{aligned}$$

similarly, the left-hand side of (6.3.54) is equal to

$$\begin{aligned} & 2\int_0^1 \left|P_n(w^{(\alpha+m/2,\beta)},m;1-2x^2)\right|^{m-1}\left\{\operatorname{sgn}\left[P_n(w^{(\alpha+m/2,\beta)},m;1-2x^2)\right]\right\} \\ & \times q(1-2x^2)x^m u^{(\alpha,\beta)}(x)dx \\ = \ & 2^{-\alpha-\beta-m/2-1}\int_{-1}^1 \left|P_n(w^{(\alpha+m/2,\beta)},m;t)\right|^{m-1}\left\{\operatorname{sgn}\left[P_n(w^{(\alpha+m/2,\beta)},m;t)\right]\right\} \\ & \times q(t)w^{(\alpha+m/2,\beta)}(t)dt = 0. \quad \square \end{aligned}$$

Since $u^{(-1/2,\alpha)} = w^{(\alpha,\alpha)}$, as a direct consequence of Theorem 6.3.5 by (6.2.18) we have

Corollary 6.3.3. *We have*

$$\begin{aligned} P_{2n}(w^{(\alpha,\alpha)},m;x) &= 2^{-n}P_n(w^{(\alpha,-1/2)},m;2x^2-1) \\ &= (-2)^{-n}P_n(w^{(-1/2,\alpha)},m;1-2x^2), \end{aligned} \tag{6.3.55}$$

and

$$\begin{aligned} P_{2n+1}(w^{(\alpha,\alpha)},m;x) &= 2^{-n}xP_n(w^{(\alpha,(m-1)/2)},m;2x^2-1) \\ &= (-2)^{-n}xP_n(w^{((m-1)/2,\alpha)},m;1-2x^2). \end{aligned} \tag{6.3.56}$$

Theorem 6.3.6. *We have*

$$P_n(v_m,m;x) = 2^{-n}V_n(x), \tag{6.3.57}$$

$$P_n(w_m,m;x) = 2^{-n}W_n(x), \tag{6.3.58}$$

and

$$\begin{cases} x_{kn}(w_m,m) = -x_{n+1-k,n}(v_m,m), \\ \ell_{kn}(w_m,m;-x) = \ell_{n+1-k,n}(v_m,m;x), \end{cases} \quad k = 1,2,\ldots,n. \tag{6.3.59}$$

Proof. Setting $\alpha = (m-1)/2$ in (6.3.55) and using (6.3.4) and (6.2.16), we obtain

$$P_n(w_m, m; 2x^2-1) = 2^n P_{2n}(u_m, m; x) = 2^{-n} U_{2n}(x) = 2^{-n} W_n(2x^2-1),$$

which means (6.3.58). To prove (6.3.57) by (6.2.18), (6.3.58), and (6.2.17)

$$P_n(v_m, m; x) = (-1)^n P_n(w_m, m; -x) = (-2)^{-n} W_n(-x) = 2^{-n} V_n(x).$$

With the help of (6.3.57), (6.3.58), and (6.2.17) we obtain (6.3.59). □

Remark 6.3.1. Equations (6.3.57) and (6.3.58) may also be obtained by setting $\alpha = -1/2$ in (6.3.56) .

Remark 6.3.2. Using (6.2.14) Theorem 6.3.5 can not provide m orthogonal polynomials explicitly with respect to weights other than (6.2.14).

In fact, by (6.3.52) the m orthogonal polynomials $P_n(u^{(\alpha,\beta)}, m)$ may be represented explicitly if and only if both $P_n(w^{(\alpha,\beta)}, m)$ and $P_n(w^{(\alpha+m/2,\beta)}, m)$ can be represented explicitly. But by (6.2.14) it is equivalent to either the condition $\alpha = \beta = -1/2$ or the condition $\alpha = -1/2$ and $\beta = (m-1)/2$, which is just either the case $u^{(-1/2,-1/2)} = w^{(-1/2,-1/2)}$ or the case $u^{(-1/2,(m-1)/2)} = w^{((m-1)/2,(m-1)/2)}$, respectively. That is to say, they belong to the weights (6.2.14).

Theorem 6.3.7. *With the notations of (1.4.1) and (1.4.7) we have*

$$\begin{cases} P_{2n}(h, m; x) & = P_n(\ell^{(-1/2)}, m; x^2), \\ P_{2n+1}(h, m; x) & = x P_n(\ell^{((m-1)/2)}, m; x^2). \end{cases} \tag{6.3.60}$$

Proof. By (3.2.6) it is enough to show that for $N \leq n$

$$\int_{\mathbf{R}} |P_{2n}(h, m; x)|^{m-1} [\text{sgn}\, P_{2n}(h, m; x)] P_{2N}(h, m; x) h(x) dx = \delta_{nN}$$

and

$$\int_{\mathbf{R}} |P_{2n+1}(h, m; x)|^{m-1} [\text{sgn}\, P_{2n+1}(h, m; x)] P_{2N+1}(h, m; x) h(x) dx = \delta_{nN}.$$

By making the change of variable $x = t^{1/2}$ it follows from (3.2.6) that

$$\begin{aligned} \delta_{nN} &= \int_0^\infty |P_n(\ell^{(-1/2)}, m; t)|^{m-1} [\text{sgn}\, P_n(\ell^{(-1/2)}, m; t)] \\ &\quad \times P_N(\ell^{(-1/2)}, m; t) \ell^{(-1/2)}(t) dt \\ &= 2\int_0^\infty |P_n(\ell^{(-1/2)}, m; x^2)|^{m-1} [\text{sgn}\, P_n(\ell^{(-1/2)}, m; x^2)] \\ &\quad \times P_N(\ell^{(-1/2)}, m; x^2) h(x) dx \\ &= \int_{\mathbf{R}} |P_n(\ell^{(-1/2)}, m; x^2)|^{m-1} [\text{sgn}\, P_n(\ell^{(-1/2)}, m; x^2)] \\ &\quad \times P_N(\ell^{(-1/2)}, m; x^2) h(x) dx. \end{aligned}$$

This proves the first equation of (6.3.60). Similarly we can prove the second. □

6.4 A Recurrence Approach

In this section we derive explicit expressions of Gauss-Turán quadrature formulas by a recurrence approach. To state the first main result in this section we need some notations. Let $n \in \mathbf{N}$ and $m, p, s, i \in \mathbf{N}_0$. We use the symbols in Section 1.1, say, $[x], \{x\}$, etc. and put

$$w(x) = (1-x^2)^{-\{(p+1)/2\}},$$

$$x_{kn} = \cos\frac{(2n-2k+1)\pi}{2n}, \quad k = 1,2,\ldots,n,$$

$$y_{kn} = \cos\frac{(n-k)\pi}{n}, \quad k = 0,1,\ldots,n,$$

$$S_n(x) = \operatorname{sgn}\,[T_n(x)^m U_{n-1}(x)^p],$$

$$p_k = [n_k(p-2)], \quad n_k = \begin{cases} 1, & 1 \le k \le n-1, \\ \frac{1}{2}, & k = 0, n, \end{cases}$$

$$\ell_k(x) = \frac{T_n(x)}{T_n'(x_k)(x-x_k)}, \quad k = 1,2,\ldots,n, \tag{6.4.1}$$

$$I_{kms}(x) = \ell_k(x)^m [U_{n-1}(x)/U_{n-1}(x_k)]^s, \quad k = 1,2,\ldots,n, \tag{6.4.2}$$

$$\alpha_{kims} = \frac{1}{i!}\left[I_{kms}(x)^{-1}\right]^{(i)}_{x=x_k}, \quad k = 1,2,\ldots,n, \tag{6.4.3}$$

$$\alpha^*_{kim} = \frac{1}{i!}\left[I_{km1}(x)^{-1}U_{n-1}(x)^{-1}U'_{n-1}(x)\right]^{(i)}_{x=x_k}, \quad k = 1,2,\ldots,n, \tag{6.4.4}$$

$$D_p(x) = (1-x^2)^{[p/2]} U_{n-1}(x)^p, \tag{6.4.5}$$

$$d_{kp} = D_p^{([n_k p])}(y_k) = \begin{cases} (-1)^{p(n-k)} p! n^p (1-y_k^2)^{-[(p+1)/2]}, & 1 \le k \le n-1, \\ (-1)^{p(n-1)} 2^{[p/2]} ([\frac{p}{2}])! n^p, & k = 0, \\ (-2)^{[p/2]} ([\frac{p}{2}])! n^p, & k = n, \end{cases} \tag{6.4.6}$$

$$L_{kps}(x) = \frac{([n_k p])! D_p(x) T_n(x)^s}{d_{kp}(x-y_k)^{[n_k p]} T_n(y_k)^s}, \quad k = 0,1,\ldots,n, \tag{6.4.7}$$

$$\beta_{kips} = \frac{1}{i!}\left[L_{kps}(x)^{-1}\right]^{(i)}_{x=y_k}, \quad k = 0,1,\ldots,n, \tag{6.4.8}$$

$$\beta^*_{kip} = \frac{1}{i!}\left[L_{kp1}(x)^{-1} T_n(x)^{-1} T_n'(x)\right]^{(i)}_{x=y_k}, \quad k = 0,1,\ldots,n, \tag{6.4.9}$$

$$\gamma_{kips} = \frac{1}{i!}\left\{ \sum_{\nu \in \{0,n\}\setminus\{k\}} [2(y_\nu - x) L_{kps}(x)]^{-1} \right\}^{(i)}_{x=y_k}, \quad k = 0,1,\ldots,n. \tag{6.4.10}$$

In the quadrature formula of the following theorem we prefer to use the notations different from (4.1.3).

Theorem 6.4.1. *Let $m, p \in \mathbf{N}_0$ and $m+p \geq 1$. Then the quadrature formula*

$$\int_{-1}^{1} f(x)S_n(x)w(x)dx = \sum_{k=1}^{n}\sum_{j=0}^{m-2} C_{kjmp} f^{(j)}(x_k) + \sum_{k=0}^{n}\sum_{j=0}^{p_k} D_{kjmp} f^{(j)}(y_k) \tag{6.4.11}$$

is exact for all $f \in \mathbf{P}_{(m+p)n-2\{p/2\}-1}$. Here for $m, p \geq 2$ we have the recurrence relations

$$C_{kjmp} = \frac{m-1}{m+p-2} C_{k,j,m-2,p} + \frac{p-1}{m+p-2} C_{k,j,m,p-2}, \quad k = 1,2,\ldots,n, \quad j = 0,1,\ldots,m-2, \tag{6.4.12}$$

and

$$D_{kjmp} = \frac{m-1}{m+p-2} D_{k,j,m-2,p} + \frac{p-1}{m+p-2} D_{k,j,m,p-2}, \quad k = 0,1,\ldots,n, \quad j = 0,1,\ldots,p_k. \tag{6.4.13}$$

Meanwhile, the initial values of C_{kjmp}, $1 \leq k \leq n$, $0 \leq j \leq m-2$, are given by

$$C_{k,m-2,m,p} = \frac{(-1)^{(m+p)(n-k)} B(\frac{m-1}{2}, \frac{p+1}{2})(1-x_k^2)^{(m-2)/2+\{p/2\}}}{(m-2)!n^{m-1}}, \tag{6.4.14}$$

$$C_{kjm0} = C_{k,j,m-2,0} + \frac{j(m-3)!}{j!} C_{k,m-2,m,0}\alpha_{k,m-2-j,m-2,0}, \quad m \geq 3, \tag{6.4.15}$$

$$C_{kjm1} = C_{k,j,m-2,1} + \frac{(m-3)!}{j!} C_{k,m-2,m,1}\left(j\alpha_{k,m-2-j,m-2,1} - \alpha^*_{k,m-3-j,m-2}\right), \quad m \geq 3, \tag{6.4.16}$$

and the initial values of D_{kjmp}, $0 \leq k \leq n$, $0 \leq j \leq p_k$, are given by

$$D_{k,p_k,m,p} = \frac{(-1)^{m(n-k+1)} n_k B((m+1)/2,(p-1)/2)}{d_{k,p-2}n}, \tag{6.4.17}$$

$$D_{kj0p} = D_{k,j,0,p-2} + \frac{p_k! D_{k,p_k,0,p}}{j! n_k(p-2)}\Big((j+\{n_k(p-2)\})\beta_{k,p_k-j,p-2,0} - \frac{1}{2}[1+(-1)^{p+1}]\gamma_{k,p_k-j-1,p-2,0}\Big), \tag{6.4.18}$$

$$D_{kj1p} = D_{k,j,1,p-2} + \frac{p_k! D_{k,p_k,1,p}}{j! n_k(p-2)}\Big((j+\{n_k(p-2)\})\beta_{k,p_k-j,p-2,1} - \beta^*_{k,p_k-j-1,p-2} - \frac{1}{2}[1+(-1)^{p+1}]\gamma_{k,p_k-j-1,p-2,1}\Big). \tag{6.4.19}$$

Remark 6.4.1. Here we agree that $C_{kjmp} = 0$ if $j > m-2$ and $D_{kjmp} = 0$ if $j > p_k$. In particular,

$$C_{kj0p} = C_{kj1p} = C_{k,j,j+1,p} = D_{kjm0} = D_{kjm1} = D_{k,j,m,j+1} = 0.$$

Using the recurrence relation (6.4.12) by induction determination of C_{kjmp} may be reduced to calculations of $C_{k,j,j+1,q}, C_{k,j,j+2,q}, C_{k,j,r,0}$, or $C_{k,j,r,1}$. We see that $C_{k,j,j+1,q} = 0$; the terms $C_{k,j,j+2,p}, C_{k,j,r,0}$, or $C_{k,j,r,1}$ may be calculated by (6.4.14), (6.4.15), or (6.4.16), respectively. Similarly D_{kjmp} may be calculated by (6.4.13) and (6.4.17)–(6.4.19).

Proof of Theorem 6.4.1. Applying Theorem 6.3.1 and using the same argument as that of Theorem 4.1.1 we conclude that the quadrature formula (6.4.11) with certain numbers C_{kjmp} and D_{kjmp} holds for all $f \in \mathbf{P}_{(m+p)n-2\{p/2\}-1}$. Now let us determine these numbers. To this end we begin by proving (6.4.14)–(6.4.19). We give the proofs of (6.4.14)–(6.4.16) only, the others being similar. It suffices to show these formulas for a fixed index k, $1 \le k \le n$.

To prove (6.4.14) we use the formula

$$C_{k,m-2,m,p} = \frac{1}{(m-2)!} \times \int_{-1}^{1} \frac{T_n(x)^{m-1}(1-x^2)^{[p/2]}U_{n-1}(x)^{p-1}}{T_n'(x_k)^{m-1}(1-x_k^2)^{[p/2]}U_{n-1}(x_k)^{p-1}(x-x_k)} S_n(x)w(x)dx, \tag{6.4.20}$$

which is obtained from (6.4.11). Expand $T_n(x)/(x-x_k)$ as

$$\frac{T_n(x)}{x-x_k} = U_{n-1}(x) + \sum_{j=0}^{n-2} a_j U_j(x) \tag{6.4.21}$$

with certain constants a_j. Applying (6.3.2) with replacing m by $m-2$ yields

$$\int_{-1}^{1} [U_j(x)T_n(x)]T_n(x)^{m-3}U_{n-1}(x)^{p-1}(1-x^2)^{[p/2]}S_n(x)w(x)dx = 0, \quad j \le n-2.$$

Hence (6.4.20) becomes

$$C_{k,m-2,m,p} = \frac{\int_{-1}^{1} |T_n(x)|^{m-2}|U_{n-1}(x)|^p(1-x^2)^{(p-1)/2}dx}{(m-2)!T_n'(x_k)^{m-1}(1-x_k^2)^{[p/2]}U_{n-1}(x_k)^{p-1}},$$

which by (6.2.13) gives

$$C_{k,m-2,m,p} = \frac{B(\frac{m-1}{2}, \frac{p+1}{2})}{(m-2)!(1-x_k^2)^{[p/2]}T_n'(x_k)^{m-1}U_{n-1}(x_k)^{p-1}}. \tag{6.4.22}$$

Using the well known formulas

$$T_n'(x) = nU_{n-1}(x) \tag{6.4.23}$$

and

$$T_n'(x_k) = (-1)^{n-k}n(1-x_k^2)^{-1/2}, \quad k = 1,2,\ldots,n, \tag{6.4.24}$$

we get (6.4.14).

Let us prove (6.4.15). For $r \in \mathbf{N}$ let A_{kjr}, $0 \le j \le r-1$, be the fundamental polynomials for Hermite interpolation with $m_k \equiv r$ on the zeros x_k's given by (2.2.9). Then

$$A_{kjr}(x) = \frac{1}{j!}\sum_{i=0}^{r-1-j} \alpha_{kjr0}(x-x_k)^{i+j}\ell_k(x)^r, \quad 0 \le j \le r-1. \tag{6.4.25}$$

Formula (6.4.22) with $p=0$ by (6.4.23) becomes

$$C_{k,m-2,m,0} = \frac{B(\frac{m-1}{2},\frac{1}{2})}{n(m-2)!T_n'(x_k)^{m-2}}. \tag{6.4.26}$$

For $0 \le j \le m-3$ applying (6.4.11) with $p=0$ twice (replacing m by $m-2$ for the first time), we obtain

$$\begin{aligned} C_{k,j,m-2,0} &= \int_{-1}^{1} A_{k,j,m-2}(x)[\text{sgn } T_n(x)^m]\frac{dx}{\sqrt{1-x^2}} \\ &= C_{k,j,m,0} + \sum_{\nu=1}^{n} C_{\nu,m-2,m,0}A_{k,j,m-2}^{(m-2)}(x_\nu). \end{aligned} \tag{6.4.27}$$

To calculate the sum in the above formula we distinguish two cases.

Case 1. $\nu = k$. Using (6.4.25), (6.4.2), (6.4.3), and (2.2.5) and applying Newton-Leibniz rule twice we obtain

$$\begin{aligned} & A_{k,j,m-2}^{(m-2)}(x_k) \\ = & \frac{1}{j!}\sum_{i=0}^{m-3-j} \alpha_{k,i,m-2,0}[(x-x_k)^{i+j}\ell_k(x)^{m-2}]_{x=x_k}^{(m-2)} \\ = & \frac{(m-2)!}{j!}\sum_{i=0}^{m-3-j} \alpha_{k,i,m-2,0}\frac{[\ell_k(x)^{m-2}]_{x=x_k}^{(m-2-i-j)}}{(m-2-i-j)!} \\ = & \binom{m-2}{j}\sum_{i=0}^{m-3-j}\binom{m-2-j}{i}[\ell_k(x)^{-(m-2)}]_{x=x_k}^{(i)}[\ell_k(x)^{m-2}]_{x=x_k}^{(m-2-i-j)} \\ = & \binom{m-2}{j}\left\{[1]_{x=x_k}^{(m-2-j)} - (m-2-j)!\alpha_{k,m-2-j,m-2,0}\right\} \\ = & -\frac{(m-2)!\alpha_{k,m-2-j,m-2,0}}{j!}. \end{aligned}$$

Case 2. $\nu \ne k$. Using (6.4.25) and applying Newton-Leibniz rule again we have

$$\begin{aligned} & \sum_{\nu \ne k} C_{\nu,m-2,m,0}A_{k,j,m-2}^{(m-2)}(x_\nu) \\ = & \frac{1}{j!}\sum_{\nu \ne k} C_{\nu,m-2,m,0}\sum_{i=0}^{m-3-j} \alpha_{k,i,m-2,0}[(x-x_k)^{i+j}\ell_k(x)^{m-2}]_{x=x_\nu}^{(m-2)} \\ = & \frac{1}{j!}\sum_{i=0}^{m-3-j} \alpha_{k,i,m-2,0}\sum_{\nu \ne k} C_{\nu,m-2,m,0}[(x-x_k)^{i+j}\ell_k(x)^{m-2}]_{x=x_\nu}^{(m-2)} \\ = & \frac{(m-2)!}{j!}\sum_{i=0}^{m-3-j} \alpha_{k,i,m-2,0}\sum_{\nu \ne k} C_{\nu,m-2,m,0}(x_\nu - x_k)^{i+j}\ell_k'(x_\nu)^{m-2}. \end{aligned}$$

Substituting the formulas (6.4.26) and

$$\ell_k'(x_\nu) = \frac{T_n'(x_\nu)}{(x_\nu - x_k)T_n'(x_k)}, \quad \nu \neq k \tag{6.4.28}$$

into the above relation, we get

$$\begin{aligned}
&\sum_{\nu\neq k} C_{\nu,m-2,m,0} A^{(m-2)}_{k,j,m-2}(x_\nu) \\
= \; & \frac{B(\frac{m-1}{2},\frac{1}{2})}{j!n} \sum_{i=0}^{m-3-j} \alpha_{k,i,m-2,0} \sum_{\nu\neq k} \frac{(x_\nu - x_k)^{i+j}\ell_k'(x_\nu)^{m-2}}{T_n'(x_\nu)^{m-2}} \\
= \; & \frac{(m-2)!C_{k,m-2,m,0}}{j!} \sum_{i=0}^{m-3-j} \alpha_{k,i,m-2,0} \sum_{\nu\neq k} \frac{1}{(x_\nu - x_k)^{m-2-i-j}}.
\end{aligned}$$

By means of (2.2.12) we can obtain a simple formula

$$\sum_{\nu\neq k} C_{\nu,m-2,m,0} A^{(m-2)}_{k,j,m-2}(x_\nu) = \frac{(m-2-j)(m-3)!C_{k,m-2,m,0}\alpha_{k,m-2-j,m-2,0}}{j!}. \tag{6.4.29}$$

At last, from (6.4.27) we get the recurrence relations (6.4.15) with respect to the index m.

Now let us show (6.4.16). Formula (6.4.22) with $p = 1$ reads

$$C_{\nu,m-2,m,1} = \frac{B((m-1)/2,1)}{(m-2)!T_n'(x_\nu)^{m-1}}, \quad \nu = 1,2,\ldots,n. \tag{6.4.30}$$

To determine C_{kjm1}, $0 \le j \le m-3$, we consider the function

$$f(x) = \frac{1}{j!} \sum_{i=0}^{m-3-j} \alpha_{k,i,m-2,1}(x - x_k)^{i+j} I_{k,m-2,1}(x). \tag{6.4.31}$$

It is easy to check that f satisfies the interpolatory conditions

$$\begin{cases} f^{(\mu)}(x_\nu) = \delta_{j\mu}\delta_{k\nu}, & \nu = 1,2,\ldots,n, \quad \mu = 0,1,\ldots,m-3, \\ f(y_\nu) = 0, & \nu = 1,2,\ldots,n-1. \end{cases}$$

Substituting f into (6.4.11) with $p = 1$ twice yields (replacing m by $m-2$ for the first time)

$$C_{k,j,m-2,1} = \int_{-1}^{1} f(x)\text{sgn}\,[T_n(x)^{m-2}U_{n-1}(x)]dx = C_{kjm1} + \sum_{\nu=1}^{n} C_{\nu,m-2,m,1} f^{(m-2)}(x_\nu). \tag{6.4.32}$$

To calculate the sum in the above formula we distinguish the cases when $\nu = k$ and $\nu \neq k$.

Case I. $\nu = k$. Using (6.4.31) and applying Newton-Leibniz rule twice gives

$$\begin{aligned}
f^{(m-2)}(x_k) \; &= \; \frac{1}{j!} \sum_{i=0}^{m-3-j} \alpha_{k,i,m-2,1} [(x - x_k)^{i+j} I_{k,m-2,1}(x)]^{(m-2)}_{x=x_k} \\
&= \; \frac{(m-2)!}{j!} \sum_{i=0}^{m-3-j} \alpha_{k,i,m-2,1} \frac{I^{(m-2-i-j)}_{k,m-2,1}(x_k)}{(m-2-i-j)!} \\
&= \; \binom{m-2}{j} \sum_{i=0}^{m-3-j} \binom{m-2-j}{i} [I_{k,m-2,1}(x)^{-1}]^{(i)}_{x=x_k} I^{(m-2-i-j)}_{k,m-2,1}(x_k) \\
&= \; \binom{m-2}{j} \left\{ [1]^{(m-2-j)}_{x=x_k} - (m-2-j)!\alpha_{k,m-2-j,m-2,1} \right\}.
\end{aligned}$$

At last we obtain

$$f^{(m-2)}(x_k) = -\frac{(m-2)!\alpha_{k,m-2-j,m-2,1}}{j!}. \tag{6.4.33}$$

Case II. $\nu \neq k$. Using (6.4.33) and applying Newton-Leibniz rule again we have

$$\begin{aligned}
&\sum_{\nu\neq k} C_{\nu,m-2,m,1} f^{(m-2)}(x_\nu) \\
&= \frac{1}{j!}\sum_{\nu\neq k} C_{\nu,m-2,m,1} \sum_{i=0}^{m-3-j} \alpha_{k,i,m-2,1}[(x-x_k)^{i+j} I_{k,m-2,1}(x)]^{(m-2)}_{x=x_\nu} \\
&= \frac{1}{j!}\sum_{i=0}^{m-3-j} \alpha_{k,i,m-2,1} \sum_{\nu\neq k} C_{\nu,m-2,m,1}(x_\nu-x_k)^{i+j} I^{(m-2)}_{k,m-2,1}(x_\nu) \\
&= \frac{(m-2)!}{j!}\sum_{i=0}^{m-3-j} \alpha_{k,i,m-2,1} \sum_{\nu\neq k} C_{\nu,m-2,m,1}\frac{(x_\nu-x_k)^{i+j} U_{n-1}(x_\nu)\ell_k'(x_\nu)^{m-2}}{U_{n-1}(x_k)}.
\end{aligned}$$

Using (6.4.23), (6.4.24), (6.4.30), and (6.4.28) the above formula becomes

$$\begin{aligned}
&\sum_{\nu\neq k} C_{\nu,m-2,m,1} f^{(m-2)}(x_\nu) \\
&= \frac{(m-2)!C_{k,m-2,m,1}}{j!}\sum_{i=0}^{m-3-j} \alpha_{k,i,m-2,1}\sum_{\nu\neq k}\frac{1}{(x_\nu-x_k)^{m-2-i-j}}.
\end{aligned}$$

Since (see Theorem 2.2.2 and its proof)

$$\sum_{\nu\neq k}\frac{1}{(x_\nu-x_k)^{m-2-i-j}} = -\frac{1}{(m-3-i-j)!}\left[\frac{l_k'(x)}{l_k(x)}\right]^{(m-3-i-j)}_{x=x_k},$$

applying Newton-Leibniz rule we have

$$\begin{aligned}
&\sum_{\nu\neq k} C_{\nu,m-2,m,1} f^{(m-2)}(x_\nu) \\
&= -\frac{(m-2)!C_{k,m-2,m,1}}{j!(m-3-j)!} \\
&\quad \times \sum_{i=0}^{m-3-j}\binom{m-3-j}{i}[I_{k,m-2,1}(x)^{-1}]^{(i)}_{x=x_k}\left[\frac{\ell_k'(x)}{\ell_k(x)}\right]^{(m-3-i-j)}_{x=x_k} \\
&= -\frac{(m-2)!C_{k,m-2,m,1}}{j!(m-3-j)!}\left[I_{k,m-2,1}(x)^{-1}\ell_k(x)^{-1}\ell_k'(x)\right]^{(m-3-j)}_{x=x_k}.
\end{aligned}$$

Using the relation

$$[I_{k,m-2,1}(x)^{-1}]' = -I_{k,m-2,1}(x)^{-1}[(m-2)\ell_k(x)^{-1}\ell_k'(x) + U_{n-1}(x)^{-1}U_{n-1}'(x)]$$

we obtain

$$\begin{aligned}
&-I_{k,m-2,1}(x)^{-1}\ell_k(x)^{-1}\ell_k'(x) \\
&= \frac{1}{m-2}\left\{[I_{k,m-2,1}(x)^{-1}]' - U_{n-1}(x)^{-1}U_{n-1}'(x)\right\}
\end{aligned}$$

and hence

$$\sum_{\nu\neq k} C_{\nu,m-2,m,1}f^{(m-2)}(x_\nu) = \frac{(m-3)!}{j!}C_{k,m-2,m,1}$$
$$\times[(m-2-j)\alpha_{k,m-2-j,m-2,1}+\alpha^*_{k,m-3-j,m-2}]. \tag{6.4.34}$$

Combining (6.4.31)–(6.4.34) gives (6.4.16).

We give the proof of (6.4.12) only, that of (6.4.13) being similar. Relation (6.4.17) yields

$$D_{k,p_k,m,p} = \frac{m-1}{m+p-2}D_{k,p_k,m-2,p}, \quad k=0,1,\ldots,n. \tag{6.4.35}$$

We consider the function $f \in \mathbf{P}_{(m+p-3)n-2\{p/2\}-1}$ which satisfies the interpolation conditions

$$\begin{cases} f^{(\mu)}(x_\nu) = \delta_{j\mu}\delta_{k\nu}, & \nu=1,2,\ldots,n, \quad \mu=0,1,\ldots,m-2, \\ f^{(\mu)}(y_\nu)=0, & \nu=0,1,\ldots,n, \quad \mu=0,1,\ldots,p_\nu-1. \end{cases}$$

Substituting f into (6.4.11) with p replaced by $p-2$ leads to

$$C_{k,j,m,p-2} = \int_{-1}^{1} f(x)S_n(x)w(x)dx. \tag{6.4.36}$$

On the other hand, first applying (6.4.11) to f and using (6.4.36) yields

$$C_{k,j,m,p-2} = C_{kjmp} + \sum_{i=0}^{n} D_{i,p_i,m,p}f^{(p_i)}(y_i). \tag{6.4.37}$$

Next, inserting f into (6.4.11) with m replaced by $m-2$ and using (6.4.36) gives

$$C_{k,j,m,p-2} = C_{k,j,m-2,p} + \sum_{i=0}^{n} D_{i,p_i,m-2,p}f^{(p_i)}(y_i). \tag{6.4.38}$$

Substituting (6.4.35) into (6.4.37) and canceling the term

$$\sum_{i=0}^{n} D_{i,p_i,m-2,p}f^{(p_i)}(y_i)$$

from the two equations (6.4.37) and (6.4.38) we obtain the recurrence relations (6.4.12) with respect to the index m. □

Since Theorem 6.4.1 is very general and complicated, for convenience of use we prefer to state some special cases and to use the notation $b_{kir} = \alpha_{kir0}$. By definition

$$\lambda_{kj}(\nu,m) = C_{kjm0}.$$

Theorem 6.4.2. *Let $m \in \mathbf{N}$. Then the generalized Gaussian quadrature formula*

$$\int_{-1}^{1} f(x)[sgn\ T_n(x)^m]\frac{dx}{\sqrt{1-x^2}} = \sum_{k=1}^{n}\sum_{j=0}^{m-2}\lambda_{kj}(\nu,m)f^{(j)}(x_k) \tag{6.4.39}$$

holds for all $f \in \mathbf{P}_{mn-1}$, *where for each* k, $1 \le k \le n$,

$$\lambda_{k0}(v,m) = \begin{cases} 0, & m \in \mathbf{N}_1, \\ \frac{\pi}{n}, & m \in \mathbf{N}_2, \end{cases} \tag{6.4.40}$$

and

$$\begin{aligned} \lambda_{kj}(v,m) &= \frac{(-1)^{m(n-k)}\tau_m}{(j-1)!} \sum_{i=1}^{[\frac{m-j}{2}]} \frac{(m-2i-1)!b_{k,m-2i-j,m-2i}(1-x_k^2)^{(m-2i)/2}}{[(m-2i)!!]^2 n^{m-2i+1}}, \\ & \quad j = 1,2,\ldots,m-2, \ m \ge 3. \end{aligned} \tag{6.4.41}$$

Here

$$\tau_m = \begin{cases} 2, & m \in \mathbf{N}_1, \\ \pi, & m \in \mathbf{N}_2. \end{cases}$$

According to the formulas (6.4.41) and (2.2.5) we obtain the following corollaries.

Corollary 6.4.1. *Let* $m \in \mathbf{N}_2$. *Then for each* k, $1 \le k \le n$,

$$\lambda_{k,m-2}(v,m) = \frac{\pi(1-x_k^2)^{(m-2)/2}}{[(m-2)!!]^2 n^{m-1}}, \tag{6.4.42}$$

$$\lambda_{k,m-3}(v,m) = -\frac{\pi(m-3)x_k(1-x_k^2)^{(m-4)/2}}{2(m-2)!!(m-4)!!n^{m-1}}, \ m \ge 4, \tag{6.4.43}$$

$$\begin{cases} \lambda_{k0}(v,4) = \frac{\pi}{n}, \\ \lambda_{k1}(v,4) = -\frac{\pi x_k}{4n^3}, \\ \lambda_{k2}(v,4) = \frac{\pi(1-x_k^2)}{4n^3}, \end{cases} \tag{6.4.44}$$

$$\begin{cases} \lambda_{k0}(v,6) = \frac{\pi}{n}, \\ \lambda_{k1}(v,6) = -\frac{\pi x_k(20n^2-1)}{64n^5}, \\ \lambda_{k2}(v,6) = \frac{\pi[(20n^2-7)(1-x_k^2)+3]}{64n^5}, \\ \lambda_{k3}(v,6) = -\frac{3\pi x_k(1-x_k^2)}{32n^5}, \\ \lambda_{k4}(v,6) = \frac{\pi(1-x_k^2)^2}{64n^5}, \end{cases} \tag{6.4.45}$$

$$\begin{cases} \lambda_{k0}(v,8) = \frac{\pi}{n}, \\ \lambda_{k1}(v,8) = -\frac{\pi x_k(28n^2-1)^2}{2304n^7}, \\ \lambda_{k2}(v,8) = \frac{\pi[(784n^4-392n^2+31)(1-x_k^2)+168n^2-15]}{2304n^7}, \\ \lambda_{k3}(v,8) = -\frac{\pi x_k[(112n^2-30)(1-x_k^2)+5]}{768n^7}, \\ \lambda_{k4}(v,8) = \frac{\pi(1-x_k^2)[(56n^2-65)(1-x_k^2)+45]}{2304n^7}, \\ \lambda_{k5}(v,8) = -\frac{5\pi x_k(1-x_k^2)^2}{768n^7}, \\ \lambda_{k6}(v,8) = \frac{\pi(1-x_k^2)^3}{2304n^7}. \end{cases} \tag{6.4.46}$$

Corollary 6.4.2. *Let $m \in \mathbf{N}_1$ and $m \geq 3$. Then for each k, $1 \leq k \leq n$,*

$$\lambda_{k,m-2}(\nu,m) = \frac{(-1)^{n-k}2(1-x_k^2)^{(m-2)/2}}{[(m-2)!!]^2 n^{m-1}}, \tag{6.4.47}$$

$$\lambda_{k,m-3}(\nu,m) = \frac{(-1)^{n-k+1}(m-2)(m-3)x_k(1-x_k^2)^{(m-4)/2}}{[(m-2)!!]^2 n^{m-1}}, \tag{6.4.48}$$

$$\begin{cases} \lambda_{k0}(\nu,3) = 0, \\ \lambda_{k1}(\nu,3) = \frac{(-1)^{n-k}2(1-x_k^2)^{1/2}}{n^2}, \end{cases} \tag{6.4.49}$$

$$\begin{cases} \lambda_{k0}(\nu,5) = 0, \\ \lambda_{k1}(\nu,5) = \frac{(-1)^{n-k}2(10n^2-1)(1-x_k^2)^{1/2}}{9n^4}, \\ \lambda_{k2}(\nu,5) = \frac{(-1)^{n-k+1}2x_k(1-x_k^2)^{1/2}}{3n^4}, \\ \lambda_{k3}(\nu,5) = \frac{(-1)^{n-k}2(1-x_k^2)^{3/2}}{9n^4}, \end{cases} \tag{6.4.50}$$

$$\begin{cases} \lambda_{k0}(\nu,7) = 0, \\ \lambda_{k1}(\nu,7) = \frac{(-1)^{n-k}2(259n^4-35n^2+1)(1-x_k^2)^{1/2}}{225n^6}, \\ \lambda_{k2}(\nu,7) = \frac{(-1)^{n-k+1}2(7n^2-1)x_k(1-x_k^2)^{1/2}}{15n^6}, \\ \lambda_{k3}(\nu,7) = \frac{(-1)^{n-k}2(1-x_k^2)^{1/2}[(7n^2-5)(1-x_k^2)+3]}{45n^6}, \\ \lambda_{k4}(\nu,7) = \frac{(-1)^{n-k+1}4x_k(1-x_k^2)^{3/2}}{45n^6}, \\ \lambda_{k5}(\nu,7) = \frac{(-1)^{n-k}2(1-x_k^2)^{5/2}}{225n^6}. \end{cases} \tag{6.4.51}$$

Theorem 6.4.3. *Let $p \in \mathbf{N}$. Then the quadrature formula*

$$\int_{-1}^{1} f(x)[sgn\, U_{n-1}(x)^p](1-x^2)^{-\{(p+1)/2\}}dx = \sum_{k=0}^{n}\sum_{j=0}^{p_k} D_{kj0p} f^{(j)}(y_k) \tag{6.4.52}$$

is exact for all $f \in \mathbf{P}_{pn+2[\frac{p}{2}]-1}$, where for each j, $1 \leq j \leq p_k$, and for each k, $0 \leq k \leq n$,

$$\begin{cases} D_{k,p_k,0,p} = \frac{n_k \tau_p (p-2)!}{d_{k,p-2}[(p-2)!!]^2 n}, & p \geq 2, \\ D_{k,p_k+1,0,p} = 0, \end{cases} \tag{6.4.53}$$

and $D_{k,j,0,p}$ is given by (6.4.18).

In order to state following results we introduce the notation ($x_k = x_{kn}(\nu_m,m)$, $k = 1,2,\ldots,n$):

$$-1 = x_0 < x_1 < \cdots < x_n < x_{n+1} = 1,$$

$$n_k = \begin{cases} \frac{1}{2}, & k = 0, n+1, \\ 1, & 1 \leq k \leq n, \end{cases}$$

$$\Pi_m(x) = (1+x)^{m/2} V_n(x)^m, \tag{6.4.54}$$

$$d_{km} = \Pi_m^{(n_k m)}(x_k) = \begin{cases} (m/2)! V_n(-1)^m, & k = 0, \\ m!(1+x_k)^{m/2} V_n'(x_k)^m, & 1 \leq k \leq n, \end{cases} \tag{6.4.55}$$

$$L_{km}(x) = \frac{(n_k m)!\Pi_m(x)}{d_{km}(x-x_k)^{n_k m}}, \quad k=0,1,\ldots,n, \tag{6.4.56}$$

$$b_{kim} = \frac{1}{i!}\left[L_{km}(x)^{-1}\right]^{(i)}_{x=x_k}, \quad k=0,1,\ldots,n, \quad i=0,1,\ldots. \tag{6.4.57}$$

Lemma 6.4.1. *Let $m \in \mathbf{N}_2$ and for $1 \le k \le n$, $0 \le j \le m-2$,*

$$c_{kjm}(v_m) = \int_{-1}^{1} [2^{-n}V_n(x)]^j \ell_k(x) v_m(x)dx, \tag{6.4.58}$$

where

$$\ell_k(x) = \frac{V_n(x)}{V_n'(x_k)(x-x_k)}.$$

Then

$$c_{kjm}(v_m) = \begin{cases} \frac{\pi j!(1+x_k)^{(m-j)/2}}{2^{jn-j/2-1}(j!!)^2(2n+1)} + \frac{\pi(1+x_k)^{(m-j-4)/2}}{2^{jn+j/2}V_n'(x_k)} \\ \times \sum\limits_{i=0}^{j/2} \sum\limits_{r=i(2n+1)+n}^{(m-j-4)/2} \frac{1}{2^r(1+x_k)^r}\binom{j+1}{j/2-i}\binom{2r+1}{r-i(2n+1)-n}, & j \in \mathbf{M}_2, \\ \frac{\pi(1+x_k)^{(m-j-3)/2}}{2^{jn+j/2-1/2}V_n'(x_k)} \\ \times \sum\limits_{i=0}^{(j+1)/2} \sum\limits_{r=i(2n+1)}^{(m-j-3)/2} \frac{1}{2^r(1+x_k)^r}\binom{j+1}{(j+1)/2-i}\binom{2r}{r-i(2n+1)}, & j \in \mathbf{M}_1. \end{cases} \tag{6.4.59}$$

Proof. By making the change of variable $x = 2t^2 - 1$ and applying (6.2.15)

$$\begin{aligned} c_{kjm}(v_m) &= 2^{(m+2)/2-jn}\int_0^1 V_n(2t^2-1)^j \ell_k(2t^2-1)t^m v(t)dt \\ &= 2^{m/2-jn}\int_{-1}^1 V_n(2t^2-1)^j \ell_k(2t^2-1)t^m v(t)dt \\ &= 2^{m/2-jn}\int_{-1}^1 \ell_k(2t^2-1)t^{m-j}T_{2n+1}(t)^j v(t)dt. \end{aligned}$$

We distinguish the cases when $j \in \mathbf{M}_2$ and $j \in \mathbf{M}_1$.
Case 1. $j \in \mathbf{M}_2$. In this case

$$\begin{aligned} c_{kjm}(v_m) &= 2^{m/2-jn}\left[\int_{-1}^1 \ell_k(2t^2-1)t^2(t^{m-j-2}-t_k^{m-j-2})T_{2n+1}(t)^j v(t)dt \right. \\ &\quad \left. + t_k^{m-j-2}\int_{-1}^1 \ell_k(2t^2-1)t^2 T_{2n+1}(t)^j v(t)dt\right] \\ &= 2^{m/2-jn}[S_1+S_2]. \end{aligned}$$

By (6.2.3) and (6.2.15)

$$\begin{aligned}
S_1 &= \int_{-1}^{1} \frac{t(t^{m-j-2}-t_k^{m-j-2})}{2V_n'(x_k)(t^2-t_k^2)} T_{2n+1}(t)^{j+1} v(t)dt \\
&= \frac{1}{2V_n'(x_k)} \sum_{r=0}^{(m-j-4)/2} t_k^{m-j-4-2r} \int_{-1}^{1} t^{2r+1} T_{2n+1}(t)^{j+1} v(t)dt \\
&= \frac{\pi}{2^{j+2}V_n'(x_k)} \sum_{r=0}^{(m-j-4)/2} t_k^{m-j-4-2r} \\
&\quad \times \sum_{i=0}^{j/2} \frac{1}{2^{2r}} \binom{j+1}{j/2-i} \binom{2r+1}{(2r+1-(2i+1)(2n+1))/2} \\
&= \frac{\pi(1+x_k)^{(m-j)/2-2}}{2^{(m+j)/2}V_n'(x_k)} \sum_{i=0}^{j/2} \sum_{r=i(2n+1)+n}^{(m-j-4)/2} \frac{1}{2^r(1+x_k)^r} \\
&\quad \times \binom{j+1}{j/2-i} \binom{2r+1}{r-i(2n+1)-n}.
\end{aligned}$$

By (6.2.12)

$$S_2 = \frac{2(j!)\pi t_k^{m-j}}{(j!!)^2(2n+1)} = \frac{\pi(j!)(1+x_k)^{(m-j)/2}}{2^{(m-j)/2-1}(j!!)^2(2n+1)}.$$

Case 2. $j \in \mathbf{M}_1$. In this case by (6.2.12) and (6.2.15)

$$\begin{aligned}
c_{kjm}(v_m) &= 2^{m/2-jn} \left[\int_{-1}^{1} \ell_k(2t^2-1)t(t^{m-j-1}-t_k^{m-j-1}) T_{2n+1}(t)^j v(t)dt \right. \\
&\quad \left. + t_k^{m-j-1} \int_{-1}^{1} \ell_k(2t^2-1)t T_{2n+1}(t)^j v(t)dt \right] \\
&= 2^{m/2-jn} \left[\int_{-1}^{1} \ell_k(2t^2-1)t(t^{m-j-1}-t_k^{m-j-1}) T_{2n+1}(t)^j v(t)dt \right. \\
&= \frac{2^{m/2-jn-1}}{V_n'(x_k)} \int_{-1}^{1} \frac{t^{m-j-1}-t_k^{m-j-1}}{t^2-t_k^2} T_{2n+1}(t)^{j+1} v(t)dt \\
&= \frac{2^{m/2-jn-1}}{V_n'(x_k)} \sum_{r=0}^{(m-j-3)/2} \int_{-1}^{1} t^{2r} T_{2n+1}(t)^{j+1} v(t)dt.
\end{aligned}$$

By (6.2.3)

$$\begin{aligned}
c_{kjm}(v_m) &= \frac{2^{m/2-jn-j-1}\pi}{V_n'(x_k)} \sum_{r=0}^{(m-j-3)/2} t_k^{m-j-3-2r} 2^{-2r} \\
&\quad \times \sum_{i=0} (j+1)/2' \binom{j+1}{(j+1)/2-i} \binom{2r}{r-i(2n+1)} \\
&= \frac{\pi(1+x_k)^{(m-j-3)/2}}{2^{jn+j/2-1/2}V_n'(x_k)} \sum_{i=0} (j+1)/2' \sum_{r=i(2n+1)}^{(m-j-3)/2} \frac{1}{2^r(1+x_k)^r} \\
&\quad \times \binom{j+1}{(j+1)/2-i} \binom{2r}{r-i(2n+1)}. \qquad \square
\end{aligned}$$

Theorem 6.4.4. *Let $m \in \mathbf{N}_2$, $m_k = n_k(m-2)$, and let x_k's be the zeros of $V_n(x)$. Then the Gaussian quadrature formula*

$$\int_{-1}^{1} f(x)v(x)dx = \sum_{k=0}^{n}\sum_{j=0}^{m_k} \lambda_{kjm} f^{(j)}(x_k) \tag{6.4.60}$$

holds for all $f \in \mathbf{P}_{mn+m/2-1}$, where for each j, $0 \le j \le m_k$, and each k, $0 \le k \le n$,

$$\begin{cases} \lambda_{k,m_k,m} = \frac{2^{m/2}(m-2)!\pi n_k}{[(m-2)!!]^2(2n+1)d_{k,m-2}}, \\ \lambda_{k,m_k+1,m} = 0, \end{cases} \tag{6.4.61}$$

$$\lambda_{kjm} = \lambda_{k,j,m-2} + \frac{(m_k-1)!\lambda_{k,m_k,m} b_{k,m_k-j,m-2}}{(j-1)!}, \quad m \ge 4. \tag{6.4.62}$$

Proof. Let k, $0 \le k \le n$, be fixed.

First let us prove (6.4.61), since (6.4.60) is obvious. To this end put $\Omega_{n+1}(x) = (1+x)V_n(x)$. It is easy to check that the polynomial of degree $(m-1)n + m/2 - 1$

$$f(x) = \frac{\Pi_{m-2}(x)\Omega_{n+1}(x)}{d_{k,m-2}\Omega'_{n+1}(x_k)(x-x_k)}$$

satisfies the interpolatory conditions:

$$f^{(\mu)}(x_\nu) = \delta_{\mu,m_k}\delta_{\nu,k}, \quad \mu = 0,1,\ldots,m_k, \quad \nu = 0,1,\ldots,n.$$

In fact, it is sufficient to show

$$f^{(m_k)}(x_k) = 1.$$

By Newton-Leibniz rule and (6.4.55) this is indeed the case:

$$f^{(m_k)}(x_k) = \left[\frac{\Pi_{m-2}(x)\Omega_{n+1}(x)}{d_{k,m-2}\Omega'_{n+1}(x_k)(x-x_k)}\right]^{(m_k)}_{x=x_k} = \frac{1}{d_{k,m-2}}\Pi^{(m_k)}_{m-2}(x_k) = 1.$$

Substituting f into (6.4.60) gives

$$\lambda_{k,m_k,m} = \int_{-1}^{1} \frac{\Pi_{m-2}(x)\Omega_{n+1}(x)}{d_{k,m-2}\Omega'_{n+1}(x_k)(x-x_k)} v(x)dx, \quad k = 0,1,\ldots,n. \tag{6.4.63}$$

We distinguish the cases when $1 \le k \le n$ and $k = 0$.

Case 1. $1 \le k \le n$. By means of (6.4.63), (6.4.58), and (6.4.59), we have

$$\begin{aligned} \lambda_{k,m_k,m} &= \frac{1}{d_{k,m-2}(1+x_k)} \int_{-1}^{1} V_n(x)^{m-2}\ell_k(x)v_m(x)dx \\ &= \frac{2^{(m-2)n}c_{k,m-2,m}(v_m)}{d_{k,m-2}(1+x_k)} = \frac{2^{m/2}(m-2)!\pi}{[(m-2)!!]^2(2n+1)d_{k,m-2}}. \end{aligned}$$

Case 2. $k=0$. By making the change of variable $x=2t^2-1$, it follows from (6.4.63), (6.2.15), and (6.2.18) that

$$\begin{aligned}
\lambda_{0,(m-2)/2,m} &= \frac{1}{d_{0,m-2}V_n(-1)}\int_{-1}^{1} V_n(x)^{m-1}v_{m-2}(x)dx \\
&= \frac{2^{(m-2)/2}}{d_{0,m-2}V_n(-1)}\int_{-1}^{1} V_n(2t^2-1)^{m-1}t^{m-2}v(t)dt \\
&= \frac{2^{(m-2)/2}}{d_{0,m-2}V_n(-1)}\int_{-1}^{1} V_n(2t^2-1)T_{2n+1}(t)^{m-2}v(t)dt \\
&= \frac{2^{(m-2)/2}}{d_{0,m-2}V_n(-1)}\cdot\frac{(m-2)!\pi V_n(-1)}{[(m-2)!!]^2(2n+1)} \\
&= \frac{2^{m/2}(m-2)!\pi n_0}{[(m-2)!!]^2(2n+1)d_{0,m-2}}.
\end{aligned}$$

Next let us prove (6.4.62). By (2.2.9) we conclude that the polynomial of degree $(m-2)n+m/2-2$

$$f(x)=\frac{1}{j!}\sum_{i=0}^{m_k-j-1} b_{k,i,m-2}(x-x_k)^{i+j}L_{k,m-2}(x) \tag{6.4.64}$$

satisfies the interpolatory conditions:

$$f^{(\mu)}(x_\nu)=\delta_{j\mu}\delta_{k\nu},\quad \mu=0,1,\ldots,m_k-1,\quad \nu=0,1,\ldots,n.$$

Substituting f into (6.4.60) twice (replacing m by $m-2$ for the first time), we have

$$\lambda_{k,j,m-2}=\int_{-1}^{1} f(x)v(x)dx=\lambda_{kjm}+\sum_{\nu=0}^{n}\lambda_{\nu,m_\nu,m}f^{(m_\nu)}(x_\nu). \tag{6.4.65}$$

We separate the cases when $\nu=k$ and $\nu\neq k$.

Case 1. $\nu=k$. By (6.4.55) and (6.4.56) we have $L_{k,m-2}(x_k)=1$. Using (6.4.64) and (6.4.57), and applying Newton-Leibniz rule twice, we obtain

$$\begin{aligned}
f^{(m_k)}(x_k) &= \frac{1}{j!}\sum_{i=0}^{m_k-j-1} b_{k,i,m-2}\left[(x-x_k)^{i+j}L_{k,m-2}(x)\right]_{x=x_k}^{(m_k)} \\
&= \frac{(m_k)!}{j!}\sum_{i=0}^{m_k-j-1} b_{k,i,m-2}\frac{L_{k,m-2}^{(m_k-j-i)}(x_k)}{(m_k-j-i)!} \\
&= \binom{m_k}{j}\sum_{i=0}^{m_k-j-1}\binom{m_k-j}{i}\left[L_{k,m-2}(x)^{-1}\right]_{x=x_k}^{(i)}L_{k,m-2}^{(m_k-j-i)}(x_k) \\
&= \binom{m_k}{j}\left\{[1]_{x=x_k}^{(m_k-j)}-(m_k-j)!b_{k,m_k-j,m-2}\right\} \\
&= -\frac{(m_k)!b_{k,m_k-j,m-2}}{j!}.
\end{aligned} \tag{6.4.66}$$

Case 2. $\nu \neq k$. Using (6.4.64) and applying Newton-Leibniz rule again we have

$$\begin{aligned}
&\sum_{\nu\neq k} \lambda_{\nu,m_\nu,m} f^{(m_\nu)}(x_\nu) \\
= &\ \frac{1}{j!}\sum_{\nu\neq k}\lambda_{\nu,m_\nu,m}\sum_{i=0}^{m_k-j-1} b_{k,i,m-2}\left[(x-x_k)^{i+j}L_{k,m-2}(x)\right]_{x=x_\nu}^{(m_\nu)} \\
= &\ \frac{1}{j!}\sum_{i=0}^{m_k-j-1} b_{k,i,m-2}\sum_{\nu\neq k}\lambda_{\nu,m_\nu,m}\left[(x-x_k)^{i+j}L_{k,m-2}(x)\right]_{x=x_\nu}^{(m_\nu)} \\
= &\ \frac{1}{j!}\sum_{i=0}^{m_k-j-1} b_{k,i,m-2}\sum_{\nu\neq k}\lambda_{\nu,m_\nu,m}(x_\nu-x_k)^{i+j}L_{k,m-2}^{(m_\nu)}(x_\nu). \qquad (6.4.67)
\end{aligned}$$

According to (6.4.55) and (6.4.56) we see

$$L_{k,m-2}^{(m_\nu)}(x_\nu) = \frac{d_{\nu,m-2}m_k!}{d_{k,m-2}(x_\nu - x_k)^{m_k}}, \quad \nu \neq k. \qquad (6.4.68)$$

Substituting (6.4.61) and (6.4.68) into (6.4.67) and applying Theorem 2.2.2, we obtain

$$\begin{aligned}
&\sum_{\nu\neq k} \lambda_{\nu,m_\nu,m} f^{(m_\nu)}(x_\nu) \\
= &\ \frac{2^{m/2}(m-2)!\pi m_k!}{j!(m-2)[(m-2)!!]^2(2n+1)d_{k,m-2}}\sum_{i=0}^{m_k-j-1} b_{k,i,m-2}\sum_{\nu\neq k}\frac{m_\nu}{(x_\nu-x_k)^{m_k-i-j}} \\
= &\ \frac{(m_k-1)!\lambda_{k,m_k,m}}{j!}\sum_{i=0}^{m_k-j-1} b_{k,i,m-2}a_{k,m_k-j-i} \\
= &\ \frac{(m_k-1)!(m_k-j)\lambda_{k,m_k,m}b_{k,m_k-j,m-2}}{j!},
\end{aligned}$$

which, coupled with (6.4.65) and (6.4.66), gives (6.4.62). □

Theorem 6.4.5. *Let $m \in \mathbf{N}_2$ and let x_k's be the zeros of $V_n(x)$. Then the Gaussian quadrature formula*

$$\int_{-1}^{1} f(x)v_m(x)dx = \sum_{k=1}^{n}\sum_{j=0}^{m-2}\lambda_{kj}(v_m,m)f^{(j)}(x_k) \qquad (6.4.69)$$

holds for all $f \in \mathbf{P}_{mn-1}$, where for each j, $0 \le j \le m-2$, and for each k, $1 \le k \le n$,

$$\lambda_{kj}(v_m,m) = \sum_{i=j}^{m-2}\binom{i}{j}\frac{(m/2)!}{(m/2+j-i)!}\lambda_{kim}(1+x_k)^{m/2+j-i}. \qquad (6.4.70)$$

Proof. Let $A_{kj} \in \mathbf{P}_{mn-1}$, $0 \le j \le m-1$, $1 \le k \le n$, be the fundamental polynomials based on the zeros x_k's. Inserting $f = A_{kj}$ into (6.4.69) and using (6.4.60) yields

$$\begin{aligned}\lambda_{kj}(v_m, m) &= \int_{-1}^{1} A_{kj}(x)v_m(x)dx \\ &= \int_{-1}^{1} A_{kj}(x)(1+x)^{m/2}v(x)dx \\ &= \sum_{i=0}^{m-2} \lambda_{kim}[A_{kj}(x)(1+x)^{m/2}]_{x=x_k}^{(i)} \\ &= \sum_{i=j}^{m-2} \binom{i}{j} \frac{(m/2)!}{(m/2+j-i)!}\lambda_{kim}(1+x_k)^{m/2+j-i}.\end{aligned}$$

This proves (6.4.70). □

Theorem 6.4.6. *Let $m \in \mathbf{N}_2$, $m_k = n_k(m-2)$, and let x_k's be the zeros of $W_n(x)$. We have the Gaussian quadrature formulas*

$$\int_{-1}^{1} f(x)v(x)dx = \sum_{k=1}^{n+1}\sum_{j=0}^{m_k} \mu_{kjm} f^{(j)}(x_k), \quad f \in \mathbf{P}_{mn+m/2-1}, \tag{6.4.71}$$

and

$$\int_{-1}^{1} f(x)w_m(x)dx = \sum_{k=1}^{n}\sum_{j=0}^{m-2} \lambda_{kj}(w_m, m) f^{(j)}(x_k), \quad f \in \mathbf{P}_{mn-1}, \tag{6.4.72}$$

where

$$\mu_{kjm} = (-1)^j \lambda_{n+1-k,j,m}, \quad 0 \le j \le m_k, \quad 1 \le k \le n+1, \tag{6.4.73}$$

$$\lambda_{kj}(w_m, m) = (-1)^j \lambda_{n+1-k,j}(v_m, m), \quad 0 \le j \le m-2, \quad 1 \le k \le n. \tag{6.4.74}$$

Proof. Equations (6.4.71) and (6.4.72) are obvious. Now let us prove (6.4.74) only, since the proof of (6.4.73) may be derived in a similar way.

Let $A_{kj}(w_m), A_{n+1-k,j}(v_m) \in \mathbf{P}_{mn-1}$, $0 \le j \le m-1$, $1 \le k \le n$, satisfy the interpolatory conditions:

$$A_{kj}^{(\mu)}(w_m; x_\nu(w_m, m)) = A_{n+1-k,j}^{(\mu)}(v_m; x_{n+1-\nu}(v_m, m)) = \delta_{j\mu}\delta_{k\nu},$$
$$\mu = 0, 1, \ldots, m-1, \quad \nu = 1, 2, \ldots, n.$$

Then by (6.3.59)

$$\begin{aligned}[A_{n+1-k,j}(v_m; -x)]_{x=x_\nu(w_m,m)}^{(\mu)} &= (-1)^\mu A_{n+1-k,j}^{(\mu)}(v_m; x_{n+1-\nu}(v_m, m)) \\ &= (-1)^j \delta_{j\mu}\delta_{k\nu}, \quad \mu = 0, 1, \ldots, m-1, \quad \nu = 1, 2, \ldots, n,\end{aligned}$$

which means

$$A_{kj}(w_m; x) = (-1)^j A_{n+1-k,j}(v_m; -x). \tag{6.4.75}$$

By making the change of variable $x = -t$, it follows from (6.4.72), (6.4.75), and (6.4.69) that

$$\begin{aligned}\lambda_{kj}(w_m, m) &= \int_{-1}^{1} A_{kj}(w_m; x)w_m(x)dx = (-1)^j \int_{-1}^{1} A_{n+1-k,j}(v_m; -x)w_m(x)dx \\ &= (-1)^j \int_{-1}^{1} A_{n+1-k,j}(v_m; t)v_m(t)dt = (-1)^j \lambda_{n+1-k,j}(v_m, m).\end{aligned}$$ □

6.5 An Approach of Divided Differences

In this section we give a *divided difference quadrature formula* and use the notations of Chapters 2 and 4.

Theorem 6.5.1. *Let $f \in \mathbf{P}_N$ and let $(x_1,\dots,x_n)$ be the solution of (3.1.1). We have the generalized Gaussian quadrature formulas*

$$\int_{\mathbf{R}} f(x)S_n(x)d\mu(x) = \sum_{k=1}^{n}\sum_{j=0}^{m_k-2} \mu_{kj} f[x_1^{m_1-1},\dots,x_{k-1}^{m_{k-1}-1},x_k^{j+1}] \tag{6.5.1}$$

and

$$\int_{\mathbf{R}} f(x)S_n(x)d\mu(x) = \sum_{j=0}^{m-2}\sum_{m_k\geq j+2} \nu_{kj} f[x_1^{\min\{j+1,m_1-1\}},\dots, x_k^{\min\{j+1,m_k-1\}},x_{k+1}^{\min\{j,m_{k+1}-1\}},\dots,x_n^{\min\{j,m_n-1\}}], \tag{6.5.2}$$

where

$$\mu_{kj} = \int_{\mathbf{R}} \prod_{i=1}^{k-1}(x-x_i)^{m_i-1}(x-x_k)^j S_n(x)d\mu(x) \tag{6.5.3}$$

and

$$\nu_{kj} = \int_{\mathbf{R}} \prod_{i=1}^{k-1}(x-x_i)^{\min\{j+1,m_i-1\}}\prod_{i=k}^{n}(x-x_i)^{\min\{j,m_i-1\}} S_n(x)d\mu(x). \tag{6.5.4}$$

In particular, if $m_1 = \cdots = m_n = m$ then

$$\int_{\mathbf{R}} f(x)S_n(x)d\mu(x) = \sum_{k=1}^{n}\sum_{j=0}^{m-2} \mu_{kj} f[x_1^{m-1},\dots,x_{k-1}^{m-1},x_k^{j+1}] \tag{6.5.5}$$

and

$$\int_{\mathbf{R}} f(x)S_n(x)d\mu(x) = \sum_{k=1}^{n}\sum_{j=0}^{m-2} \nu_{kj} f[x_1^{j+1},\dots,x_k^{j+1},x_{k+1}^{j},\dots,x_n^{j}], \tag{6.5.6}$$

where

$$\mu_{kj} = \int_{\mathbf{R}} \prod_{i=1}^{k-1}(x-x_i)^{m-1}(x-x_k)^j S_n(x)d\mu(x) \tag{6.5.7}$$

and

$$\nu_{kj} = \int_{\mathbf{R}} \omega_n(x)^j \prod_{i=1}^{k-1}(x-x_i) S_n(x)d\mu(x). \tag{6.5.8}$$

Meanwhile, we have the generalized Gaussian quadrature formula

$$\int_{\mathbf{R}} f(x)S_n(x)d\mu(x) = \sum_{k=1}^{n}\sum_{j=0}^{m-2} \rho_{kj} f[x_1^{j},\dots,x_n^{j},x_k], \tag{6.5.9}$$

where

$$\rho_{kj} = \int_{\mathbf{R}} \omega_n(x)^j \ell_k(x) S_n(x)d\mu(x). \tag{6.5.10}$$

Proof. Multiplying (2.3.8)–(2.3.10), (2.3.6), and (2.3.7) by $S_n(x)d\mu(x)$, and then integrating them from -1 to 1, by orthogonality we obtain (6.5.1)–(6.5.10). □

For the Chebyshev weight we can get a simpler formula. To this end we introduce the symbol

$$\mathcal{L}_j f = \begin{cases} \sum_{k=1}^n f(x_k), & j=0, \\ f'[x_1^j,\ldots,x_n^j], & j \ge 1. \end{cases} \tag{6.5.11}$$

Then we have

Theorem 6.5.2. *Let x_i's be the zeros of T_n. Let $m \in \mathbf{N}$, $k \in \mathbf{N}_0$, $f \in \mathbf{P}_{(m+k)n-1}$, and*

$$\mathcal{A}_{kn}(f) = \frac{2}{\pi}\int_{-1}^{1} f(x)T_{kn}(x)[sgn\ T_n(x)^m]\frac{dx}{\sqrt{1-x^2}}. \tag{6.5.12}$$

Then

$$\mathcal{A}_{kn}(f) = \frac{2}{n}\left\{ \sum_{\substack{k+m\in\mathbf{N}_2 \\ k\notin\mathbf{N}_2}} \binom{0}{k/2}\mathcal{L}_0 f + \sum_{\substack{1\le j\le m+k-2 \\ j+k+m\in\mathbf{N}_2 \\ k-j\notin\mathbf{N}_2}} \frac{1}{2^{jn}j}\binom{j}{(j+k)/2}\mathcal{L}_j f \right\}. \tag{6.5.13}$$

In particular for $m \in \mathbf{N}_2$

$$\mathcal{A}_{kn}(f) = \begin{cases} \frac{2}{n}\left\{ \mathcal{L}_0 f + \sum_{j=1}^{(m-2)/2} \frac{(2j-1)!}{2^{2jn}(j!)^2}\mathcal{L}_{2j} f \right\}, & k=0, \\ \frac{2}{n}\sum_{j=0}^{(m-2)/2} \frac{(k+2j-1)!}{2^{(k+2j)n}j!(k+j)!}\mathcal{L}_{k+2j} f, & k>0, \end{cases} \tag{6.5.14}$$

and for $m \in \mathbf{N}_1$

$$\mathcal{A}_{kn}(f) = \begin{cases} \frac{2}{n}\sum_{j=1}^{(m-3)/2} \frac{(2j-2)!}{2^{(2j-1)n}\Gamma((2j+1)/2)^2}\mathcal{L}_{2j-1} f, & k=0, \\ \frac{2}{n}\left\{ \frac{1-(-1)^k}{2}\binom{0}{k/2}\mathcal{L}_0 f + \sum_{j=-[k/2]}^{(m-3)/2} \frac{\binom{k+2j+1}{(2j+1)/2}}{2^{(k+2j+1)n}(k+2j+1)}\mathcal{L}_{k+2j+1} f \right\}, & k>0. \end{cases} \tag{6.5.15}$$

Proof. For this polynomial f by (2.3.1) and (2.3.7) we have

$$\begin{aligned} f(x) &= \sum_{i=1}^{n}\sum_{j=0}^{m+k-2} f[x_1^j,\ldots,x_n^j,x_i][2^{1-n}T_n(x)]^j\ell_i(x) \\ &\quad + f[x_1^{m+k-1},\ldots,x_n^{m+k-1},x][2^{1-n}T_n(x)]^{m+k-1}. \end{aligned}$$

Multiplying this with

$$\frac{2}{\pi}T_{kn}(x)[\mathrm{sgn}\ T_n(x)^m]\frac{dx}{\sqrt{1-x^2}}$$

and then integrating between -1 and 1, we obtain

$$\mathcal{A}_{kn}(x) = S_1 + S_2,$$

where

$$S_1 = \frac{2}{\pi}\sum_{i=1}^{n}\sum_{j=0}^{m+k-2} 2^{j(1-n)} f[x_1^j,\ldots,x_n^j,x_i] \times \int_{-1}^{1} T_n(x)^j T_{kn}(x)\ell_i(x)[\operatorname{sgn} T_n(x)^m]\frac{dx}{\sqrt{1-x^2}}$$

and

$$S_2 = \frac{2^{(m+k-1)(1-n)+1}}{\pi} \times \int_{-1}^{1} f[x_1^{m+k-1},\ldots,x_n^{m+k-1},x]T_n(x)^{m+k-1}T_{kn}(x)[\operatorname{sgn} T_n(x)^m]\frac{dx}{\sqrt{1-x^2}}.$$

It is particularly simple to calculate S_2; since $f[x_1^{m+k-1},\ldots,x_n^{m+k-1},\cdot] \in \mathbf{P}_{n-1}$, by (6.2.7) $S_2 = 0$. To determine S_1 we use (6.2.7) to get

$$S_1 = \sum_{\substack{0\le j\le m+k-2\\ j+k+m\in\mathbf{N}_2\\ k-j\notin\mathbf{N}_2}} \frac{1}{2^{jn-1}}\binom{j}{(j+k)/2}\sum_{i=1}^{n} f[x_1^j,\ldots,x_n^j,x_i],$$

which coupled with (2.3.15) gives (6.5.13). Formulas (6.5.14) and (6.5.15) follow directly from (6.5.13). □

Remark 6.5.1. With the help of (2.3.14) we can transform a divided difference quadrature formula (6.5.13) into a Gauss-Turán quadrature formula. For example, applying (6.5.14) and (2.3.14) and recalling the notation (4.5.5)

$$Q_{n,m-2}(v;f) = \sum_{k=1}^{n}\sum_{j=0}^{m-2}\lambda_{kj}(v,m)f^{(j)}(x_k) \tag{6.5.16}$$

we have for $m \in \mathbf{N}_2$

$$\begin{aligned} &Q_{n,m-2}(v;f)\\ =\ & \frac{\pi}{n}\left\{\sum_{k=1}^{n} f(x_k) + \sum_{j=1}^{m/2-1}\frac{(2j-1)!}{4^{jn}(j!)^2}f'[x_1^{2j},\ldots,x_n^{2j}]\right\}\\ =\ & \frac{\pi}{n}\sum_{k=1}^{n}\left\{f(x_k) + \sum_{j=1}^{m/2-1}\frac{1}{4^j(j!)^2T_n'(x_k)^{2j}}[\ell_k(x)^{-2j}f'(x)]_{x=x_k}^{(2j-1)}\right\}. \end{aligned} \tag{6.5.17}$$

Using the notation of (6.4.57) we see

$$[\ell_k(x)^{-2j}f'(x)]_{x=x_k}^{(2j-1)} = \sum_{i=0}^{2j-1}\frac{(2j-1)!}{i!}b_{k,2j-i-1,2j}f^{(i+1)}(x_k).$$

Substituting the above formula and (6.4.24) into (6.5.15), we obtain

$$\begin{aligned}
& Q_{n,m-2}(v;f) \\
= & \frac{\pi}{n}\sum_{k=1}^{n}\left\{f(x_k)+\sum_{j=1}^{m/2-1}\sum_{i=0}^{2j-1}\frac{(2j-1)!(1-x_k^2)^j}{i![(2j)!!]^2n^{2j}}b_{k,2j-i-1,2j}f^{(i+1)}(x_k)\right\} \\
= & \frac{\pi}{n}\sum_{k=1}^{n}\left\{f(x_k)+\sum_{j=1}^{m/2-1}\sum_{i=1}^{2j}\frac{(2j-1)!(1-x_k^2)^j}{(i-1)![(2j)!!]^2n^{2j}}b_{k,2j-i,2j}f^{(i)}(x_k)\right\} \\
= & \frac{\pi}{n}\sum_{k=1}^{n}\left\{f(x_k)+\sum_{j=1}^{m-2}\frac{1}{(j-1)!}\sum_{i=[(j+1)/2]}^{m/2-1}\frac{(2i-1)!(1-x_k^2)^i}{[(2i)!!]^2n^{2i}}b_{k,2i-j,2i}f^{(j)}(x_k)\right\}.
\end{aligned}$$

It is easy to examine that this formula is equivalent to (6.4.41) with $m \in \mathbf{N}_2$.

6.6 A Trigonometric Gaussian Quadrature Formula

To derive a *trigonometric Gaussian quadrature formula* at the zeros

$$\theta_k = \frac{2n-2k+1}{2n}\pi, \quad k=1,2,\ldots,n, \tag{6.6.1}$$

we introduce the numbers c_{ki}, $i=0,1,\ldots,k$, $k\in\mathbf{N}$, which are defined by the *generating function*

$$\prod_{i=0}^{k-1}(x+i^2)=\sum_{i=1}^{k}c_{ki}x^i. \tag{6.6.2}$$

Using these numbers we give an explicit expression for the linear functional $\mathcal{L}_j f$ based on the zeros $x_k=\cos\theta_k$ of T_n.

Theorem 6.6.1. *Let* $g(\theta)=f(\cos\theta)$. *Then*

$$\mathcal{L}_{2j}f=\begin{cases}\mathcal{D}_0(g), & j=0,\\ \frac{2^{2jn}}{(2j-1)!}\sum_{i=1}^{j}\frac{c_{ji}}{(2n)^{2i}}\mathcal{D}_{2i}(g), & j>0,\end{cases} \tag{6.6.3}$$

where

$$\mathcal{D}_i(g)=\sum_{k=1}^{n}g^{(i)}(\theta_k). \tag{6.6.4}$$

Proof. The equation (6.6.3) with $j=0$ is trivial. Let $j\geq 1$. In is not hard to check that the linear functional $\mathcal{L}_{2j}f$ may be uniquely determined by the following conditions:

(a) We have

$$\mathcal{L}_{2j}f=\sum_{k=1}^{n}\sum_{i=0}^{2j}\rho_{kji}f^{(i)}(x_k)$$

for certain numbers ρ_{kji}.

(b) The equation $\mathcal{L}_{2j}f=0$ holds for all $f\in\mathbf{P}_{2jn-1}$.

(c) The equation $\mathcal{L}_{2j}T_{2jn}=2jn2^{2jn-1}$ holds.

On the other hand, by differentiation the right-hand side of (6.6.3) must satisfy Condition (a). To examine Conditions (b) and (c) we choose $f = T_N$. Then

$$D_{2i}(g) = (-1)^i N^{2i} \sum_{k=1}^{n} \cos N\theta_k = \begin{cases} 0, & N \neq 2pn, \\ (-1)^{i+p} n N^{2i}, & N = 2pn. \end{cases}$$

Meanwhile, by (6.6.2)

$$\begin{aligned} \sum_{i=1}^{j} (-1)^i c_{ji} p^{2i} &= \sum_{i=1}^{j} c_{ji} (-p^2)^i = \prod_{i=0}^{j-1} (i^2 - p^2) \\ &= \begin{cases} 0, & 1 \leq p \leq j-1, \\ (-1)^j \prod_{i=0}^{j-1} (j+i)(j-i) = (-1)^j j (2j-1)!, & p = j. \end{cases} \end{aligned}$$

Denote by S the right-hand side of (6.6.3). If $N \neq 2pn$ then $S = 0$; if $N = 2pn$ then

$$\begin{aligned} S &= \frac{(-1)^p 2^{2jn} n}{(2j-1)!} \sum_{i=1}^{j} (-1)^i c_{ji} p^{2i} \\ &= \begin{cases} 0, & 1 \leq p \leq j-1, \\ \frac{2^{2jn} n j (2j-1)!}{(2j-1)!} = 2jn 2^{2jn-1}, & p = j. \end{cases} \end{aligned}$$

Thus Conditions (b) and (c) are satisfied. □

Let

$$P_r(x) = \begin{cases} 1, & r = 1, \\ \prod_{i=1}^{r-1} (x + i^2), & r > 1. \end{cases} \tag{6.6.5}$$

The following theorem gives an explicit expression of a trigonometric Gaussian quadrature formula at the zeros (6.6.1).

Theorem 6.6.2. *Let $m \in \mathbf{N}_2$, $g(\theta) = f(\cos\theta)$ and $\mathcal{D}g = g'$. Then*

$$Q_{n,m-2}(v; f) = \frac{\pi}{[(m/2-1)!]^2 n} \sum_{k=1}^{n} P_{m/2} \left(\frac{\mathcal{D}^2}{4n^2} \right) g(\theta_k). \tag{6.6.6}$$

Proof. By (6.5.14) and (6.6.3)

$$\begin{aligned} Q_{n,m-2}(v; f) &= \frac{\pi}{n} \left\{ \mathcal{L}_0 f + \sum_{j=1}^{(m-2)/2} \frac{(2j-1)!}{4^{jn} (j!)^2} \mathcal{L}_{2j} f \right\} \\ &= \frac{\pi}{n} \left\{ \mathcal{D}_0(g) + \sum_{j=1}^{(m-2)/2} \frac{1}{(j!)^2} \sum_{i=1}^{j} \frac{c_{ji}}{(2n)^{2i}} \mathcal{D}_{2i}(g) \right\} \\ &= \frac{\pi}{n} \sum_{k=1}^{n} \left\{ g(\theta_k) + \sum_{j=1}^{(m-2)/2} \frac{1}{(j!)^2} \sum_{i=1}^{j} c_{ji} \left(\frac{\mathcal{D}^2}{4n^2} \right)^i g(\theta_k) \right\}, \end{aligned}$$

which, coupled with (6.6.2) and (6.6.5), gives

$$\begin{aligned} & Q_{n,m-2}(v;f) \\ = \ & \frac{\pi}{n}\sum_{k=1}^{n}\left\{g(\theta_k)+\sum_{j=1}^{(m-2)/2}\frac{1}{(j!)^2}\left(\frac{\mathcal{D}^2}{4n^2}\right)\prod_{i=1}^{j-1}\left(\frac{\mathcal{D}^2}{4n^2}+i^2\right)g(\theta_k)\right\} \\ = \ & \frac{\pi}{n}\sum_{k=1}^{n}\left\{g(\theta_k)+\sum_{j=1}^{(m-2)/2}\frac{1}{(j!)^2}\left(\frac{\mathcal{D}^2}{4n^2}\right)P_j\left(\frac{\mathcal{D}^2}{4n^2}\right)g(\theta_k)\right\}. \end{aligned} \tag{6.6.7}$$

We can prove from (6.6.2) the identity by induction

$$1+x\sum_{j=1}^{r-1}\frac{1}{(j!)^2}P_j(x)=\frac{1}{[(r-1)!]^2}P_r(x), \quad r\geq 1. \tag{6.6.8}$$

Then equation (6.6.7) by (6.6.8) gives (6.6.6). □

The following theorem provides an approach to transform the quadrature formula (6.6.6) into a Gauss-Turán quadrature formula, although this approach is not explicit.

Theorem 6.6.3. *Let* $m\in\mathbf{N}_2$, $g(\theta)=f(\cos\theta)$, *and*

$$g^{(2r)}(\theta)=\sum_{i=1}^{2r}e_{ri}(\cos\theta)f^{(i)}(\cos\theta), \quad r\geq 0. \tag{6.6.9}$$

Then

$$\lambda_{kj}(v,m)=\begin{cases}\frac{\pi}{n}, & j=0,\\ \frac{\pi}{n[(m/2-1)!]^2}\sum_{i=[(j+1)/2]}^{(m-2)/2}\frac{c_{m/2,i+1}}{(2n)^{2i}}e_{ij}(x_k), & 1\leq j\leq m-2.\end{cases} \tag{6.6.10}$$

Proof. By (6.6.5), (6.6.2), and (6.6.9) for a fixed index k, $1\leq k\leq n$,

$$\begin{aligned} & P_{m/2}\left(\frac{\mathcal{D}^2}{4n^2}\right)g(\theta_k)=\prod_{i=1}^{m/2-1}\left(\frac{\mathcal{D}^2}{4n^2}+i^2\right)g(\theta_k) \\ = \ & \sum_{i=1}^{m/2}c_{m/2,i}\left(\frac{\mathcal{D}^2}{4n^2}\right)^{i-1}g(\theta_k)=c_{m/2,1}g(\theta_k)+\sum_{i=1}^{m/2-1}\frac{c_{m/2,i+1}}{(2n)^{2i}}g^{(2i)}(\theta_k) \\ = \ & c_{m/2,1}f(x_k)+\sum_{i=1}^{m/2-1}\frac{c_{m/2,i+1}}{(2n)^{2i}}\sum_{p=1}^{2i}e_{ip}(x_k)f^{(p)}(x_k). \end{aligned}$$

Substituting this into (6.6.6) yields

$$\begin{aligned} Q_{n,m-2}(v;f) &= \frac{\pi}{n[(m/2-1)!]^2} \\ &\times\sum_{k=1}^{n}\left[c_{m/2,1}f(x_k)+\sum_{i=1}^{m/2-1}\frac{c_{m/2,i+1}}{(2n)^{2i}}\sum_{p=1}^{2i}e_{ip}(x_k)f^{(p)}(x_k)\right]. \end{aligned} \tag{6.6.11}$$

Noticing that $c_{m/2,1}=[(m/2-1)!]^2$ by (6.6.2) and rearranging terms from (6.6.11) we obtain (6.6.10). □

6.7 The Gori-Micchelli Weight Class

In [36] Gori and Micchelli introduced for each n a class of weights for which explicit Gaussian quadrature formulas of all orders can be found.

Definition 6.7.1. *Let $n \in \mathbf{N}$ be fixed. A weight class $\mathbf{W}_n$ is called the Gori-Micchelli weight class if $w \in \mathbf{W}_n$ has an expansion*

$$w(x)\sqrt{1-x^2} = \sum_{k=0} \infty' \rho_k T_{kn}(x), \tag{6.7.1}$$

where the prime indicates the term corresponding to $k = 0$ is to be halved.

We introduce the set of indices

$$\mathbf{I}_m = \{i \in \mathbf{N}_0 : \ i \in \mathbf{N}_{02} \text{ or } i - m \in \mathbf{N}_1\}. \tag{6.7.2}$$

Theorem 6.7.1. *Let the weight $w \in \mathbf{W}_n$ satisfy the condition*

$$\rho_k = 0, \quad k \in \mathbf{N}_0 \setminus \mathbf{I}_m. \tag{6.7.3}$$

Then

$$P_n(w, m; x) = 2^{1-n} T_n(x) \tag{6.7.4}$$

and for $f \in C[-1, 1]$

$$\int_{-1} {}^1 f(x)[sgn\ T_n(x)^m] w(x) dx = \frac{\pi}{2} \sum_{k \in \mathbf{I}_m} {}' \rho_k \mathcal{A}_{kn}(f), \tag{6.7.5}$$

where $\mathcal{A}_{kn}(f)$ is given by (6.5.12) and (6.5.13).

Moreover, for $m \in \mathbf{N}_2$ the quadrature formula

$$\int_{-1}^{1} f(x) w(x) dx = \frac{\pi}{2} \sum_{k=0} m - 1' \rho_k \mathcal{A}_{kn}(f) \tag{6.7.6}$$

holds for all $f \in \mathbf{P}_{mn-1}$.

Proof. By (6.7.1), (6.7.3), and (6.2.7) for $q \in \mathbf{P}_{n-1}$

$$\begin{aligned}
&\int_{-1}^{1} T_n(x)^{m-1} [\text{sgn } T_n(x)^m] q(x) w(x) dx \\
&= \sum_{k=0}^{\infty}{}' \rho_k \int_{-1}^{1} T_n(x)^{m-1} T_{kn}(x) [\text{sgn } T_n(x)^m] q(x) \frac{dx}{\sqrt{1-x^2}} \\
&= \sum_{k \in \mathbf{K}_m} {}' \rho_k \int_{-1}^{1} T_n(x)^{m-1} T_{kn}(x) [\text{sgn } T_n(x) m] q(x) \frac{dx}{\sqrt{1-x^2}} = 0.
\end{aligned}$$

According to Theorem 3.2.2 we obtain (6.7.4).

Equation (6.7.5) follows directly from (6.7.1) and (6.7.3).

Let us prove the last part of the theorem and let $f \in \mathbf{P}_{mn-1}$. By (6.5.14) for $k \geq m$, we have $\mathcal{L}_{k+2j} f = 0$ and hence $\mathcal{A}_{kn}(f) = 0$. Then formula (6.7.6) follows from (6.7.5). □

As an immediate consequence of Theorem 6.7.1 we formulate the following

Corollary 6.7.1. *Let $m \in \mathbf{N}_2$ and let the weight $w \in \mathbf{W}_n$ satisfy*

$$\rho_k = 0, \quad k \in \mathbf{N}_1. \tag{6.7.7}$$

Then the quadrature formula

$$\int_{-1}^{1} f(x)w(x)dx = \frac{\pi}{2} \sum_{k=0} m/2 - 1' \rho_{2k} \mathcal{A}_{2kn}(f) \tag{6.7.8}$$

holds for all $f \in \mathbf{P}_{mn-1}$.

Remark 6.7.1. The original definition of $\mathbf{W}_n$ given by Gori and Micchelli has restriction (6.7.7) and their original expression for $\mathcal{A}_{2kn}(f)$ is implicitly determined [36].

Theorem 6.7.2. *Let $w \in \mathbf{W}_n$ with $\rho_1 = 0$ and $u(x) = (1+x)w(x)$. Then*

$$P_{n-1}(u;x) = 2^{1-n} V_{n-1}(x) \tag{6.7.9}$$

and the formula

$$\int_{-1}^{1} f(x)u(x)dx = \frac{1}{2}\rho_0 \int_{-1}^{1} f(x)\sqrt{\frac{1+x}{1-x}}dx \tag{6.7.10}$$

holds for all $f \in \mathbf{P}_{2n-2}$.

Proof. By (6.7.1) and (6.2.7) for $f \in \mathbf{P}_{2n-2}$

$$\begin{aligned} & \int_{-1}^{1} f(x)u(x)dx \\ = & \int_{-1}^{1} f(x)(1+x)w(x)dx \\ = & \sum_{k=0} \infty' \rho_k \int_{-1}^{1} T_{kn}(x) f(x)(1+x)\frac{dx}{\sqrt{1-x^2}} \\ = & \frac{1}{2}\rho_0 \int_{-1}^{1} f(x)\sqrt{\frac{1+x}{1-x}}dx + \sum_{k=2}^{\infty} \rho_k \int_{-1}^{1} T_{kn}(x) f(x)(1+x)\frac{dx}{\sqrt{1-x^2}} \\ = & \frac{1}{2}\rho_0 \int_{-1}^{1} f(x)\sqrt{\frac{1+x}{1-x}}dx. \end{aligned} \tag{6.7.11}$$

This proves (6.7.10). In particular, if we choose $q \in \mathbf{P}_{n-2}$ and $f = V_{n-1}q$, then by orthogonality from (6.7.11) we obtain

$$\int_{-1}^{1} V_{n-1}(x)q(x)u(x)dx = 0.$$

According to Theorem 3.2.2 we obtain (6.7.9). □

Similarly, we can get

Theorem 6.7.3. *Let $w \in \mathbf{W}_n$ with $\rho_1 = 0$ and $u(x) = (1-x^2)w(x)$. Then*

$$P_{n-1}(u;x) = 2^{1-n} U_{n-1}(x) \tag{6.7.12}$$

and the formula

$$\int_{-1}^{1} f(x)u(x)dx = \frac{1}{2}\rho_0 \int_{-1}^{1} f(x)\sqrt{1-x^2}dx \tag{6.7.13}$$

holds for all $f \in \mathbf{P}_{2n-3}$.

6.8 Notes

6.8.1. Theorems 6.3.1–6.3.4 appear in [100, 101, 103]. Formulas (6.3.4), (6.3.57), and (6.3.58) were found by Ossicini and Rosati [71].

6.8.2. The explicit expressions of Gaussian quadrature formulas except (6.4.59) and the recurrence approach in Section 6.4 are due to Shi [87, 88, 91, 100, 105]. Shi [105] established an incomplete formula of the integral (6.4.58) with the restriction $n \geq m/2-1$. The complete form (6.4.59) was given by Yang and Wang [157].

6.8.3. In 1972 Micchelli and Rivlin [58] established the remarkable quadrature formula in terms of divided differences at the zeros of T_n, i.e., formula (6.5.14) with $k=0$. In 1996 Bojanov [8] gave a simple approach to establish general quadrature formulas of this type at arbitrary system of nodes. In 2003 Yang and Wang [156] established the interesting formula (6.5.14) and found a simple method to transform a divided difference quadrature formula into an explicit Gauss-Turán quadrature formula (Remark 6.5.1). The further results we refer to the dissertation given by Yang [153].

6.8.4. The trigonometric Gaussian quadrature formula (6.6.6) was first given by Kis [45] in 1957. The explicit formula (6.6.3) for the functional $\mathcal{L}_j f$ was established and used to prove (6.6.6) by Micchelli and Sharma [59] in 1983. Theorem 6.6.3 is also due to Micchelli and Sharma [59].

6.8.5. The Gori-Micchelli weight class $\mathbf{W}_n$ was introduced by Gori and Micchelli [36] in 1996. This class $\mathbf{W}_n$ includes certain generalized Jacobi weights $w_{n,\alpha}(x) = |U_{n-1}(x)|^{2\alpha+1}(1-x^2)^{\alpha}$, $\alpha > -1$. For $n=2$ this weight reduce to the weight $w_{2,\alpha}(x) = |x|^{2\alpha+1}(1-x^2)^{\alpha}$.

6.8.6. In this chapter many results, say, Lemma 6.4.1 and Theorems 6.4.4–6.4.6, 6.6.2, 6.6.3, etc., deal with the case when $m \in \mathbf{N}_2$ only. It would be interesting to treat the case when $m \in \mathbf{N}_1$ here.

Chapter 7

Zeros of Power Orthogonal Polynomials

7.1 General Properties

In this chapter we discuss the zeros of power orthogonal polynomials and use the notations of Chapter 3.

Theorem 7.1.1. *Let the assumptions of Theorem 3.2.1 prevail. If* $\int_c^d d\mu(x) = 0$*, then the power orthogonal polynomial* $\omega_n(d\mu, \mathbf{m})$ *has at most one zero in the interval* $\langle c, d\rangle$.

Proof. Suppose to the contrary that $x_k, x_{k+1} \in \langle c, d\rangle$, where $x_i = x_{in}(d\mu, \mathbf{m})$, $i = k, k+1$. Choose the polynomial

$$q(x) = \frac{\omega_n(d\mu, \mathbf{m}; x)}{(x - x_k)(x - x_{k+1})}.$$

We observe that the inequality

$$\omega_n(d\mu, \mathbf{m}; x) q(x) = \frac{\omega_n(d\mu, \mathbf{m}; x)^2}{(x - x_k)(x - x_{k+1})} \geq 0$$

holds for all $x \in \langle a, b\rangle \setminus \langle c, d\rangle$. By (3.2.6)

$$\begin{aligned} 0 &= \int_a^b \frac{\Omega_n(d\mu, \mathbf{m}; x)}{\omega_n(d\mu, \mathbf{m}; x)} q(x) d\mu(x) \\ &= \int_a^b \frac{\Omega_n(d\mu, \mathbf{m}; x)}{\omega_n(d\mu, \mathbf{m}; x)^2} \omega_n(d\mu, \mathbf{m}; x) q(x) d\mu(x) \\ &= \int_{\langle a,b\rangle \setminus \langle c,d\rangle} \frac{\Omega_n(d\mu, \mathbf{m}; x)}{\omega_n(d\mu, \mathbf{m}; x)^2} \omega_n(d\mu, \mathbf{m}; x) q(x) d\mu(x) > 0, \end{aligned}$$

a contradiction. □

Formula (4.1.3) with $m_k \equiv m$ becomes

$$\int_a^b f(x) S_n(x) d\mu(x) = \sum_{k=1}^{n} \sum_{j=0}^{m-2} \lambda_{kj} f^{(j)}(x_k), \tag{7.1.1}$$

where $x_k = x_k(d\mu,m)$, $\lambda_{kj} = \lambda_{kj}(d\mu,m)$, and $S_n(x) = \operatorname{sgn} P_n(d\mu,m;x)^m$.

A. Markov proved an important result concerning the dependence of the zeros of $P_n(w,2;x)$ on a parameter τ which appears in the weight function $w(x) = w(\tau;x)$ (see, say, [137, Theorem 6.12.1, p. 111]). Using his idea we can deduce an analogue for m orthogonal polynomials. But for our applications later we relax the original assumption about continuity of $w(\tau;x)$ and $w'_\tau(\tau;x)$ with respect to x and τ and assume continuity of $w(\tau;x)$ and $w'_\tau(\tau;x)$ with respect to τ only. That is the following

Theorem 7.1.2. *Let $w(\tau;x)$ be a weight on $[a,b]$ depending on a parameter τ such that $w(\tau;x)$ is positive and continuous with respect to τ for $x \in (a,b)$, $\tau \in (\tau_1,\tau_2)$. Also, assume the existence and continuity (with respect to τ) of the partial derivative $w'_\tau(\tau;x)$ for $x \in (a,b)$, $\tau \in (\tau_1,\tau_2)$, and the convergence of the integrals*

$$\int_a^b x^j w'_\tau(\tau;x)dx, \quad j = 0,1,\ldots,mn-1,$$

uniformly in every closed interval $[\tau',\tau''] \subset (\tau_1,\tau_2)$. If w'_τ/w is an increasing function of x in (a,b), then the kth zero $x_k(\tau) = x_k(w(\tau),m)$ (for a fixed index k) is an increasing function of τ.

Proof. Differentiating the formula (7.1.1) with $d\mu(x) = w(\tau;x)dx$ with respect to τ, we obtain

$$\begin{aligned}&\int_a^b f(x)[\operatorname{sgn} P_n(w(\tau),m;x)^m]w'_\tau(\tau;x)dx \\ &= \sum_{k=1}^n \sum_{j=0}^{m-2} [\lambda_{kj}(\tau) f^{(j+1)}(x_k(\tau))x'_k(\tau) + \lambda'_{kj}(\tau) f^{(j)}(x_k(\tau))],\end{aligned} \tag{7.1.2}$$

where

$$x'_k(\tau) = \frac{\partial x_k(w(\tau);m)}{\partial\tau}, \quad \lambda'_{kj}(\tau) = \frac{\partial\lambda_{kj}(w(\tau),m)}{\partial\tau}.$$

Substituting

$$f(x) = \frac{P_n(w(\tau),m;x)^m}{(m-1)!(x-x_k(\tau))}$$

into (7.1.2) and observing $f^{(m-1)}(x_k(\tau)) = P'_n(w(\tau),m;x_k(\tau))^m$, here

$$P'_n(w(\tau),m;x_k(\tau)) = \frac{d}{dx} P_n(w(\tau),m;x) \,|_{x=x_k(\tau)},$$

we get

$$\int_a^b \frac{|P_n(w(\tau),m;x)|^m}{(m-1)!(x-x_k(\tau))} w'_\tau(\tau;x)dx = \lambda_{k,m-2}(\tau)P'_n(w(\tau),m;x_k(\tau))^m x'_k(\tau).$$

By (3.2.6) we see

$$
\begin{aligned}
&\int_a^b \frac{|P_n(w(\tau),m;x)|^m}{x-x_k(\tau)} w'_\tau(\tau;x)dx \\
= &\int_a^b \frac{|P_n(w(\tau),m;x)|^m}{x-x_k(\tau)} \left[w'_\tau(\tau;x) - \frac{w'_\tau(\tau;x_k(\tau))}{w(\tau;x_k(\tau))} w(\tau;x) \right] dx \\
= &\int_a^b \frac{|P_n(w(\tau),m;x)|^m}{x-x_k(\tau)} \left[\frac{w'_\tau(\tau;x)}{w(\tau;x)} - \frac{w'_\tau(\tau;x_k(\tau))}{w(\tau;x_k(\tau))} \right] w(\tau;x)dx \\
> &\ 0,
\end{aligned}
$$

because the difference in the square brackets has the same sign as $x - x_k(\tau)$ by the assumption. This proves $x'_k(\tau) > 0$, since $\lambda_{k,m-2}(\tau)P'_n(w_\tau, m; x_k(\tau))^m > 0$ by virtue of (4.1.6). □

As consequences of this lemma we state two lemmas which are extensions of [137, Theorems 6.12.2 and 6.21.1, pp. 112–117] and may be derived by the same arguments as that of those theorems in [137].

Theorem 7.1.3. *Let $w(x)$ and $u(x)$ be weights on $[a,b]$, both positive in (a,b). Let $u(x)/w(x)$ be increasing. Then*

$$x_k(w,m) < x_k(u,m), \quad k = 1,2,\ldots,n.$$

Theorem 7.1.4. *Let $w(x)$ be a weight on $[-1,1]$ and $\sigma = \pm 1$. Let*

$$w(\tau;x) = w(x)(1+\sigma x)^\tau \tag{7.1.3}$$

be a weight on $[-1,1]$ with a parameter τ for $\tau \in (\tau_1, \tau_2)$. Assume that the convergence of the integrals

$$\int_{-1}^1 x^j \ln(1+\sigma x) w(\tau;x)dx, \quad j = 0,1,\ldots,mn-1,$$

holds uniformly in every closed interval $[\tau', \tau''] \subset (\tau_1, \tau_2)$. Then

$$\sigma \frac{\partial x_k(w(\tau),m)}{\partial \tau} > 0, \quad \sigma \frac{\partial \theta_k(w(\tau),m)}{\partial \tau} < 0, \quad k = 1,2,\ldots,n, \tag{7.1.4}$$

where $x_k(w(\tau),m) = \cos[\theta_k(w(\tau),m)]$, $0 \le \theta_k(w(\tau),m) \le \pi$.

7.2 Zeros of Special Power Orthogonal Polynomials

In this section we enumerate the zeros of m orthogonal polynomial $x_{kn} = \cos\theta_{kn}$ in increasing order

$$
\begin{cases}
-1 = x_{0n} < x_{1n} < \cdots < x_{nn} \le x_{n+1,n} = 1, \\
\pi = \theta_{0n} > \theta_{1n} > \cdots > \theta_{nn} > \theta_{n+1,n} = 0.
\end{cases} \tag{7.2.1}
$$

Let

$$x_{kn}^{(\alpha,\beta)} = x_{kn}(w^{(\alpha,\beta)}, m),\ x_{kn}^{(\alpha,\beta)} = \cos\theta_{kn}^{(\alpha,\beta)},\ \alpha,\beta \ge -1,\ k = 1,2,\ldots,n.$$

Applying Theorem 7.1.4 to the weight $w(\tau;x)$ with $w(x) = (1+x)^\beta$, $\tau = \alpha$, $\sigma = -1$, or $w(x) = (1-x)^\alpha$, $\tau = \beta$, $\sigma = 1$, we obtain

Theorem 7.2.1. *We have*

$$\frac{\partial x_k^{(\alpha,\beta)}}{\partial\alpha} < 0, \quad \frac{\partial x_k^{(\alpha,\beta)}}{\partial\beta} > 0, \quad k = 1,2,\ldots,n, \tag{7.2.2}$$

and

$$\frac{\partial\theta_k^{(\alpha,\beta)}}{\partial\alpha} > 0, \quad \frac{\partial\theta_k^{(\alpha,\beta)}}{\partial\beta} < 0, \quad k = 1,2,\ldots,n. \tag{7.2.3}$$

In particular, for the case when $\alpha=\beta$

$$\frac{\partial x_k^{(\alpha,\alpha)}}{\partial\alpha} < 0, \quad \frac{\partial\theta_k^{(\alpha,\alpha)}}{\partial\alpha} > 0, \quad k = 1,2,\ldots,[n/2]. \tag{7.2.4}$$

Theorem 7.2.2. *With the notations of (7.1.3) assume that*

$$w \sim w^{(\alpha,\beta)}, \ \alpha,\beta > -1. \tag{7.2.5}$$

If

$$\begin{cases} \beta = 0, & \sigma = 1, \\ \alpha = 0, & \sigma = -1, \end{cases} \tag{7.2.6}$$

then

$$\begin{aligned} \omega_n(w(-1),m;x) &:= \lim_{\tau\to-1+0} \omega_n(w(\tau),m;x) \\ &= (x+\sigma)\omega_{n-1}(w(m-1),m;x). \end{aligned} \tag{7.2.7}$$

Proof. According to Theorem 7.1.4 the limit

$$\lim_{\tau\to-1+0} \omega_n(w(\tau),m;x)$$

exists and the convergence is uniform with respect to x in $[-1,1]$. By the definition of $\omega_n(w(\tau),m;x)$ for $-1<\tau\le 0$

$$\begin{aligned} \int_{-1}^{1} |\omega_n(w(\tau),m;x)|^m w(\tau;x)dx &\le \int_{-1}^{1} |(x+\sigma)^n|^m w(\tau;x)dx \\ &\le c\int_{-1}^{1} |x+\sigma|^{mn-1} w(x)dx \le c < \infty. \end{aligned}$$

Thus

$$\begin{cases} \lim_{\tau\to-1+0} x_1(w(\tau)) = -1, & \sigma = 1, \\ \lim_{\tau\to-1+0} x_n(w(\tau)) = 1, & \sigma = -1. \end{cases} \tag{7.2.8}$$

Meanwhile by (3.2.6)

$$\int_{-1}^{1} \omega_n(w(\tau),m;x)^{m-1}[\text{sgn}\ \omega_n(w(\tau),m;x)^m]q(x)w(\tau;x)dx = 0, \quad \forall q\in\mathbf{P}_{n-1}.$$

For an arbitrary $Q\in\mathbf{P}_{n-2}$ substituting $q(x) = (x+\sigma)Q(x)$ into the above equation, we obtain

$$\int_{-1}^{1} \omega_n(w(\tau),m;x)^{m-1}[\text{sgn}\ \omega_n(w(\tau),m;x)^m]Q(x)w(\tau+1;x)dx = 0.$$

As $\tau \to -1+0$ the above equation gives

$$\int_{-1}^{1} \omega_n(w(-1),m;x)^{m-1}[\text{sgn}\ \omega_n(w(-1),m;x)^m]Q(x)w(0,x)dx = 0,$$

or by (7.2.8) equivalently,

$$\int_{-1}^{1} \left[\frac{\omega_n(w(-1),m;x)}{x+\sigma}\right]^{m-1} \left[\text{sgn}\ \left(\frac{\omega_n(w(-1),m;x)}{x+\sigma}\right)^m\right] Q(x)w(m-1,x)dx = 0,$$

which by (3.2.6) means

$$\omega_{n-1}(w(m-1),m;x) = \frac{\omega_n(w(-1),m;x)}{x+\sigma}.$$

Thus relation (7.2.7) follows. □

Theorem 7.2.3. *We have for each k, $1 \le k \le n$,*

$$\theta_{kn}^{(-1,\beta)} = \theta_{k,n-1}^{(m-1,\beta)}, \tag{7.2.9}$$

$$\theta_{kn}^{(\alpha,-1)} = \theta_{k-1,n-1}^{(\alpha,m-1)}, \tag{7.2.10}$$

$$\theta_{kn}^{(\alpha+\gamma,\beta)} \ge \theta_{k,n-1}^{(m-1+\gamma,\beta)}, \quad \mu+\gamma > -1,\ m+\gamma > -2, \tag{7.2.11}$$

and

$$\theta_{kn}^{(\alpha,\beta+\gamma)} \le \theta_{k-1,n-1}^{(\alpha,m-1+\gamma)}, \quad \beta+\gamma > -1,\ m+\gamma > -2. \tag{7.2.12}$$

Proof. Formulas (7.2.9) and (7.2.10) follow directly from (7.2.7) with $\sigma = -1$ and $\sigma = 1$, respectively.

To prove (7.2.11) we introduce a weight with a parameter τ

$$w^{(\alpha,\beta)}(\tau;x) = (1-\tau x)^\gamma w^{(\alpha,\beta)}(x), \quad 0 < \tau < 1.$$

For each fixed τ by the same argument as that of Theorem 7.2.1 we obtain

$$\begin{aligned} \omega_n(w^{(-1,\beta)}(\tau),m;x) &= \lim_{\alpha\to-1+0} \omega_n(w^{(\alpha,\beta)}(\tau),m;x) \\ &= (x-1)\omega_{n-1}(w^{(m-1,\beta)}(\tau),m;x). \end{aligned}$$

Hence by Theorem 7.1.4

$$\theta_{kn}(w^{(\alpha,\beta)}(\tau),m) > \theta_{k,n-1}(w^{(m-1,\beta)}(\tau),m),\ k = 1,2,\ldots,n.$$

As $\tau \to 1$ the above inequalities become

$$\theta_{kn}(w^{(\alpha,\beta)}(1),m) \ge \theta_{k,n-1}(w^{(m-1,\beta)}(1),m),\ k = 1,2,\ldots,n.$$

That is

$$\theta_{kn}^{(\alpha+\gamma,\beta)} \ge \theta_{k,n-1}^{(m-1+\gamma,\beta)},\ k = 1,2,\ldots,n.$$

This proves (7.2.11). Similarly we can prove (7.2.12). □

Theorem 7.2.4. *Let*

$$p = -\left[-\frac{\alpha+1/2}{m-1}\right], \quad q = -\left[-\frac{\beta+1/2}{m-1}\right]. \tag{7.2.13}$$

Then for each k, $1 \le k \le n$,

$$\theta_{k+1,n}^{(-1/2,\beta)} \le \theta_{kn}^{(\alpha,\beta)} \le \theta_{kn}^{(-1/2,\beta)}, \quad \mu \le -\frac{1}{2}, \tag{7.2.14}$$

$$\theta_{kn}^{(-1/2,\beta)} \le \theta_{kn}^{(\alpha,\beta)} \le \theta_{k,n+p}^{(-1/2,\beta)}, \quad \mu > -\frac{1}{2}, \tag{7.2.15}$$

$$\theta_{kn}^{(\alpha,-1/2)} \le \theta_{kn}^{(\alpha,\beta)} \le \theta_{k-1,n}^{(\alpha,-1/2)}, \quad \beta \le -\frac{1}{2}, \tag{7.2.16}$$

and

$$\theta_{k+q,n+q}^{(\alpha,-1/2)} \le \theta_{kn}^{(\alpha,\beta)} \le \theta_{kn}^{(\alpha,-1/2)}, \quad \beta > -\frac{1}{2}. \tag{7.2.17}$$

Proof. To prove (7.2.14) it is sufficient to show the relations

$$\theta_{k+1,n}^{(-1/2,\beta)} \le \theta_{k,n-1}^{(-1/2,\beta)} \le \theta_{k,n-1}^{(m-1,\beta)} = \theta_{kn}^{(-1,\beta)} \le \theta_{kn}^{(\alpha,\beta)} \le \theta_{kn}^{(-1/2,\beta)}.$$

In fact, the equality follows from (7.2.9); the first inequality follows from Theorem 3.3.1 and the others from (7.2.3).

We prove (7.2.15). The first inequality in it follows from (7.2.3). To prove the second inequality we observe that the first equality of (7.2.13) implies $p \ge 1$ and $\alpha \le p(m-1) - 1/2$. Hence by (7.2.3)

$$\theta_{kn}^{(\alpha,\beta)} \le \theta_{kn}^{(p(m-1)-1/2,\beta)}. \tag{7.2.18}$$

Applying (7.2.11) p times, here for the ith time ($i = 1, \ldots, p$) we put $\alpha = 0$, $\gamma = (i-1)(m-1) - 1/2$, by induction we obtain

$$\theta_{k,n+p}^{(-1/2,\beta)} \ge \theta_{k,n+p-1}^{(m-1-1/2,\beta)} \ge \theta_{k,n+p-2}^{(2(m-1)-1/2,\beta)} \ge \cdots \ge \theta_{kn}^{(p(m-1)-1/2,\beta)},$$

which together with (7.2.18) gives the second inequality of (7.2.15).

Similarly, inequalities (7.2.16) and (7.2.17) follow from (7.2.3), (7.2.10), (7.2.12), (7.2.13), and Theorem 3.3.1. □

As a immediate consequence of Theorem 7.2.4 we formulate the following

Theorem 7.2.5. *With the notations of (7.2.13) we have for each* k, $1 \le k \le n$,

$$\theta_{k+q,n+q}^{(-1/2,-1/2)} \le \theta_{kn}^{(\alpha,\beta)} \le \theta_{k,n+p}^{(-1/2,-1/2)}, \ \alpha, \beta > -\frac{1}{2}, \tag{7.2.19}$$

$$\theta_{k+1,n}^{(-1/2,-1/2)} \le \theta_{kn}^{(\alpha,\beta)} \le \theta_{k-1,n}^{(-1/2,-1/2)}, \ \alpha, \beta \le -\frac{1}{2}, \tag{7.2.20}$$

$$\theta_{kn}^{(-1/2,-1/2)} \le \theta_{kn}^{(\alpha,\beta)} \le \theta_{k-1,n+p}^{(-1/2,-1/2)}, \ \alpha > -\frac{1}{2}, \ \beta \le -\frac{1}{2}, \tag{7.2.21}$$

and

$$\theta_{k+q+1,n+q}^{(-1/2,-1/2)} \le \theta_{kn}^{(\alpha,\beta)} \le \theta_{kn}^{(-1/2,-1/2)}, \ \alpha \le -\frac{1}{2}, \ \beta > -\frac{1}{2}. \tag{7.2.22}$$

In particular,

$$\theta_{k+p,n+p}^{(-1/2,-1/2)} \le \theta_{kn}^{(\alpha,\alpha)} \le \theta_{k,n+p}^{(-1/2,-1/2)}, \ \alpha > -\frac{1}{2} \tag{7.2.23}$$

and

$$\theta_{k+1,n}^{(-1/2,-1/2)} \le \theta_{kn}^{(\alpha,\alpha)} \le \theta_{k-1,n}^{(-1/2,-1/2)}, \ \alpha \le -\frac{1}{2}. \tag{7.2.24}$$

Theorem 7.2.5 implies the following

Corollary 7.2.1. *With the notations of (7.2.13) we have for each k, $1 \le k \le n$,*

$$\theta_{k+2q+2,n+2q}^{(-1/2,-1/2)} \le \theta_{kn}^{(\alpha,\beta)} \le \theta_{k-2,n+2p}^{(-1/2,-1/2)}, \quad \alpha,\beta > -1. \tag{7.2.25}$$

As a consequence of Theorems 7.2.1 and 6.3.6 we formulate an extension of [137, Theorems 6.21.2 and 6.21.3, p. 118].

Theorem 7.2.6. *If*

$$-\frac{1}{2} \le \alpha,\beta \le \frac{m-1}{2}, \tag{7.2.26}$$

then

$$\frac{2n-2k+1}{2n+1}\pi \le \theta_{kn}^{(\alpha,\beta)} \le \frac{2n-2k+2}{2n+1}\pi, \quad k=1,2,\ldots,n, \tag{7.2.27}$$

with equality only in the special cases when $\alpha = -1/2$, $\beta = (m-1)/2$ and $\alpha = (m-1)/2$, $\beta = -1/2$, respectively.

Proof. By means of (7.2.3) the maximum and minimum of $\theta_k^{(\alpha,\beta)}$ are attained in the special cases mentioned above. It remains to apply (6.3.57), (6.3.58), (1.3.10), and (1.3.11). □

Theorem 7.2.7. *If*

$$-\frac{1}{2} \le \alpha = \beta \le \frac{m-1}{2}, \tag{7.2.28}$$

then

$$\frac{n-k+1}{n+1}\pi \le \theta_{kn}^{(\alpha,\alpha)} \le \frac{2n-2k+1}{2n}\pi, \quad k=1,2,\ldots,[n/2], \tag{7.2.29}$$

with equality only in the special cases when $\alpha=\beta=-1/2$ and $\alpha=\beta=(m-1)/2$, respectively.

Proof. By (7.2.4) the maximum and minimum of $\theta_k^{(\alpha,\alpha)}$ are attained in the special cases mentioned above. It remains to use (6.3.3) and (6.3.4). □

As a consequence of Theorems 7.2.1 and 6.3.7 we formulate the following

Theorem 7.2.8. *With the notations of (1.4.1) and (1.4.7) if*

$$-\frac{1}{2} \le \alpha \le \frac{m-1}{2}, \tag{7.2.30}$$

then

$$x_{k,2n}(h,m)^2 \le x_{kn}(\ell^{(\alpha)},m) \le x_{k,2n+1}(h,m)^2, \quad k=1,2,\ldots,n. \tag{7.2.31}$$

Lemma 7.2.1. [21, Theorem IV, p. 718] *Let for each* $k,\ 1 \le k \le n,$

$$\ell_{kn}(x) = \prod_{i \neq k} \left(\frac{x - x_{in}}{x_{kn} - x_{in}} \right), \tag{7.2.32}$$

where $-1 \le x_{1n} < x_{2n} < \cdots < x_{nn} \le 1$. *If the inequality*

$$\left| \int_{-1}^{1} \ell_{kn}(x)dx \right| \le cn^{\rho}, \quad \rho > 0, \tag{7.2.33}$$

holds for all $k,\ 1 \le k \le n$, *and* $n \in \mathbf{N}$ *then inequality*

$$\theta_{kn} - \theta_{k+1,n} \le \frac{c\ln(n+1)}{n} \tag{7.2.34}$$

holds for all $k,\ 0 \le k \le n$, *and* $n \in \mathbf{N}$.

Corollary 7.2.2. *If the inequality*

$$\|\ell_{kn}\| = \max_{1 \le k \le n} |\ell_{kn}(x)| \le cn^{\rho}, \quad \rho > 0, \tag{7.2.35}$$

holds for all $k,\ 1 \le k \le n$, *and* $n \in \mathbf{N}$, *then inequality (7.2.34) holds for all* $k,\ 0 \le k \le n$, *and* $n \in \mathbf{N}$.

Proof. Since inequality (7.2.35) implies (7.2.33), we can obtain the corollary by Lemma 7.2.1. □

Lemma 7.2.2. [78] *If* $P \in \mathbf{P}_n$ *attains its absolute maximum in* $[-1,1]$ *at* $y = \cos\tau$ *then the interval* $(\tau - \pi/(2n), \tau + \pi/(2n))$ *contains no any root of* $P(\cos\theta)$.

Lemma 7.2.3. *Let* $P_k \in \mathbf{P}_n,\ k = 1,2,\ldots,M$, *and* $-1 \le y_1 < y_2 < \cdots < y_M \le 1$. *If*

$$\left\| \sum_{k=1}^{M} |(x - y_k)P_k(x)| \right\| = \mu_n \tag{7.2.36}$$

and

$$\sum_{k=1}^{M} |P_k(y_j)| \le \nu_n, \qquad j = 1,2,\ldots,M, \tag{7.2.37}$$

then

$$\left\| \sum_{k=1}^{M} |P_k| \right\| \le 2(n^2\mu_n + \nu_n). \tag{7.2.38}$$

In particular, if $M = 1$ *and* $P_1(y_1) = 0,\ |y_1| < 1$, *then*

$$\|P_1\| \le \frac{4n\mu_n}{(1 - y_1^2)^{1/2}}. \tag{7.2.39}$$

Proof. Let

$$\sum_{k=1}^{M}|P_k(\xi)| = \rho_n := \left\|\sum_{k=1}^{M}|P_k|\right\|, \quad \xi\in[-1,1].$$

If $|\xi-y_k|\geq 1/(2n^2)$ holds for every k then

$$\mu_n \geq \sum_{k=1}^{M}|(\xi-y_k)P_k(\xi)| \geq \frac{\rho_n}{2n^2},$$

which implies (7.2.38).

If $|\xi-y_j|<1/(2n^2)$ holds for some j then with the notation

$$P(x)=\sum_{k=1}^{M}[\text{sgn}\,P_k(\xi)]P_k(x)$$

by the mean value theorem for the derivative and Markov inequality

$$\rho_n-\nu_n\leq P(\xi)-P(y_j)=P'(\eta)(\xi-y_j)\leq n^2\|P\||\xi-y_j|\leq n^2\rho_n\cdot\frac{1}{2n^2}=\frac{1}{2}\rho_n.$$

Hence $\rho_n\leq 2\nu_n$, which implies (7.2.38).

To prove the second conclusion let $y_1=\cos\theta$, $\xi=\cos\tau$ and choose $t=\theta+\varepsilon\pi/(2n)$, where $\varepsilon=\text{sgn}\,(\tau-\theta)$. It is enough to establish

$$|\xi-y_1|\geq\frac{\sin\theta}{4n}, \tag{7.2.40}$$

because by (7.2.36) and (7.2.40)

$$\mu_n\geq|(\xi-y_1)P_1(\xi)|\geq\frac{\sin\theta}{4n}\|P_1\|=\frac{(1-y_1^2)^{1/2}}{4n}\|P_1\|,$$

which is equivalent to (7.2.39). Let us prove (7.2.40). By Lemma 7.2.2 we have $|\tau-\theta|\geq\pi/(2n)$ and hence

$$\begin{aligned}
|\xi-y_1| &\geq |\cos t-\cos\theta| = \left|2\sin\frac{t+\theta}{2}\sin\frac{t-\theta}{2}\right| \\
&= \left|2\sin\left(\theta+\frac{\varepsilon\pi}{4n}\right)\sin\frac{\pi}{4n}\right| \geq \left|\frac{1}{n}\sin\left(\theta+\frac{\varepsilon\pi}{4n}\right)\right| \\
&= \frac{1}{n}\left|\sin\theta\cos\frac{\pi}{4n}+\varepsilon\cos\theta\sin\frac{\pi}{4n}\right| \\
&\geq \frac{1}{n}\left[\frac{2^{1/2}}{2}\sin\theta-\sin\frac{\pi}{4n}\right]\geq\frac{1}{n}\left[\frac{2^{1/2}}{2}\sin\theta-\frac{\pi}{4n}\right].
\end{aligned}$$

If

$$\sin\theta\geq\frac{\pi}{4n(\frac{2^{1/2}}{2}-\frac{1}{4})},$$

i.e.,

$$\frac{2^{1/2}}{2}\sin\theta-\frac{\pi}{4n}\geq\frac{1}{4}\sin\theta,$$

then (7.2.40) follows. If

$$\sin\theta < \frac{\pi}{4n(\frac{2^{1/2}}{2} - \frac{1}{4})},$$

then in this case $\sin\theta \leq 2/n$ and hence

$$|\xi - y_1| \geq |\cos t - \cos\theta| \geq |1 - \cos(t-\theta)| \geq \left|1 - \cos\frac{\pi}{2n}\right| = 2\sin^2\frac{\pi}{4n} \geq \frac{1}{2n^2} \geq \frac{\sin\theta}{4n}. \qquad \square$$

Theorem 7.2.9. *If*

$$w(x) \geq Cw^{(\alpha,\beta)}(x), \tag{7.2.41}$$

where $C > 0$ is a constant, then inequality (7.2.35) holds for all k, $1 \leq k \leq n$, and $n \in \mathbf{N}$, and inequality (7.2.34) holds for all k, $0 \leq k \leq n$, and $n \in \mathbf{N}$.

Proof. By (7.1.1) with $-a = b = 1$ and (4.4.9) for $1 \leq k \leq n$

$$\begin{aligned}
&\int_{-1}^{1}\left|(x-x_k)^{m-2}\ell_k(x)^m\right| w^{(\alpha,\beta)}(x)dx \leq C^{-1}\int_{-1}^{1}\left|(x-x_k)^{m-2}\ell_k(x)^m\right| w(x)dx \\
= \;& C^{-1}[\operatorname{sgn} P_n'(w,m;x_k)^m]\int_{-1}^{1}(x-x_k)^{m-2}\ell_k(x)^m S_n(x)w(x)dx \\
= \;& C^{-1}(m-2)![\operatorname{sgn} P_n'(w,m;x_k)^m]\lambda_{k,m-2}(w,m) \leq cd_k^{m-2}\int_{x_{k-1}}^{x_{k+1}} w(x)dx \leq c.
\end{aligned}$$

Applying (5.3.22) we have

$$\max_{|x|\leq 1}\left|(x-x_k)^{m-2}\ell_k(x)^m\widehat{W}_n(x)\right| \leq cn\int_{-1}^{1}\left|(x-x_k)^{m-2}\ell_k(x)^m\right| w^{(\alpha,\beta)}(x)dx \leq cn.$$

Hence

$$\max_{|x|\leq 1}\left|(x-x_k)^{m-2}\ell_k(x)^m\right| \leq cn^{\gamma}, \tag{7.2.42}$$

where

$$\gamma = 1 + (2\alpha+1)_+ + (2\beta+1)_+. \tag{7.2.43}$$

Applying repeatedly Lemma 7.2.3, we obtain

$$\|\ell_k\| \leq cn^{[2(m-2)+\gamma]/m}, \quad k = 1, 2, \ldots, n.$$

Then using Corollary 7.2.2 we get (7.2.34). $\square$

Theorem 7.2.10. *Let the assumptions of Theorem 7.2.9 prevail and let $0 \leq c < d \leq \pi$. Then*

$$card\,\{k:\; c \leq \theta_{kn} \leq d\} = \frac{d-c}{\pi}n + O(\ln^2 n). \tag{7.2.44}$$

Proof. This theorem follows directly from Theorem 7.2.9 and a result given by Erdős [18] which says that if inequality (7.2.35) is valid for all k, $0 \leq k \leq n$, and $n \in \mathbf{N}$ then the estimation (7.2.44) holds. $\square$

Lemma 7.2.4. *Let $b > a > 0$ and $d = (b-a)/h > 2$. Then*

$$(b-h)^p - (a+h)^p \geq c(b^p - a^p), \tag{7.2.45}$$

where

$$c = \begin{cases} 1, & p \leq 0, \\ \frac{(d-1)^{p}-1}{d}, & 0 < p < 1, \\ \frac{(d-2)(d-1)^p}{d^p}, & p \geq 1. \end{cases} \tag{7.2.46}$$

Proof. Rewrite (7.2.45) in the form with $\delta = h/a$

$$[1+(d-1)\delta]^p - (1+\delta)^p \geq c[(1+d\delta)^p - 1]. \tag{7.2.47}$$

For $p \leq 0$ we have $[1+(d-1)\delta]^p \geq (1+d\delta)^p$ and $(1+\delta)^p \leq 1$. Hence (7.2.47) with $c = 1$ is valid.

For $p > 0$ we consider the problem to minimize the function

$$g(\delta, c) = c$$

subject to the condition

$$g_1(\delta, c) = [1+(d-1)\delta]^p - (1+\delta)^p - c[(1+d\delta)^p - 1] \geq 0.$$

If a pair $\{\delta, c\}$ is a solution of this problem then according to Theorem 3.4 in [3] there is a pair $\{\lambda_0, \lambda_1\}$, $\lambda_0, \lambda_1 \geq 0$, $\lambda_0 + \lambda_1 > 0$, such that

$$-\lambda_1 \frac{\partial g_1}{\partial \delta} = 0, \tag{7.2.48}$$

$$\lambda_0 - \lambda_1 \frac{\partial g_1}{\partial c} = 0, \tag{7.2.49}$$

and

$$\lambda_1 g_1 = 0. \tag{7.2.50}$$

The equation (7.2.49) shows that $\lambda_1 > 0$, for otherwise it would lead to $\lambda_0 = \lambda_1 = 0$, a contradiction. Thus equation (7.2.48) yields

$$\frac{\partial g_1}{\partial \delta} = 0,$$

that is,

$$(d-1)[1+(d-1)\delta]^{p-1} - (1+\delta)^{p-1} - cd(1+d\delta)^{p-1} = 0.$$

This gives

$$c = \frac{(d-1)[1+(d-1)\delta]^{p-1} - (1+\delta)^{p-1}}{d(1+d\delta)^{p-1}}.$$

If $0 < p < 1$ then

$$\begin{aligned} c &\geq \frac{(d-1)[(d-1)+(d-1)\delta]^{p-1} - (1+\delta)^{p-1}}{d(1+d\delta)^{p-1}} \\ &= \frac{(d-1)^p - 1}{d}\left(\frac{1+\delta}{1+d\delta}\right)^{p-1} \geq \frac{(d-1)^p - 1}{d}; \end{aligned}$$

if $p \geq 1$ then

$$c \geq \frac{d-2}{d}\left(\frac{1+(d-1)\delta}{1+d\delta}\right)^{p-1} \geq \frac{d-2}{d}\left(\frac{d-1}{d}\right)^{p-1} = \frac{(d-2)(d-1)^{p-1}}{d^p}. \quad \square$$

Lemma 7.2.5. *Let $-1 \leq a < b \leq 1$ and $d = (b-a)/h > 4$. Let (5.3.2) prevail. Then*

$$\int_{a+h}^{b-h} u(t)dt \geq c(u,d)\int_a^b u(t)dt. \tag{7.2.51}$$

Proof. Put

$$\delta = \frac{d-2}{2d}\min_{1\leq i\leq r-1}(t_{i+1}-t_i) \tag{7.2.52}$$

and

$$c_1 = \left[\int_{-1}^1 u(t)dt\right]^{-1}\inf_{-1\leq\tau\leq 1-\delta}\int_\tau^{\tau+\delta} u(t)dt > 0. \tag{7.2.53}$$

We separate the cases when $b-a-2h \geq \delta$ and $b-a-2h < \delta$.

Case 1. $b-a-2h \geq \delta$. In this case by (7.2.53)

$$\begin{aligned}\int_{a+h}^{b-h} u(t)dt &\geq \inf_{-1\leq\tau\leq 1-\delta}\int_\tau^{\tau+\delta} u(t)dt \\ &\geq c_1\int_{-1}^1 u(t)dt \geq c_1\int_a^b u(t)dt.\end{aligned}$$

Case 2. $b-a-2h < \delta$. In this case since

$$b-a-2h = b-a-\frac{2(b-a)}{d} = \frac{(d-2)(b-a)}{d},$$

by (7.2.52)

$$b-a < \frac{1}{2}\min_{1\leq i\leq r-1}(t_{i+1}-t_i).$$

This shows that the interval $[a,b]$ contains at most one point of t_i's. Assume that the index i, $1 \leq i \leq r$, satisfies

$$\min_{t\in[a,b]}|t-t_i| = \min_{1\leq j\leq r}\min_{t\in[a,b]}|t-t_j|.$$

Again we separate the cases when $t_i \in [a,b]$ and $t_i \notin [a,b]$.

Case 2.1. $t_i \in [a,b]$. By a calculation

$$\begin{aligned}&\int_{a+h}^{b-h} u(t)dt \geq c\int_{a+h}^{b-h}|t-t_i|^{p_i}dt \\ &= \begin{cases}\frac{c}{p_i+1}\left[(b-t_i-h)^{p_i+1}+(t_i-a-h)^{p_i+1}\right], & t_i \in (a+h,b-h), \\ \frac{c}{p_i+1}\left[(b-t_i-h)^{p_i+1}-(a-t_i+h)^{p_i+1}\right], & t_i \leq a+h, \\ \frac{c}{p_i+1}\left[(t_i-a-h)^{p_i+1}-(t_i-b+h)^{p_i+1}\right], & t_i \geq b-h.\end{cases}\end{aligned}$$

On the other hand, we have

$$\int_a^b u(t)dt \leq c\int_a^b |t-t_i|^{p_i}dt = \frac{c}{p_i+1}\left[(b-t_i)^{p_i+1}+(t_i-a)^{p_i+1}\right].$$

For $t_i \in (a+h, b-h)$, using inequality (5.3.6), we see

$$(b-t_i-h)^{p_i+1}+(t_i-a-h)^{p_i+1} \geq \left(\frac{b-a-2h}{2}\right)^{p_i+1}$$
$$= \left[\left(\frac{d-2}{2d}\right)(b-a)\right]^{p_i+1} \geq \frac{1}{2}\left(\frac{d-2}{2d}\right)^{p_i+1}\left[(b-t_i)^{p_i+1}+(t_i-a)^{p_i+1}\right].$$

For $(a \leq) t_i \leq a+h$, we have

$$\frac{1}{h}\left[(b-t_i)-(t_i-a)\right] \geq \frac{1}{h}\left[(b-a-h)-h\right] = d-2 > 2,$$

$$b-t_i \geq b-a-h = (d-1)h \geq (d-1)(t_i-a),$$

and hence by (7.2.45)

$$\begin{aligned} & (b-t_i-h)^{p_i+1}-(a-t_i+h)^{p_i+1} \\ \geq\ & (b-t_i-h)^{p_i+1}-(t_i-a+h)^{p_i+1} \\ \geq\ & c\left[(b-t_i)^{p_i+1}-(t_i-a)^{p_i+1}\right] \\ \geq\ & \frac{c[(d-1)^{p_i+1}-1]}{(d-1)^{p_i+1}+1}\left[(b-t_i)^{p_i+1}+(t_i-a)^{p_i+1}\right]. \end{aligned}$$

Similarly, if $t_i \geq b-h$, then

$$\begin{aligned} & (t_i-a-h)^{p_i+1}-(t_i-b+h)^{p_i+1} \\ \geq\ & \frac{c[(d-1)^{p_i+1}-1]}{(d-1)^{p_i+1}+1}\left[(b-t_i)^{p_i+1}+(t_i-a)^{p_i+1}\right]. \end{aligned}$$

Thus in all the cases inequality (7.2.51) follows.

Case 2.2. $t_i \notin [a,b]$. Suppose without loss of generality that $t_i < a$. Then

$$\begin{aligned} & \int_{a+h}^{b-h} u(t)dt \geq c\int_{a+h}^{b-h} |t-t_i|^{p_i}dt \\ =\ & \frac{c}{p_i+1}\left[(b-t_i-h)^{p_i+1}-(a-t_i+h)^{p_i+1}\right] \end{aligned}$$

and

$$\int_a^b u(t)dt \leq c\int_a^b |t-t_i|^{p_i}dt = \frac{c}{p_i+1}\left[(b-t_i)^{p_i+1}-(a-t_i)^{p_i+1}\right].$$

Applying (7.2.45) we get (7.2.51). □

Lemma 7.2.6. *If* $p \geq 0$, $B > A > 0$, *and* $\sigma = \pm 1$, *then*

$$\frac{(B+\sigma A)(B^p+A^p)}{B^{p+1}+\sigma A^{p+1}} \leq 2. \tag{7.2.54}$$

Proof. In the case when $\sigma = -1$

$$\begin{aligned} & 2(B^{p+1} - A^{p+1}) - (B-A)(B^p + A^p) = B^{p+1} - A^{p+1} + AB^p - A^pB \\ = & (B+A)(B^p - A^p) \geq 0, \end{aligned}$$

which is equivalent to our assertion. The inequality (7.2.54) with $\sigma = 1$ may be proved similarly. □

Lemma 7.2.7. *Let $p \geq 0$, $B_n \geq A_n \geq 0$, $\sigma = \pm 1$, and*

$$B_n^{p+1} + \sigma A_n^{p+1} \leq \frac{C}{n}\left[\left(B_n + \frac{1}{n}\right)^p + \left(A_n + \frac{1}{n}\right)^p\right]. \tag{7.2.55}$$

Then

$$B_n + \sigma A_n \leq \frac{c(C,p)}{n}. \tag{7.2.56}$$

Proof. If $A_n \leq 1/n$ then by (7.2.55) and (5.3.6)

$$\begin{aligned} B_n^{p+1} & \leq \frac{A_n^p}{n} + \frac{C}{n}\left[\left(B_n + \frac{1}{n}\right)^p + \left(A_n + \frac{1}{n}\right)^p\right] \\ & \leq \frac{2C+1}{n}\left(B_n + \frac{1}{n}\right)^p \leq \frac{(2C+1)2^p}{n}\left[B_n^p + \left(\frac{1}{n}\right)^p\right] \end{aligned}$$

and hence

$$B_n + \sigma A_n \leq 2B_n \leq \frac{c}{n}.$$

If $A_n > 1/n$ then by (7.2.55)

$$B_n^{p+1} + \sigma A_n^{p+1} \leq \frac{c}{n}(B_n^p + A_n^p)$$

and hence by (7.2.54)

$$B_n + \sigma A_n \leq \frac{c}{n} \cdot \frac{(B_n + \sigma A_n)(B_n^p + A_n^p)}{B_n^{p+1} + \sigma A_n^{p+1}} \leq \frac{c}{n}. \quad \square$$

Remark 7.2.1. Unfortunately, Lemma 7.2.7 is not true in general for the case when $-1 < p < 0$, although we need such a result later. For example, let $A_n = n^{-3/4}$, $\varepsilon_n = n^{-1/4}$, and $B_n = (1+\varepsilon_n)^2 A_n$. Then

$$\begin{aligned} B_n^{1/2} - A_n^{1/2} & = \varepsilon_n A_n^{1/2} = \varepsilon_n(1+\varepsilon_n)A_nB_n^{-1/2} \\ & \leq 2n^{-1}B_n^{-1/2} \leq n^{-1}(B_n^{-1/2} + A_n^{-1/2}) \\ & \leq 2n^{-1}\left[\left(B_n + \frac{1}{n}\right)^{-1/2} + \left(A_n + \frac{1}{n}\right)^{-1/2}\right] \end{aligned}$$

which shows that inequality (7.2.55) is true. But inequality (7.2.56) is violated, because

$$B_n - A_n = 2\varepsilon_n A_n + \varepsilon_n^2 \geq n^{-1/2}.$$

Theorem 7.2.11. *Let $x_{kn} = x_{kn}(u,m)$ and let relation (5.3.2) with*

$$p_i \geq 0, \quad i = 2,3,\ldots,r-1, \tag{7.2.57}$$

prevail. Then the relation (4.5.30) is valid.

Proof. The proof follows and properly modifies the ideas of Nevai in [65, pp. 164–167]. Meanwhile according to Lemma 4.5.6 it is enough to prove (4.5.31). We use the notation $\ell_j = \ell_{jn}(u,m)$ of (5.2.32).

First, we prove

$$x_{k+1} - x_k \leq c\Delta_n(x_k), \quad k = 0,1,\ldots,n. \tag{7.2.58}$$

To this end choose n_0 so large that for $n \geq n_0$

$$\max_{0\leq k\leq n} (x_{k+1,n} - x_{kn}) \leq \frac{1}{2} \min_{1\leq j\leq r-1} (t_{j+1} - t_j).$$

For a fixed index k, $0 \leq k \leq n$, assume that an index i, $1 \leq i \leq r$, satisfies

$$\min_{t\in[x_k,x_{k+1}]} |t - t_i| = \min_{1\leq j\leq r} \min_{t\in[x_k,x_{k+1}]} |t - t_j|.$$

So the interval $[x_k, x_{k+1}]$ contains no any point of t_j's except t_i.

By Theorem 5.2.3 it follows from (7.1.1) that

$$\begin{aligned}
&\int_{-1}^{1} |A_{m-2}(\ell_j,x_j;t)|(1+t)u(t)dt \\
&= \int_{-1}^{1} A_{m-2}(\ell_j,x_j;t)(1+t)\text{sgn}\,[(t-x_j)\ell_j(u,m;t)]^m u(t)dt \\
&= [\text{sgn}\,P_n'(u,m;x_j)^m] \int_{-1}^{1} A_{m-2}(\ell_j,x_j;t)(1+t)[\text{sgn}\,P_n(u,m;t)^m]u(t)dt \\
&= [\text{sgn}\,P_n'(u,m;x_j)^m]\lambda_{j,m-2,n}(u,m)(1+x_j) \\
&= \lambda_{m-2,n}(u,m;x_j)(1+x_j).
\end{aligned} \tag{7.2.59}$$

By (4.3.26) and (5.3.6)

$$\ell_k(t)^m + \ell_{k+1}(t)^m \geq 2^{1-m}[\ell_k(t) + \ell_{k+1}(t)]^m \geq 2^{1-m}, \quad t \in [x_k, x_{k+1}],$$
$$k = 0,1,\ldots,n.$$

Thus by (5.2.22) and (7.2.51)

$$\begin{aligned}
&\int_{-1}^{1} |A_{m-2}(\ell_k,x_k;t)|(1+t)u(t)dt + \int_{-1}^{1} |A_{m-2}(\ell_{k+1},x_{k+1};t)|(1+t)u(t)dt \\
&\geq \frac{1}{(m-2)!}\left(\frac{x_{k+1}-x_k}{5}\right)^{m-2} \\
&\quad \times \int_{x_k+(x_{k+1}-x_k)/5}^{x_{k+1}-(x_{k+1}-x_k)/5} [\ell_k(t)^m + \ell_{k+1}(t)^m](1+t)u(t)dt \\
&\geq c(x_{k+1}-x_k)^{m-2} \int_{x_k+(x_{k+1}-x_k)/5}^{x_{k+1}-(x_{k+1}-x_k)/5} (1+t)u(t)dt \\
&\geq c(x_{k+1}-x_k)^{m-2} \int_{x_k}^{x_{k+1}} (1+t)u(t)dt,
\end{aligned}$$

which, coupled with (7.2.59), gives

$$\begin{aligned}&(x_{k+1}-x_k)^{m-2}\int_{x_k}^{x_{k+1}}(1+t)u(t)dt\\ \leq\ & c[\lambda_{m-2,n}(u,m;x_k)(1+x_k)+\lambda_{m-2,n}(u,m;x_{k+1})(1+x_{k+1})].\end{aligned} \tag{7.2.60}$$

If $x_{k+1}-x_k\leq\Delta_n(x_k)+\Delta_n(x_{k+1})$, then

$$\begin{aligned}&\left|\left[(1-x_{k+1}^2)^{1/2}\right]^2-\left[(1-x_k^2)^{1/2}\right]^2\right|=|x_{k+1}^2-x_k^2|\leq 2(x_{k+1}-x_k)\\ \leq\ & \frac{c}{n}\left\{\left[(1-x_{k+1}^2)^{1/2}+\frac{1}{n}\right]+\left[(1-x_k^2)^{1/2}+\frac{1}{n}\right]\right\}\end{aligned}$$

and hence by Lemma 7.2.7

$$|\Delta_n(x_{k+1})-\Delta_n(x_k)|=\frac{1}{n}\left|(1-x_{k+1}^2)^{1/2}-(1-x_k^2)^{1/2}\right|\leq\frac{c}{n^2}.$$

So $\Delta_n(x_{k+1})\leq c\Delta_n(x_k)$ and

$$x_{k+1}-x_k\leq c\Delta_n(x_k).$$

If $x_{k+1}-x_k>\Delta_n(x_k)+\Delta_n(x_{k+1})$, then using (5.3.26) inequality (7.2.60) gives

$$\int_{x_k}^{x_{k+1}}(1+t)u(t)dt\leq\frac{c}{n}\left[\widehat{W}_n(x_k)(1+x_k)+\widehat{W}_n(x_{k+1})(1+x_{k+1})\right]$$

or

$$\int_{x_k}^{x_{k+1}}|t-t_i|^{p_i}(1+t)dt\leq\frac{c}{n}\left[\widehat{W}_n(x_k)(1+x_k)+\widehat{W}_n(x_{k+1})(1+x_{k+1})\right]. \tag{7.2.61}$$

We distinguish the cases when $i\in\{1,r\}$ and $2\leq i\leq r-1$.

Case 1. $i\in\{1,r\}$. It is enough to treat the case when $i=1$. In this case using (5.3.26) inequality (7.2.61) yields

$$\begin{aligned}&(1+x_{k+1})^{p_1+2}-(1+x_k)^{p_1+2}\\ \leq\ & \frac{c}{n}\left\{\left[(1+x_{k+1})^{1/2}+\frac{1}{n}\right]^{2p_1+3}+\left[(1+x_k)^{1/2}+\frac{1}{n}\right]^{2p_1+3}\right\}.\end{aligned}$$

Again by Lemma 7.2.7 we have

$$(1+x_{k+1})^{1/2}-(1+x_k)^{1/2}\leq\frac{c}{n}.$$

Thus

$$\begin{aligned}x_{k+1}-x_k &= \left[(1+x_{k+1})^{1/2}+(1+x_k)^{1/2}\right]\left[(1+x_{k+1})^{1/2}-(1+x_k)^{1/2}\right]\\ &\leq \frac{c}{n}\left[(1+x_{k+1})^{1/2}+(1+x_k)^{1/2}\right]\leq\frac{c}{n}\left[(1+x_k)^{1/2}+\frac{1}{n}\right]\\ &\leq \frac{c}{n}\left[(1-x_k^2)^{1/2}+\frac{1}{n}\right]=c\Delta_n(x_k).\end{aligned}$$

Case 2. $2 \le i \le r-1$. In this case by (5.3.26) inequality (7.2.61) gives

$$\begin{aligned}&\frac{c}{n}\left[\left(|t_i - x_k| + \frac{1}{n}\right)^{p_i} + \left(|t_i - x_{k+1}| + \frac{1}{n}\right)^{p_i}\right]\\ \ge\; & \begin{cases} |t_i - x_k|^{p_i+1} + |t_i - x_{k+1}|^{p_i+1}, & t_i \in [x_k, x_{k+1}],\\ \left|\,|t_i - x_k|^{p_i+1} - |t_i - x_{k+1}|^{p_i+1}\right|, & t_i \notin [x_k, x_{k+1}].\end{cases}\end{aligned} \tag{7.2.62}$$

If $t_i \in [x_k, x_{k+1}]$ then by Lemma 7.2.7 it follows from (7.2.62) that

$$x_{k+1} - x_k = |t_i - x_k| + |t_i - x_{k+1}| \le \frac{c}{n} \le c\Delta_n(x_k).$$

If $t_i \notin [x_k, x_{k+1}]$ then by Lemma 7.2.7 it follows from (7.2.62) that

$$x_{k+1} - x_k = \left|\,|t_i - x_k| - |t_i - x_{k+1}|\,\right| \le \frac{c}{n} \le c\Delta_n(x_k).$$

Next, we prove $x_{k+1} - x_k \ge c\Delta_n(x_k)$, $k = 0, 1, \ldots, n$. To this end applying Lemma 5.3.3 several times and using (5.3.26), we obtain

$$\begin{aligned}&|A_{m-2}^{(m-1)}(\ell_k, x_k; x)|\delta_{mn}(x)^{m-1}\widehat{W}_{mn}(x)\\ \le\;& cmn\int_{-1}^{1} |A_{m-2}^{(m-1)}(\ell_k, x_k; t)|(1-t^2)^{(m-1)/2}u(t)dt\\ \le\;& c(mn)^m \int_{-1}^{1} |A_{m-2}(\ell_k, x_k; t)|u(t)dt\\ =\;& c(mn)^m \lambda_{m-2,n}(u, m; x_k)\\ \le\;& cn^{m-1}\widehat{W}_n(x_k)\Delta_n(x_k)^{m-2}.\end{aligned}$$

Thus by Lemma 5.3.4 it follows from (7.2.58) that

$$\begin{aligned}|A_{m-2}^{(m-1)}(\ell_k, x_k; x)| &\le cn^{m-1}\widehat{W}_n(x_k)\Delta_n(x_k)^{m-2}\delta_{mn}(x)^{1-m}\widehat{W}_{mn}(x)^{-1}\\ &\le c\Delta_n(x_k)^{-1}, \quad x \in [x_k, x_{k+1}].\end{aligned}$$

But by the mean value theorem for the derivatives for some point $\xi \in [x_k, x_{k+1}]$

$$\begin{aligned}1 &= A_{m-2}^{(m-2)}(\ell_k, x_k; x_k) - A_{m-2}^{(m-2)}(\ell_k, x_k; x_{k+1})\\ &= (x_k - x_{k+1})A_{m-2}^{(m-1)}(\ell_k, x_k; \xi) \le c(x_{k+1} - x_k)\Delta_n(x_k)^{-1}.\end{aligned}$$

Then

$$x_{k+1} - x_k \ge c\Delta_n(x_k). \quad \square$$

As a consequence of Theorem 7.2.11 we state the following

Corollary 7.2.3. *Let the assumptions of Theorem 7.2.11 prevail. Then with the constants associated with the symbol $\sim$ depending on u and m*

$$\theta \sim \theta_{kn}(u, m), \quad \theta \in [\theta_{k+1,n}(u, m), \theta_{kn}(u, m)], \quad k = 1, 2, \ldots, n, \tag{7.2.63}$$

and

$$x_{k+1,n}(u, m) - x_{kn}(u, m) \sim \Delta_n(x_{kn}(u, m)), \quad k = 0, 1, \ldots, n. \tag{7.2.64}$$

As a consequence of Theorems 5.2.6, 5.3.3 and 7.2.11 we state the following

Corollary 7.2.4. *Let the assumptions of Theorem 7.2.11 prevail. Then for each index* $j \in \mathbf{M}_1 \setminus \{0\}$

$$|\lambda_{jn}(u,m;x)| \le c\lambda_n(u;x)\Delta_n(x)^j \le \frac{c}{n}\widehat{W}_n(x)\Delta_n(x)^j,$$
$$x \in [-1,1] \setminus [(x_{1n},x_{2n}) \cup (x_{n-1,n},x_{nn})]. \tag{7.2.65}$$

7.3 Notes

7.3.1. The main results in this chapter are due to Shi [112, 125].

7.3.2. We see that many results for orthogonal polynomials may be extended to m orthogonal polynomials. But if a statement for orthogonal polynomials is derived from the differential equations satisfied by them only, then it can not be extended to m orthogonal polynomials for $m \ne 2$ in general.

7.3.3. Theorem 9.20 in [65, pp. 164–165] and Theorem 2 in [121] are special cases of Theorem 7.2.11 when $m = 2$ and $m \in \mathbf{N}_2$ without the restriction (7.2.57), respectively. But their proofs (the latter directly cites the former) are not suitable for the case when $p_i < 0,\ 2 \le i \le r-1$. It is still open that if Theorem 7.2.11 remains true without this restriction.

Chapter 8

Bounds and Inequalities

8.1 The Exact Order of Lebesgue Function Type Sum

In this chapter let $n \geq 2$ and

$$a = x_0 \leq x_1 < \cdots < x_n \leq x_{n+1} = b. \tag{8.1.1}$$

Denote by ℓ_k's the fundamental polynomials of Lagrange interpolation on the nodes $x_1, \ldots, x_n$. Put for $s > 0$

$$L_k(x) = |(x - x_k)\ell_k(x)|^s, \quad k = 0, 1, \cdots, n+1; \ \ell_0(x) = \ell_{n+1}(x) = 0. \tag{8.1.2}$$

For a fixed n we introduce the following notations (d, $d_k \geq 0$) and adopt the convention that $[A, B] = \phi$ if $A > B$:

$$J_k = \langle x_k, x_{k+1} \rangle, \quad k = 0, 1, \ldots, n,$$

$$J_k(d) = \begin{cases} [x_k + d, x_{k+1} - d], & k = 1, 2, \ldots, n-1, \\ \langle x_0, x_1 - d], & k = 0, \\ [x_n + d, x_{n+1} \rangle, & k = n. \end{cases}$$

Define $z_k = z_k(d_k) \in J_k(d_k)$ by

$$|\omega_n(z_k)| = \min_{x \in J_k(d_k)} |\omega_n(x)|, \tag{8.1.3}$$

where $\omega_n(x) = \prod_{k=1}^{n}(x - x_k)$.

The following lemma provides a lower bound for the sum of two adjacent terms of L_k's.

Lemma 8.1.1. *Let $s \geq 1$ and $0 \leq k, r \leq n$. Then*

$$L_k(x) + L_{k+1}(x) \geq 2^{1-s} d_k^s \left| \frac{\omega_n(z_r)}{\omega_n(z_k)} \right|^s, \quad x \in J_r(d_r). \tag{8.1.4}$$

Proof. Our proof is based on Erdős-Turán inequality (4.3.26). Using the identity

$$L_i(x) = \left| \frac{\omega_n(x)}{\omega_n'(x_i)} \right|^s = \left| \frac{\omega_n(x)}{\omega_n(z_r)} \right|^s L_i(z_r)$$

and relation (8.1.3) we have

$$L_i(x) \geq L_i(z_r), \quad x \in J_r(d_r), \quad i = k, k+1.$$

So for $x \in J_r(d_r)$

$$\begin{aligned} & L_k(x) + L_{k+1}(x) \\ \geq\ & L_k(z_r) + L_{k+1}(z_r) \\ =\ & \left|\frac{\omega_n(z_r)}{\omega_n(z_k)}\right|^s (|z_k - x_k|^s |\ell_k(z_k)|^s + |z_k - x_{k+1}|^s |\ell_{k+1}(z_k)|^s) \\ \geq\ & d_k^s \left|\frac{\omega_n(z_r)}{\omega_n(z_k)}\right|^s (|\ell_k(z_k)|^s + |\ell_{k+1}(z_k)|^s) \geq 2^{1-s} d_k^s \left|\frac{\omega_n(z_r)}{\omega_n(z_k)}\right|^s. \end{aligned}$$

Here the last inequality follows from (4.3.26) and Hölder inequality. □

To state the first main result in this section we introduce some notations. For each n define $1 \leq i_n \leq j_n \leq n$ and $\Delta_n = [\xi_{i_n}, \xi_{j_n}]$ where

$$\xi_{i_n} = \begin{cases} x_{i_n}, & i_n > 1, \\ a, & i_n = 1, \end{cases}$$

$$\xi_{j_n} = \begin{cases} x_{j_n}, & j_n < n, \\ b, & j_n = n. \end{cases}$$

Let $\mathbf{K}_n = \{k : J_k \subset \Delta_n\}$ and $m_n = |\mathbf{K}_n|$. Then we have

Theorem 8.1.1. *Let $s \geq 1$ and $1 \leq i_n \leq j_n \leq n$. Then for any positive number ε_n satisfying $\varepsilon_n < |\Delta_n|$ there exists a set*

$$I_n = \bigcup_{k=i_n}^{j_n} (x_k - h_k, x_k + h'_k) \bigcap \Delta_n$$

with $h_k, h'_k > 0$ such that $|I_n| \leq \varepsilon_n$ and the inequality

$$\Lambda_n(x) = \sum_{k=i_n}^{j_n} |(x - x_k)\ell_k(x)|^s \geq \frac{\varepsilon_n^s m_n^{1-s}}{4^s} \tag{8.1.5}$$

holds for all $x \in \Delta_n \setminus I_n$.

Moreover, the order $\varepsilon_n^s m_n^{1-s}$ is the best possible and is attained by the Chebyshev nodes

$$x_k = \cos\frac{2n-2k+1}{2n}\pi, \quad k = 1, 2, \ldots, n. \tag{8.1.6}$$

Proof. First we prove the theorem for the case when $a > -\infty$ and $b < \infty$.

If $m_n = 0$ then there is nothing to prove, so assume that $m_n > 0$. For each k, $0 \leq k \leq n$, let d_k be defined by

$$d_k = \sup\{d : \text{there exists a point } x \in J_k(d) \text{ such that (8.1.5) does not hold}\}.$$

Obviously,

$$d_0 \le |J_0|,\ d_n \le |J_n|,\ 0 < d_k \le \frac{1}{2}|J_k|,\ 1 \le k \le n-1,$$

and inequality (8.1.5) holds for all points in the set $\cup_{k\in\mathbf{K}_n} J_k(d_k)$, except at most $2m_n$ isolated points. Thus it suffices to estimate the measure $e = |\Delta_n \setminus \cup_{k\in\mathbf{K}_n} J_k(d_k)|$. The proof will be complete if one can show $e \le \varepsilon_n$. To this end let $0 < \rho < 1$. By the definition of d_r's for each r satisfying $d_r > 0$ one can choose a point $u_r \in J_r(\rho d_r)$ so that (8.1.5) does not hold, i.e.,

$$\Lambda_n(u_r) < c^s \varepsilon_n^s m_n^{1-s},\ d_r > 0,\ r \in \mathbf{K}_n.$$

Thus

$$\sum_{r\in\mathbf{K}_n} d_r^s \Lambda_n(u_r) < \left(\frac{\varepsilon_n}{4}\right)^s m_n^{1-s} \sum_{r\in\mathbf{K}_n} d_r^s. \tag{8.1.7}$$

On the other hand, let $z_k = z_k(\rho d_k) \in J_k(\rho d_k)$ be defined in (8.1.3), i.e.,

$$|\omega_n(z_k)| = \min_{x\in J_k(\rho d_k)} |\omega_n(x)|. \tag{8.1.8}$$

It is easy to see that

$$\Lambda_n(x) \ge \frac{1}{2} \sum_{k\in\mathbf{K}_n} [L_k(x) + L_{k+1}(x)].$$

Then using (8.1.8) and applying Lemma 8.1.1

$$\begin{aligned} &\sum_{r\in\mathbf{K}_n} d_r^s \Lambda_n(u_r) \\ \ge\ & \frac{1}{2} \sum_{r\in\mathbf{K}_n} d_r^s \sum_{k\in\mathbf{K}_n} [L_k(u_r) + L_{k+1}(u_r)] \ge 2^{-s}\rho^s \sum_{r\in\mathbf{K}_n}\sum_{k\in\mathbf{K}_n} d_r^s d_k^s \left|\frac{\omega_n(z_r)}{\omega_n(z_k)}\right|^s \\ =\ & 2^{-1-s}\rho^s \sum_{r\in\mathbf{K}_n}\sum_{k\in\mathbf{K}_n} d_r^s d_k^s \left(\left|\frac{\omega_n(z_r)}{\omega_n(z_k)}\right|^s + \left|\frac{\omega_n(z_k)}{\omega_n(z_r)}\right|^s\right). \end{aligned} \tag{8.1.9}$$

Using the inequality $t + \frac{1}{t} \ge 2$ $(t > 0)$ relation (8.1.9) gives

$$\sum_{r\in\mathbf{K}_n} d_r^s \Lambda_n(u_r) \ge 2^{-s}\rho^s \sum_{r\in\mathbf{K}_n}\sum_{k\in\mathbf{K}_n} d_r^s d_k^s = 2^{-s}\rho^s \left(\sum_{k\in\mathbf{K}_n} d_k^s\right)^2. \tag{8.1.10}$$

Comparing inequality (8.1.10) with (8.1.7) yields

$$\sum_{k\in\mathbf{K}_n} d_k^s \le (2^{-1}\rho^{-1}\varepsilon_n)^s m_n^{1-s}. \tag{8.1.11}$$

By Hölder inequality for $s > 1$

$$e \le 2 \sum_{k\in\mathbf{K}_n} d_k \le 2m_n^{(s-1)/s} \left(\sum_{k\in\mathbf{K}_n} d_k^s\right)^{1/s},$$

which remains true for $s = 1$. By means of (8.1.11) we have $e \le \rho^{-1}\varepsilon_n = \rho^{-1}\varepsilon_n$. As $\rho \to 1$ we obtain $e \le \varepsilon_n$.

Next, we prove the theorem for the general case. Since $|\ell_1(x)| > 1$ for $x < x_1$ and $|\ell_n(x)| > 1$ for $x > x_n$, we may choose a number h so that $h > \max\{\varepsilon_n, |x_1|, |x_n|\}$ and

$$\begin{cases} |(x-x_1)\ell_1(x)| \geq \frac{\varepsilon_n}{4}, & x \leq -h, \\ |(x-x_n)\ell_n(x)| \geq \frac{\varepsilon_n}{4}, & x \geq h. \end{cases}$$

Then inequality (8.1.5) holds for all $x \in \langle a,b\rangle \setminus [-h,h]$. Applying the previous conclusion to the intervals $[-h,h]$ and $\Delta_n' = \Delta_n \cap [-h,h]$, we can find a set

$$I_n = \bigcup_{k=i_n}^{j_n} (x_k - h_k, x_k + h_k') \bigcap \Delta_n'$$

with $h_k, h_k' > 0$ such that $|I_n| \leq \varepsilon_n$ and inequality (8.1.5) holds for all $x \in \Delta_n' \setminus I_n$. Then inequality (8.1.5) holds for all $x \in \Delta_n \setminus I_n$.

For the Chebyshev nodes by (6.4.24)

$$L_k(x) = \{n^{-1}(1-x_k^2)^{\frac{1}{2}}|T_n(x)|\}^s, \quad k = 1,2,\ldots,n, \tag{8.1.12}$$

where $T_n(x)$ denotes the Chebyshev polynomial of first kind. If we assume, say, that $\frac{1}{4}n \leq k \leq \frac{3}{4}n$ and $d_k = \frac{1}{4}|J_k|$ then

$$L_k(x) \sim n^{-s}, \quad x \in \bigcup_{r=0}^{n} J_r(d_r).$$

Meanwhile, for i_n and j_n with $\frac{1}{4}n \leq i_n < j_n \leq \frac{3}{4}n$ we have $|\Delta_n| \sim \varepsilon_n \sim \frac{m_n}{n}$. Thus for $x \in \cup_{r=o}^{n} J_r(d_r) \cap \Delta_n$ one obtains

$$\Lambda_n(x) = \sum_{k=i_n}^{j_n} L_k(x) \sim \frac{m_n}{n^s} \sim \varepsilon_n^s m_n^{1-s}. \quad \square$$

Now let $-a = b = 1$. We use the notations of Chapter 3. The *Lebesgue function type sum* with a fixed positive number s is defined by

$$\Lambda_n(\mathbf{x}, s; x) = \sum_{k=1}^{n} |(x-x_k)\ell_k(x)|^s. \tag{8.1.13}$$

It turns out that the functions Λ_n will play a fundamental role in the mean convergence of Lagrange and Hermite-Fejér interpolation, and an important role in power orthogonal polynomials. Clearly, it is an interesting problem to find the optimal vector of nodes $\mathbf{x}^* \in \mathbf{Y}$ for such a sum, which is defined by

$$\|\Lambda_n(\mathbf{x}^*, s)\| = \inf_{\mathbf{x}\in\mathbf{Y}} \|\Lambda_n(\mathbf{x}, s)\|, \tag{8.1.14}$$

where $\|\cdot\|$ stands for the uniform norm on $[-1,1]$. First let us remark that instead of infimum we can use minimum. Indeed, by the same argument as that of Theorem 3.2.1 the function $\|\Lambda_n(\mathbf{x},s)\|$ attains its minimum at $\mathbf{x}^* \in \overline{\mathbf{Y}}$. But for any index k, $1 \le k \le n-1$, we have

$$\lim_{x_{k+1}\to x_k} \|\Lambda_n(\mathbf{x},s)\| = \infty.$$

So $\mathbf{x}^* \in \mathbf{Y}$. We will give the answer to this problem and prove that as the optimal Lebesgue function $\sum_{k=1}^n |\ell_k(x)|$ [135, p. 96], $\Lambda_n(\mathbf{x}^*,s;x)$ must equioscillate.

Let $z_i = z_i(\mathbf{x})$ be defined by

$$z_0 = x_0 = -1,\ z_n = x_{n+1} = 1,\ \omega_n'(\mathbf{x};z_i) = 0,\ z_i \in (x_i,x_{i+1}),\ 1 \le 1 \le n-1. \tag{8.1.15}$$

For convenience of use we also need the notations:

$$\begin{cases} g(\mathbf{x}) = \left\{\sum_{k=1}^n \frac{1}{|\omega_n'(x_k)|^s}\right\}^{1/s}, \\ f_i(\mathbf{x}) = \Lambda_n(\mathbf{x},s;z_i)^{1/s}, \quad i = 0,1,\ldots,n. \end{cases} \tag{8.1.16}$$

Lemma 8.1.2. *We have*

$$\Lambda_n(\mathbf{x}^*,s;1) = \Lambda_n(\mathbf{x}^*,s;-1) = s_n. \tag{8.1.17}$$

Here the vector $\mathbf{x}^*$ *consists of the Chebyshev nodes and* $s_n = \|\Lambda_n(\mathbf{x}^*,s)\|$.

Proof. Suppose to the contrary that, say, $\Lambda_n(\mathbf{x}^*,s;-1) < s_n$. If $\Lambda_n(\mathbf{x}^*,s;1) = s_n$ then $x_n^* < 1$, for otherwise $\Lambda_n(\mathbf{x}^*,s;1) = 0$. Now instead of $\mathbf{x}^*$ we take a new vector of nodes $\mathbf{x}^*+c = (x_1^*+c,\ldots,x_n^*+c)$, where the number $c \in (0,1-x_n^*)$ is small enough so that both

$$\Lambda_n(\mathbf{x}^*+c,s;1) < s_n \quad \text{and} \quad \Lambda_n(\mathbf{x}^*+c,s;-1) < s_n$$

are valid. Clearly, again we have $\|\Lambda_n(\mathbf{x}^*+c,s)\| \le s_n$. So we can assume, without loss of generality, that $\Lambda_n(\mathbf{x}^*,s;1) < s_n$ and $\Lambda_n(\mathbf{x}^*,s;-1) < s_n$. Now we write

$$\Lambda_n(\mathbf{x},s;x) = (|\omega_n(\mathbf{x};x)|g(\mathbf{x}))^s. \tag{8.1.18}$$

Thus

$$\Lambda_n(c\mathbf{x}^*,s;x) = c^s\Lambda_n\left(\mathbf{x}^*,s;\frac{x}{c}\right), \quad c > 0,$$

here $c\mathbf{x}^* = (cx_1^*,\ldots,cx_n^*)$. Then we can conclude $\|\Lambda_n(c\mathbf{x}^*,s)\| < \|\Lambda_n(\mathbf{x}^*,s)\|$ provided the number $1-c > 0$ is small enough. This contradicts the assumption (8.1.14). □

Lemma 8.1.3. *We have*

$$\frac{\partial f_i(\mathbf{x})}{\partial x_j} = |\omega_n(z_i)|\left[\frac{g(\mathbf{x})}{x_j - z_i} + \frac{\partial g(\mathbf{x})}{\partial x_j}\right], \quad i = 0,1,\ldots,n, \quad j = 1,2,\ldots,n, \tag{8.1.19}$$

and

$$\frac{\partial g(\mathbf{x})}{\partial x_j} = \frac{g(\mathbf{x})^{1-s}}{\omega_n'(x_j)}\sum_{k=1}^n \frac{(-1)^{n-k+1}\ell_k'(x_j)}{|\omega_n'(x_k)|^{s-1}}, \quad j = 1,2,\ldots,n. \tag{8.1.20}$$

Proof. Let j, $1 \le j \le n$, be fixed. For $1 \le i \le n-1$ using (8.1.18) and (8.1.15), we obtain

$$\begin{aligned}\frac{\partial f_i(\mathbf{x})}{\partial x_j} &= \left[|\omega_n'(z_i)|\frac{\partial z_i(\mathbf{x})}{\partial x_j} + \frac{|\omega_n(z_i)|}{x_j - z_i}\right] g(\mathbf{x}) + |\omega_n(z_i)|\frac{\partial g(\mathbf{x})}{\partial x_j} \\ &= |\omega_n(z_i)|\left[\frac{g(\mathbf{x})}{x_j - z_i} + \frac{\partial g(\mathbf{x})}{\partial x_j}\right].\end{aligned}$$

Similarly, we can prove (8.1.19) for $i = 0$ and $i = n$.

To prove (8.1.20) we see

$$\begin{aligned}\frac{\partial g(\mathbf{x})}{\partial x_j} &= \frac{1}{s} g(\mathbf{x})^{1-s}\left[-\sum_{k\neq j}\frac{s}{|\omega_n'(x_k)|^{s+1}}\cdot\frac{|\omega_n'(x_k)|}{x_j - x_k} - \sum_{k\neq j}\frac{s|\omega_n'(x_j)|}{|\omega_n'(x_j)|^{s+1}(x_j - x_k)}\right] \\ &= g(\mathbf{x})^{1-s}\sum_{k\neq j}\left[\frac{1}{|\omega_n'(x_k)|^s(x_k - x_j)} + \frac{1}{|\omega_n'(x_j)|^s(x_k - x_j)}\right].\end{aligned} \tag{8.1.21}$$

We observe that

$$\frac{1}{\omega_n'(x_k)(x_k - x_j)} = -\frac{\ell_k'(x_j)}{\omega_n'(x_j)}, \quad k \neq j, \tag{8.1.22}$$

and

$$\sum_{k\neq j}\frac{1}{x_k - x_j} = -\ell_j'(x_j). \tag{8.1.23}$$

Substituting these formulas into (8.1.21) yields (8.1.20). □

Lemma 8.1.4. *The equality*

$$\Lambda_n(\mathbf{x}^*, s; z_i^*) = s_n \tag{8.1.24}$$

holds for $i = 0, 1, \ldots, n$, *where* $z_i^* = z_i(\mathbf{x}^*)$, $i = 0, 1, \ldots, n$, *are defined by (8.1.15) with* $\mathbf{x} = \mathbf{x}^*$.

Proof. According to Lemma 8.1.2 equality (8.1.24) holds for $i = 0$ and $i = n$. Meanwhile we conclude that the vector $\mathbf{x}^*$ is also a solution of the extremal problem

$$\min_{\mathbf{x}\in\mathbf{Y}} f_0(\mathbf{x})$$

subject to the conditions

$$f_0(\mathbf{x}) - f_i(\mathbf{x}) \ge 0, \quad i = 1, 2, \ldots, n. \tag{8.1.25}$$

Let us consider necessary conditions for a solution of the above extremal problem. By Theorem 3.4 in [3], there must exist a set of numbers $\lambda_0, \ldots, \lambda_n$ such that

$$\lambda_0 \nabla f_0(\mathbf{x}^*) - \sum_{i=1}^n \lambda_i \nabla[f_0(\mathbf{x}^*) - f_i(\mathbf{x}^*)] = 0, \tag{8.1.26}$$

$$\lambda_i[f_0(\mathbf{x}^*) - f_i(\mathbf{x}^*)] = 0, \quad i = 1, 2, \ldots, n, \tag{8.1.27}$$

and

$$\lambda_i \ge 0, \quad i = 0, 1, \ldots, n, \quad \sum_{i=0}^n \lambda_i = 1. \tag{8.1.28}$$

Here, as we know, $f_i(\mathbf{x}^*) = \Lambda_n(\mathbf{x}^*, s; z_i^*)^{1/s}$, $i = 0, 1, \ldots, n$, and

$$\nabla f(\mathbf{x}^*) = \left(\frac{\partial f(\mathbf{x})}{\partial x_1}, \cdots, \frac{\partial f(\mathbf{x})}{\partial x_n} \right) \Bigg|_{\mathbf{x}=\mathbf{x}^*}.$$

In view of (8.1.27) in order to prove our lemma it will be enough to establish $\lambda_i \neq 0$, $i = 1, 2, \ldots, n$. To this end using (8.1.19) it follows from (8.1.26) that

$$\begin{aligned} &\lambda_0 |\omega_n(\mathbf{x}^*; z_0^*)| \left[\frac{g(\mathbf{x}^*)}{x_j^* - z_0^*} + \frac{\partial g(\mathbf{x})}{\partial x_j} \Bigg|_{\mathbf{x}=\mathbf{x}^*} \right] \\ - &\sum_{i=1}^{n} \lambda_i \left\{ |\omega_n(\mathbf{x}^*, z_0^*)| \left[\frac{g(\mathbf{x}^*)}{x_j^* - z_0^*} + \frac{\partial g(\mathbf{x})}{\partial x_j} \Bigg|_{\mathbf{x}=\mathbf{x}^*} \right] \right. \\ - &\left. |\omega_n(\mathbf{x}^*, z_i^*)| \left[\frac{g(\mathbf{x}^*)}{x_j^* - z_i^*} + \frac{\partial g(\mathbf{x})}{\partial x_j} \Bigg|_{\mathbf{x}=\mathbf{x}^*} \right] \right\} = 0, \ j = 1, 2, \ldots, n. \end{aligned} \tag{8.1.29}$$

But by (8.1.16), (8.1.18), and (8.1.27) the inequality $\lambda_i \neq 0$ means that $|\omega_n(\mathbf{x}^*; z_i^*)| = |\omega_n(\mathbf{x}^*; z_0^*)|$. Thus relations (8.1.29) yield

$$\lambda_0 \left[\frac{g(\mathbf{x}^*)}{x_j^* - z_0^*} + \frac{\partial g(\mathbf{x})}{\partial x_j} \Bigg|_{\mathbf{x}=\mathbf{x}^*} \right] - \sum_{i=1}^{n} \lambda_i \left[\frac{g(\mathbf{x}^*)}{x_j^* - z_0^*} - \frac{g(\mathbf{x}^*)}{x_j^* - z_i^*} \right] = 0,$$
$$j = 1, 2, \ldots, n.$$

If we introduce the symbols $\mu_0 = 2\lambda_0 - 1$ and $\mu_i = \lambda_i$, $i = 1, 2, \ldots, n$, then by (8.1.28) the above relations may be written as

$$\lambda_0 \frac{\partial g(\mathbf{x})}{\partial x_j} \Bigg|_{\mathbf{x}=\mathbf{x}^*} + g(\mathbf{x}^*) \sum_{i=0}^{n} \frac{\mu_i}{x_j^* - z_i^*} = 0, \quad j = 1, 2, \ldots, n.$$

By means of (8.1.20) these yield

$$\frac{\lambda_0 g(\mathbf{x}^*)^{1-s}}{\omega_n'(\mathbf{x}^*; x_j^*)} \sum_{i=1}^{n} \frac{(-1)^{n-i+1} \ell_i'(\mathbf{x}^*; x_j^*)}{|\omega_n'(\mathbf{x}^*; x_i^*)|^{s-1}} + g(\mathbf{x}^*) \sum_{i=0}^{n} \frac{\mu_i}{x_j^* - z_i^*} = 0,$$
$$j = 1, 2, \ldots, n.$$

Multiplying the jth equation above with $g(\mathbf{x}^*)^{s-1}((x_j^*)^2 - 1)\omega_n'(\mathbf{x}^*; x_j^*)$ gives

$$\lambda_0((x_j^*)^2 - 1) \sum_{i=1}^{n} \frac{(-1)^{n-i+1} \ell_i'(\mathbf{x}^*; x_j^*)}{|\omega_n'(\mathbf{x}^*; x_i^*)|^{s-1}} + g(\mathbf{x}^*)^s \sum_{i=0}^{n} \frac{\mu_i((x_j^*)^2 - 1)\omega_n'(\mathbf{x}^*, x_j^*)}{x_j^* - z_i^*} = 0,$$
$$j = 1, 2, \ldots, n. \tag{8.1.30}$$

Since by definition

$$\omega_n'(\mathbf{x}^*; z_i^*) = 0, \quad i = 1, 2, \ldots, n-1,$$

we have

$$\omega_n'(\mathbf{x}^*; x) = n \prod_{i=1}^{n-1} (x - z_i^*).$$

This means that the function

$$\lambda_0(x^2-1)\sum_{i=1}^{n}\frac{(-1)^{n-i+1}\ell_i'(\mathbf{x}^*;x)}{|\omega_n'(\mathbf{x}^*;x_i^*)|^{s-1}}+g(\mathbf{x}^*)^s\sum_{i=0}^{n}\frac{\mu_i(x^2-1)\omega_n'(\mathbf{x}^*;x)}{x-z_i^*}$$

is indeed a polynomial of degree at most n. Comparing this with (8.1.30) we get

$$\begin{aligned}&\lambda_0(x^2-1)\sum_{i=1}^{n}\frac{(-1)^{n-i+1}\ell_i'(\mathbf{x}^*;x)}{|\omega_n'(\mathbf{x}^*;x_i^*)|^{s-1}}+g(\mathbf{x}^*)^s\sum_{i=0}^{n}\frac{\mu_i(x^2-1)\omega_n'(\mathbf{x}^*;x)}{x-z_i^*}\\ =\;& C\omega_n(\mathbf{x}^*;x)\end{aligned} \tag{8.1.31}$$

with a constant C to be determined. Putting $x=1$ and $x=-1$ in (8.1.31) and using (8.1.15) gives

$$2\mu_n g(\mathbf{x}^*)^s\omega_n'(\mathbf{x}^*;1)=C\omega_n(\mathbf{x}^*;1)$$

and

$$-2\mu_0 g(\mathbf{x}^*)^s\omega_n'(\mathbf{x}^*;-1)=C\omega_n(\mathbf{x}^*;-1),$$

respectively. Hence we obtain $\operatorname{sgn}C=\operatorname{sgn}\mu_0=\operatorname{sgn}\mu_n$. Comparing the leading coefficients of the polynomials on both the sides of (8.1.31) we have

$$C=-\lambda_0(n-1)g(\mathbf{x}^*)^s+ng(\mathbf{x}^*)^s\sum_{i=0}^{n}\mu_i=\lambda_0 g(\mathbf{x}^*)^s.$$

So $\operatorname{sgn}C=\operatorname{sgn}\lambda_0$, which, coupled with $\operatorname{sgn}C=\operatorname{sgn}\mu_0=\operatorname{sgn}\mu_n$ and $\mu_0=2\lambda_0-1$, gives $C>0$, $\lambda_0>0$, and $\lambda_n>0$. Again, putting $x=z_j^*$, $1\le j\le n-1$, in (8.1.31) yields

$$\begin{aligned}&\lambda_0((z_j^*)^2-1)\sum_{i=1}^{n}\frac{(-1)^{n-i+1}\ell_i'(\mathbf{x}^*;z_j^*)}{|\omega_n'(\mathbf{x}^*;x_i^*)|^{s-1}}+\mu_j g(\mathbf{x}^*)^s((z_j^*)^2-1)\omega_n''(\mathbf{x}^*;z_j^*)\\ =\;& C\omega_n(\mathbf{x}^*;z_j^*).\end{aligned} \tag{8.1.32}$$

Remembering $\omega'(\mathbf{x}^*;z_j^*)=0$, $1\le j\le n-1$, we have

$$\ell_i'(\mathbf{x}^*;z_j^*)=-\frac{\omega_n(\mathbf{x}^*;z_j^*)}{\omega_n'(\mathbf{x}^*;x_i^*)(z_j^*-x_i^*)^2}.$$

Substituting this into (8.1.32) gives

$$\begin{aligned}&\mu_j g(\mathbf{x}^*)^s((z_j^*)^2-1)\omega_n''(\mathbf{x}^*;z_j^*)\\ =\;& \omega_n(\mathbf{x}^*;z_j^*)\left\{C+\sum_{i=1}^{n}\frac{\lambda_0(1-(z_j^*)^2)}{|\omega_n'(\mathbf{x}^*;x_i^*)|^s(z_j^*-x_i^*)^2}\right\}.\end{aligned}$$

The right side of the above equality is not zero, so $\lambda_j=\mu_j\neq 0$. □

Lemma 8.1.5. *Let s be a fixed positive number. Let $T_n(x)$ be the nth Chebyshev polynomial of the first kind and x_{kn}'s its zeros. Then*

$$\tau_n(s)=\left\{\frac{1}{n^{1-s}}\sum_{k=1}^{n}\frac{1}{|T_n'(x_{kn}|^s}\right\}^{1/s}=\left\{\frac{1}{n}\sum_{k=1}^{n}\sin^s\frac{(2k-1)\pi}{2n}\right\}^{1/s} \tag{8.1.33}$$

and

$$\lim_{n\to\infty} \tau_n(s) = \left\{ \frac{2^s\Gamma((s+1)/2)^2}{\pi\Gamma(s+1)} \right\}^{1/s}. \tag{8.1.34}$$

Proof. Formula (8.1.33) follows from (8.1.6) and (6.4.24). To prove (8.1.34) we use (8.1.33) to get

$$\lim_{n\to\infty} \tau_n(s) = \lim_{n\to\infty} \left\{ \frac{1}{\pi} \cdot \frac{\pi}{n} \sum_{k=1}^{n} \sin^s \frac{(2k-1)\pi}{2n} \right\}^{1/s} = \left\{ \frac{1}{\pi} \int_0^\pi \sin^s\theta\, d\theta \right\}^{1/s},$$

which, coupled with the formula [37, Formula 3.621-1, p. 369]

$$\int_0^\pi \sin^s\theta\, d\theta = \frac{2^s\Gamma((s+1)/2)^2}{\Gamma(s+1)},$$

implies (8.1.34). □

Theorem 8.1.2. *Let s be a fixed positive number. Then*

$$s_n = \|\Lambda_n(\mathbf{x}^*, s)\| = \frac{1}{n^s} \sum_{k=1}^{n} \sin^s \frac{(2k-1)\pi}{2n} \tag{8.1.35}$$

and the optimal vector of nodes is uniquely given by the Chebyshev nodes (8.1.6).

Proof. According to Lemma 8.1.4 we have

$$|\omega_n(\mathbf{x}^*; z_i^*)| = \|\omega_n(\mathbf{x}^*)\|, \quad i = 0, 1, \ldots, n.$$

It is easy to see that $\omega_n(\mathbf{x}^*; x) = 2^{1-n}T_n(x)$. Then by (8.1.6) and (8.1.33) we obtain

$$s_n = \|T_n\|^s \sum_{k=1}^{n} \frac{1}{|T_n'(x_k)|^s} = \frac{1}{n^s} \sum_{k=1}^{n} (1-x_k^2)^{s/2} = \frac{1}{n^s} \sum_{k=1}^{n} \sin^s \frac{(2k-1)\pi}{2n}. \quad \square$$

For a general Lebesgue function type sum we also have an estimation which is the following

Theorem 8.1.3. [148, 84] *Let $s, t \in \mathbf{N}_0$ and $s + t \geq 1$. Then for any positive number ε_n there exists a set $I_n \subset [-1, 1]$ such that $|I_n| \leq \varepsilon_n$ and*

$$\sum_{k=1}^{n} |x - x_k|^s |\ell_k(x)|^t \geq \begin{cases} c\varepsilon_n^s n^{-s}, & s \leq t-2, \\ c\varepsilon_n^{s+1} n^{-s} \ln n, & s = t-1, \\ c\varepsilon_n^s n^{1-t}, & s \geq t \end{cases} \tag{8.1.36}$$

holds for all $x \in [-1, 1] \setminus I_n$, where $c = c(s, t) > 0$.

Moreover, the order of n is the best possible and is attained by the Chebyshev nodes (8.1.6).

Since this general theorem is not needed in this book and its proof is rather complicated, we omit the details.

Similarly, a Lebesgue function type sum of Hermite interpolation is defined by

$$\Lambda_{jmn}(\mathbf{x},s,t;x) = \sum_{k=1}^{n} |x-x_k|^s |A_{kjm}(x)|^t, \tag{8.1.37}$$

where A_{kjm} are the fundamental polynomials of Hermite interpolation on the nodes x_k's given by (2.2.9) with $m_k \equiv m$. We state the following theorem without proof.

Theorem 8.1.4. [151, 98] *Let $s,t \in \mathbf{N}_0$ and $s+t \geq 1$. If $m-j$ is odd, then for any positive number ε_n there exists a set $I_n \subset [-1,1]$ such that $|I_n| \leq \varepsilon_n$ and*

$$\Lambda_{jmn}(\mathbf{x},s,t;x) \geq \begin{cases} c\varepsilon_n^{s+(m-1)t} n^{-s-jt}, & s \leq t-2, \\ c\varepsilon_n^{2mt} n^{-s-jt} \ln n, & s = t-1, \\ c\varepsilon_n^{s+(m-1)t} n^{1-(j+1)t}, & s \geq t \end{cases} \tag{8.1.38}$$

holds for all $x \in [-1,1] \setminus I_n$, where $c = c(m,j,s,t) > 0$.

Moreover, the order of n is the best possible and is attained by the Chebyshev nodes (8.1.6).

We define for a measure $d\mu$ on $\langle a,b \rangle$

$$\begin{aligned} \mathbf{Z}(\mu') &= \{x \in \langle a,b \rangle : \mu'(x) = 0\}, \\ \mathbb{M} &= \text{the collection of all Lebesgue measurable sets in } \mathbf{R}, \end{aligned}$$

and for $\Delta \in \mathbb{M},\ \Delta \subset \langle a,b \rangle$, and $0 \leq \delta \leq |\Delta|$

$$\sigma(\Delta;\delta) = \sigma(d\mu,\Delta;\delta) = \left(\int_\Delta d\mu(x) \right)^{-1} \inf_{\substack{\Omega \in \mathbb{M} \\ \Omega \subset \Delta \\ |\Omega| = \delta}} \int_{\Delta \setminus \Omega} d\mu(x). \tag{8.1.39}$$

We adopt the simple symbol

$$\sigma(\delta) = \sigma(\langle a,b \rangle;\delta).$$

Now we have a basic

Lemma 8.1.6. *Let $d\mu$ be a measure on $\langle a,b \rangle$ and $\Delta \in \mathbb{M}$. If $\int_\Delta d\mu(x) > 0$, then there exists a number δ, $0 < \delta < |\Delta|$ (when $|\Delta \setminus \mathbf{Z}(\mu')| > 0$ each δ satisfying $0 < \delta < |\Delta \setminus \mathbf{Z}(\mu')|$ is suitable) such that $\sigma(d\mu,\Delta;\delta) > 0$.*

In particular, there must exist a number $\delta > 0$ such that $\sigma(\delta) > 0$.

Proof. Let $\varepsilon > 0$ be an arbitrary number. Suppose to the contrary that for each $n \in \mathbf{N}$ and $\delta_n = \min\{1,|\Delta|\}/(2n)$ we have $\sigma(\Delta;\delta_n) = 0$. Then for each $n \in \mathbf{N}$ there is a set $E_n \in \mathbb{M}$ such that $E_n \subset \Delta$, $|E_n| = \delta_n$, and

$$\int_{\Delta \setminus E_n} d\mu(x) \leq 2^{-n}\varepsilon.$$

Let $E = \cap_{n\in\mathbf{N}} E_n$. Thus $|E| = 0$ and

$$\int_\Delta d\mu(x) = \int_{\Delta\setminus E} d\mu(x) = \int_{\Delta\setminus\cap_{n\in\mathbf{N}} E_n} d\mu(x)$$
$$= \int_{\cup_{n\in\mathbf{N}}(\Delta\setminus E_n)} d\mu(x) \le \sum_{n\in\mathbf{N}} \int_{\Delta\setminus E_n} d\mu(x) \le \sum_{n\in\mathbf{N}} 2^{-n}\varepsilon \le \varepsilon.$$

Since $\varepsilon > 0$ is an arbitrary number, $\int_\Delta d\mu(x) = 0$, a contradiction. This contradiction proves the first part of the lemma. The second follows from the first. □

Theorem 8.1.5. *Let $d\mu$ be a measure on $\mathbf{R}$. For $1 \le i_n \le j_n \le n$ let Δ_n be defined in Theorem 8.1.1 and $\Delta'_n = [x_{i_n}, x_{j_n}]$. If $0 < \delta < |\Delta_n|$, then*

$$\int_{\Delta_n} \sum_{k=i_n}^{j_n} |(x - x_k)\ell_k(x)| d\mu(x) \ge \frac{\delta\sigma(\Delta_n;\delta)}{4} \int_{\Delta_n} d\mu(x) \tag{8.1.40}$$

and if $0 < \delta < |\Delta'_n|$, then

$$\int_{\Delta'_n} \sum_{k=i_n}^{j_n} |(x - x_k)\ell_k(x)| d\mu(x) \ge \frac{\delta\sigma(\Delta'_n;\delta)}{4} \int_{\Delta'_n} d\mu(x). \tag{8.1.41}$$

Proof. Applying Theorem 8.1.1 with $s = 1$ and $\varepsilon_n = \delta$, we obtain I_n such that $|I_n| \le \delta$ and inequality (8.1.5) holds for all $x \in \Delta_n \setminus I_n$. Integrating (8.1.5) over $\Delta_n \setminus I_n$ with respect to $d\mu$ and using (8.1.39), we obtain

$$\int_{\Delta_n} \sum_{k=i_n}^{j_n} |(x - x_k)\ell_k(x)| d\mu(x) \ge \int_{\Delta_n\setminus I_n} \sum_{k=i_n}^{j_n} |(x - x_k)\ell_k(x)| d\mu(x)$$
$$\ge \frac{\delta}{4} \int_{\Delta_n\setminus I_n} d\mu(x) \ge \frac{\delta\sigma(\Delta_n;\delta)}{4} \int_{\Delta_n} d\mu(x).$$

This proves (8.1.40). Inequality (8.1.41) may be proved in a similar way. □

Theorem 8.1.6. *Let $d\mu$ be a measure on $\mathbf{R}$. Then there exists a number $\delta = \delta(d\mu) > 0$ (when $|\mathbf{R}\setminus\mathbf{Z}(\mu')| > 0$ each δ satisfying $0 < \delta < |\mathbf{R}\setminus\mathbf{Z}(\mu')|$ is suitable) such that*

$$\int_{\mathbf{R}} \sum_{k=1}^{n} |(x - x_k)\ell_k(x)| d\mu(x) \ge \frac{\delta\sigma(\delta)}{4} \int_{\mathbf{R}} d\mu(x) > 0. \tag{8.1.42}$$

Proof. The first inequality of (8.1.42) follows directly from (8.1.40) and the second from Lemma 8.1.6. □

8.2 Relationship Between Power Orthogonal Polynomials and Orthogonal Polynomials

We use the ideas of Lubinsky and Saff in the proof of Lemma 3.1 in [52] which gives the relationship between m orthogonal polynomials and classical orthogonal polynomials ($m = 2$).

Lemma 8.2.1. *Let $d\mu$ be a measure on $\mathbf{R}$ and let for $2 \le m < \infty$*

$$d\mu_n(x) = |P_n(d\mu, m; x)|^{m-2} d\mu(x). \tag{8.2.1}$$

Then for each $n \in \mathbf{N}_0$

$$P_n(d\mu_n; x) = P_n(d\mu, m; x), \quad \gamma_n(d\mu_n) = \gamma_n(d\mu, m), \tag{8.2.2}$$

and

$$x_{kn}(d\mu_n) = x_{kn}(d\mu, m), \quad k = 1, 2, \ldots, n. \tag{8.2.3}$$

Proof. The lemma for the case when $n = 0$ is trivial. Now let $n \ge 1$. By (3.2.6) the orthogonality relation

$$\int_{\mathbf{R}} |P_n(d\mu, m; x)|^{m-2} P_n(d\mu, m; x) q(x) d\mu(x) = 0 \tag{8.2.4}$$

holds for all $q \in \mathbf{P}_{n-1}$. By means of (8.2.1) relation (8.2.4) may be written as

$$\int_{\mathbf{R}} P_n(d\mu, m; x) q(x) d\mu_n(x) = 0. \tag{8.2.5}$$

According to the definition of $P_n(d\mu, m)$ we also have

$$\int_{\mathbf{R}} P_n(d\mu, m; x)^2 d\mu_n(x) = 1. \tag{8.2.6}$$

Clearly, relations (8.2.5) and (8.2.6) mean (8.2.2) and (8.2.3). □.

Theorem 8.2.1. *Let $d\mu$ be a measure supported in $[-1, 1]$ and $2 \le m < \infty$. Then*

$$\gamma_{n-1}(d\mu, m) \le \gamma_{n-1}(d\mu_n) \le \gamma_n(d\mu_n) = \gamma_n(d\mu, m). \tag{8.2.7}$$

Proof. To prove the first inequality of (8.2.7) we use the definition of orthogonal polynomials to obtain, recalling that the polynomial $P_{n-1}(d\mu_n; x)/\gamma_{n-1}(d\mu_n)$ is monic,

$$\begin{aligned}
& \int_{-1}^{1} P_{n-1}(d\mu, m; x)^2 d\mu_n(x) \\
= \ & \gamma_{n-1}(d\mu, m)^2 \int_{-1}^{1} \left[\frac{P_{n-1}(d\mu, m; x)}{\gamma_{n-1}(d\mu, m)}\right]^2 d\mu_n(x) \\
\ge \ & \gamma_{n-1}(d\mu, m)^2 \int_{-1}^{1} \left[\frac{P_{n-1}(d\mu_n; x)}{\gamma_{n-1}(d\mu_n)}\right]^2 d\mu_n(x) \\
= \ & \frac{\gamma_{n-1}(d\mu, m)^2}{\gamma_{n-1}(d\mu_n)^2}.
\end{aligned}$$

On the other hand, by Hölder inequality and by the definition of m orthogonal polynomials we have

$$\begin{aligned}
& \int_{-1}^{1} P_{n-1}(d\mu, m; x)^2 d\mu_n(x) \\
= \ & \int_{-1}^{1} P_{n-1}(d\mu, m; x)^2 |P_n(d\mu, m; x)|^{m-2} d\mu(x) \\
\le \ & \left[\int_{-1}^{1} |P_{n-1}(d\mu, m; x)|^m d\mu(x)\right]^{2/m} \left[\int_{-1}^{1} |P_n(d\mu, m; x)|^m d\mu(x)\right]^{(m-2)/m} \\
= \ & 1.
\end{aligned}$$

Hence the first inequality of (8.2.7) follows. The second immediately follows from a well known fact $\gamma_{n-1}(d\mu_n)/\gamma_n(d\mu_n) \le 1$. The equality of (8.2.7) may be found in (8.2.2). □

Theorem 8.2.2. *Let $d\mu$ be a measure supported in $[-1,1]$ and $m \in \mathbf{N}$ with $m \ge 2$. Then for $1 \le k \le n$ and $x_{kn} = x_{kn}(d\mu, m)$ we have*

$$\lambda_{kn}(d\mu_n) = (m-2)!P_n'(d\mu, m; x_{kn})^{m-2}\lambda_{k,m-2,n}(d\mu, m) \tag{8.2.8}$$

and

$$P_{n-1}(d\mu_n; x_{kn}) = \frac{\gamma_n(d\mu_n)}{(m-2)!\gamma_{n-1}(d\mu_n)\lambda_{k,m-2,n}(d\mu, m)P_n'(d\mu, m; x_{kn})^{m-1}}. \tag{8.2.9}$$

Proof. By definition

$$\begin{aligned}\lambda_{kn}(d\mu_n) &= \lambda_n(d\mu_n; x_{kn}) = \int_{-1}^{1} \ell_{kn}(d\mu_n; x)^2 d\mu_n(x) \\ &= \int_{-1}^{1} \ell_{kn}(d\mu_n; x)^2 |P_n(d\mu, m; x)|^{m-2} d\mu(x).\end{aligned} \tag{8.2.10}$$

On the other hand, by (5.2.31), (5.2.32), and (5.2.17)

$$\begin{aligned}&\lambda_{k,m-2,n}(d\mu, m) = [\text{sgn } P_n'(d\mu, m; x_{kn})^m]\lambda_{m-2,n}(d\mu, m; x_{kn}) \\ =& \frac{1}{(m-2)!}[\text{sgn } P_n'(d\mu, m; x_{kn})^m] \int_{-1}^{1} |\ell_{kn}(d\mu, m; x)|^m |x - x_{kn}|^{m-2} d\mu(x) \\ =& \frac{1}{(m-2)!P_n'(d\mu, m; x_{kn})^{m-2}} \int_{-1}^{1} \ell_{kn}(d\mu, m; x)^2 |P_n(d\mu, m; x)|^{m-2} d\mu(x),\end{aligned}$$

which, together with (8.2.10), yields (8.2.8).

We prove (8.2.9). Relation (1.2.14) with replacing $d\mu$ by $d\mu_n$ gives

$$\begin{aligned}&\frac{\gamma_{n-1}(d\mu_n)}{\gamma_n(d\mu_n)}\lambda_{kn}(d\mu_n)P_{n-1}(d\mu_n; x_{kn})P_n(d\mu_n; x) = (x - x_{kn})\ell_{kn}(d\mu_n; x), \\ &k = 1, 2, \ldots, n.\end{aligned} \tag{8.2.11}$$

Comparing the leading coefficients in both the sides of (8.2.11) we get

$$\frac{\gamma_{n-1}(d\mu_n)}{\gamma_n(d\mu_n)}\lambda_{kn}(d\mu_n)P_{n-1}(d\mu_n; x_{kn}) = \frac{1}{P_n'(d\mu, m; x_{kn})}, \tag{8.2.12}$$

which, coupled with (8.2.8), yields (8.2.9). □

Theorem 8.2.3. *Let $d\mu$ be a measure supported in $[-1,1]$ and $m \in \mathbf{N}$ with $m \ge 2$. Then*

$$\lambda_n(d\mu_n; x) = (m-2)! \left[\sum_{k=1}^{n} \frac{\ell_{kn}(d\mu, m; x)^2}{\lambda_{k,m-2,n}(d\mu, m)P_n'(d\mu, m; x_{kn})^{m-2}}\right]^{-1} \tag{8.2.13}$$

and

$$P_{n-1}(d\mu_n; x) = \frac{\gamma_n(d\mu_n)}{(m-2)!\gamma_{n-1}(d\mu_n)} \sum_{k=1}^{n} \frac{P_n(d\mu, m; x)}{\lambda_{k,m-2,n}(d\mu, m)P_n'(d\mu, m; x_{kn})^m (x - x_{kn})}. \tag{8.2.14}$$

Proof. The formula (8.2.13) follows from (8.2.8) and a well known formula

$$\lambda_n(d\mu;x)^{-1} = \sum_{k=1}^{n} \frac{\ell_{kn}(d\mu;x)^2}{\lambda_{kn}(d\mu)}. \tag{8.2.15}$$

The formula (8.2.14) follows from (8.2.9) and the identity

$$P_{n-1}(d\mu_n;x) = \sum_{k=1}^{n} P_{n-1}(d\mu_n;x_{kn})\ell_{kn}(d\mu_n;x). \quad \square$$

Theorem 8.2.4. *Let $d\mu$ be a measure supported in $[-1,1]$. Then*

$$\lambda_n(d\mu_n;x) \le \lambda_n(d\mu,m;x)^{2/m}, \quad 2 \le m < \infty, \tag{8.2.16}$$

and

$$\lambda_n(d\mu_n;x) \ge P_n(d\mu,m;x)^{m-2}\lambda_{mn/2}(d\mu;x), \quad m \in \mathbf{N}_2, \tag{8.2.17}$$

where $\lambda_n(d\mu,m;x)$ is defined in (5.1.7).

Proof. By (5.1.7) and Hölder inequality

$$\begin{aligned}
\lambda_n(d\mu_n;x) &= \min_{P\in\mathbf{P}_{n-1},P(x)=1} \int_{-1}^{1} |P(t)|^2 d\mu_n(t) \\
&= \min_{P\in\mathbf{P}_{n-1},P(x)=1} \int_{-1}^{1} |P(t)|^2 |P_n(d\mu,m;x)|^{m-2} d\mu(t) \\
&\le \min_{P\in\mathbf{P}_{n-1},P(x)=1} \left[\int_{-1}^{1} |P(t)|^m d\mu(t)\right]^{2/m} \left[\int_{-1}^{1} |P_n(d\mu,m;x)|^m d\mu(t)\right]^{(m-2)/m} \\
&= \left[\min_{P\in\mathbf{P}_{n-1},P(x)=1} \int_{-1}^{1} |P(t)|^m d\mu(t)\right]^{2/m} \\
&= \lambda_n(d\mu,m;x)^{2/m}.
\end{aligned}$$

This proves (8.2.16).

To prove (8.2.17) let $P \in \mathbf{P}_{n-1}$ with $P(x) = 1$ satisfy

$$\lambda_n(d\mu_n;x) = \int_{-1}^{1} |P(s)|^2 |P_n(d\mu,m;s)|^{m-2} d\mu(s).$$

By the definition of $\lambda_{mn/2}(d\mu;x)$ we have

$$\begin{aligned}
P(t)^2 P_n(d\mu,m;t)^{m-2} &\le \lambda_{mn/2}(d\mu,t)^{-1} \int_{-1}^{1} |P(s)|^2 |P_n(d\mu,m;s)|^{m-2} d\mu(s) \\
&= \lambda_{mn/2}(d\mu;t)^{-1}\lambda_n(d\mu_n;x),
\end{aligned}$$

because $PP_n(d\mu,m)^{(m-2)/2} \in \mathbf{P}_{(mn/2)-1}$. Inserting $t = x$ into the above relation yields (8.2.17). $\square$

8.3 Bounds and Inequalities for Quadrature Sum

Lemma 8.3.1. *Let $d\mu$ be a measure on* **R** *and* $\xi_{n+1} = \max\{|x_{1,n+1}|, |x_{n+1,n+1}|\}$, *where* $x_{kn} = x_{kn}(d\mu)$. *Then*

$$\sum_{k=1}^{n} \frac{\lambda_{kn}(d\mu)P_{n-1}(d\mu; x_{kn})^2}{\xi_{n+1}^2 - x_{kn}^2} \le 2\left(\frac{\gamma_n(d\mu)}{\gamma_{n-1}(d\mu)}\right)^2. \tag{8.3.1}$$

In particular, if the measure $d\mu$ is supported in $[-1,1]$ then

$$\sum_{k=1}^{n} \frac{\lambda_{kn}(d\mu)P_{n-1}(d\mu; x_{kn})^2}{1 - x_{kn}^2} \le 2\left(\frac{\gamma_n(d\mu)}{\gamma_{n-1}(d\mu)}\right)^2. \tag{8.3.2}$$

Proof. By the recurrence formula (1.2.3) we have

$$\xi_{n+1} - \alpha_n = \frac{\gamma_n P_{n+1}(\xi_{n+1})}{\gamma_{n+1} P_n(\xi_{n+1})} + \frac{\gamma_{n-1} P_{n-1}(\xi_{n+1})}{\gamma_n P_n(\xi_{n+1})}$$

in which each term of the right-hand side is positive. According to the inequality in [65, p. 19]

$$\alpha_n = \sum_{k=1}^{n+1} \lambda_{k,n+1} x_{k,n+1} P_n(x_{k,n+1})^2,$$

we obtain $|\alpha_n| \le \xi_{n+1}$. So

$$\frac{\gamma_{n-1} P_{n-1}(\xi_{n+1})}{\gamma_n P_n(\xi_{n+1})} \le \xi_{n+1} - \alpha_n \le 2\xi_{n+1}.$$

Similarly

$$0 \le -\frac{\gamma_{n-1} P_{n-1}(-\xi_{n+1})}{\gamma_n P_n(-\xi_{n+1})} \le 2\xi_{n+1}.$$

By (1.2.13) and (1.2.14)

$$\begin{aligned} P_{n-1}(\xi_{n+1}) &= \sum_{k=1}^{n} \ell_k(\xi_{n+1}) P_{n-1}(x_k) \\ &= \frac{\gamma_{n-1}}{\gamma_n} P_n(\xi_{n+1}) \sum_{k=1}^{n} \lambda_k P_{n-1}(x_k)^2 (\xi_{n+1} - x_k)^{-1} \end{aligned}$$

and

$$P_{n-1}(-\xi_{n+1}) = -\frac{\gamma_{n-1}}{\gamma_n} P_n(-\xi_{n+1}) \sum_{k=1}^{n} \lambda_k P_{n-1}(x_k)^2 (\xi_{n+1} + x_k)^{-1}.$$

Using these formulas, we obtain

$$\begin{aligned} &\sum_{k=1}^{n} \lambda_k P_{n-1}(x_k)^2 (\xi_{n+1}^2 - x_k^2)^{-1} \\ = &\ \frac{1}{2\xi_{n+1}} \left[\sum_{k=1}^{n} \lambda_k P_{n-1}(x_k)^2 (\xi_{n+1} - x_k)^{-1} + \sum_{k=1}^{n} \lambda_k P_{n-1}(x_k)^2 (\xi_{n+1} + x_k)^{-1}\right] \\ = &\ \frac{\gamma_n}{2\xi_{n+1}\gamma_{n-1}} \left[\frac{P_{n-1}(\xi_{n+1})}{P_n(\xi_{n+1})} - \frac{P_{n-1}(-\xi_{n+1})}{P_n(-\xi_{n+1})}\right] \\ \le &\ 2\left(\frac{\gamma_n}{\gamma_{n-1}}\right)^2. \end{aligned}$$

This proves (8.3.1). Inequality (8.3.2) is a direct consequence of (8.3.1). □

Remark 8.3.1. Inequality (8.3.1) fills a gap of Theorem 4 in [90].

Lemma 8.3.2. *Let the assumptions of Theorem 8.1.5 prevail. If* $2 \le m < \infty$ *and* $0 < \delta < |\Delta_n|$, *then*

$$\frac{\gamma_{n-1}(d\mu_n)}{\gamma_n(d\mu_n)} \sum_{k=i_n}^{j_n} \lambda_{kn}(d\mu_n)|P_{n-1}(d\mu_n;x_{kn})| \ge \frac{\delta\sigma(\Delta_n;\delta)}{4} \left[\int_{\Delta_n} d\mu(x)\right]^{1/m} \tag{8.3.3}$$

and if $0 < \delta < |\Delta_n'|$, *then*

$$\frac{\gamma_{n-1}(d\mu_n)}{\gamma_n(d\mu_n)} \sum_{k=i_n}^{j_n} \lambda_{kn}(d\mu_n)|P_{n-1}(d\mu_n;x_{kn})| \ge \frac{\delta\sigma(\Delta_n';\delta)}{4} \left[\int_{\Delta_n'} d\mu(x)\right]^{1/m}. \tag{8.3.4}$$

Proof. The relation (8.2.11) gives

$$\begin{aligned}&\sum_{k=i_n}^{j_n} |(x-x_{kn})\ell_{kn}(d\mu_n;x)| \\ =\;& \frac{\gamma_{n-1}(d\mu_n)}{\gamma_n(d\mu_n)} \sum_{k=i_n}^{j_n} \lambda_{kn}(d\mu_n)|P_{n-1}(d\mu_n;x_{kn})|\,|P_n(d\mu_n;x)|.\end{aligned} \tag{8.3.5}$$

By Hölder inequality it follows that

$$\begin{aligned}&\int_{\Delta_n} \sum_{k=i_n}^{j_n} |(x-x_{kn})\ell_{kn}(d\mu_n;x)|d\mu(x) \\ =\;& \frac{\gamma_{n-1}(d\mu_n)}{\gamma_n(d\mu_n)} \sum_{k=i_n}^{j_n} \lambda_{kn}(d\mu_n)|P_{n-1}(d\mu_n;x_{kn})| \int_{\Delta_n} |P_n(d\mu,m;,x)|d\mu(x) \\ \le\;& \frac{\gamma_{n-1}(d\mu_n)}{\gamma_n(d\mu_n)} \sum_{k=i_n}^{j_n} \lambda_{kn}(d\mu_n)|P_{n-1}(d\mu_n;x_{kn})| \left[\int_{\Delta_n} d\mu(x)\right]^{(m-1)/m}.\end{aligned}$$

Thus inequality (8.3.3) follows from (8.1.40). Similarly, inequality (8.3.4) follows from (8.1.41). □

As a consequence of Lemma 8.3.2 and Theorem 8.1.6 we state

Theorem 8.3.1. *Let* $d\mu$ *be a measure on* $\mathbf{R}$ *and* $2 \le m < \infty$. *Then there exists a number* $\delta = \delta(d\mu) > 0$ *(when* $|\mathbf{R} \setminus \mathbf{Z}(\mu')| > 0$ *every* δ *satisfying* $0 < \delta < |\mathbf{R} \setminus \mathbf{Z}(\mu')|$ *is suitable), such that for each* $n \in \mathbf{N}$

$$\frac{\gamma_{n-1}(d\mu_n)}{\gamma_n(d\mu_n)} \sum_{k=1}^{n} \lambda_{kn}(d\mu_n)|P_{n-1}(d\mu_n;x_{kn})| \ge \frac{\delta\sigma(\delta)}{4} \left[\int_{\mathbf{R}} d\mu(x)\right]^{1/m} > 0. \tag{8.3.6}$$

This result has many applications.

Corollary 8.3.1. *Let $d\mu$ be a measure supported in $[-1,1]$ and $2 \le m < \infty$. Then there exists a number $\delta = \delta(d\mu)$, $0 < \delta < 2$ (when $|\mathbf{Z}(\mu')| < 2$ every δ satisfying $0 < \delta < 2 - |\mathbf{Z}(\mu')|$ is suitable), such that for $n \in \mathbf{N}$*

$$0 < \frac{\delta}{4}\sigma(\delta) \le \frac{\gamma_{n-1}(d\mu_n)}{\gamma_n(d\mu_n)} \le 1. \tag{8.3.7}$$

Moreover, we have

$$c_1(d\mu,m)|P_n(d\mu,m;x)| \le \sum_{k=1}^{n} |(x-x_{kn})\ell_{kn}(d\mu,m;x)| \le c_2(d\mu,m)|P_n(d\mu,m;x)|, \tag{8.3.8}$$

$$c_1(d\mu,m) \le \sum_{k=1}^{n} \frac{1}{|P_n'(d\mu,m;x_{kn})|} \le c_2(d\mu,m), \tag{8.3.9}$$

and

$$\lim_{n\to\infty} \left\{ \sum_{k=1}^{n} \frac{1}{|P_n'(d\mu,m;x_{kn})|} \right\}^{1/n} = 1. \tag{8.3.10}$$

Proof. By Hölder inequality and the Gaussian quadrature formulas

$$\begin{aligned}
&\sum_{k=1}^{n} \lambda_{kn}(d\mu_n)|P_{n-1}(d\mu_n;x_{kn})| \\
\le\ & \left[\sum_{k=1}^{n} \lambda_{kn}(d\mu_n)\right]^{1/2} \left[\sum_{k=1}^{n} \lambda_{kn}(d\mu_n)|P_{n-1}(d\mu_n;x_{kn})|^2\right]^{1/2} \\
=\ & \left[\int_{-1}^{1} d\mu_n(x)\right]^{1/2} \\
=\ & \left[\int_{-1}^{1} |P_n(d\mu,m,x)|^{m-2} d\mu(x)\right]^{1/2} \\
\le\ & \left\{ \left[\int_{-1}^{1} d\mu(x)\right]^{2/m} \left[\int_{-1}^{1} |P_n(d\mu,m,x)|^{m} d\mu(x)\right]^{(m-2)/m} \right\}^{1/2} \\
=\ & \left[\int_{-1}^{1} d\mu(x)\right]^{1/m},
\end{aligned} \tag{8.3.11}$$

which, together with (8.3.6), yields

$$\frac{\gamma_{n-1}(d\mu_n)}{\gamma_n(d\mu_n)} \ge \frac{\delta}{4}\sigma(\delta) > 0.$$

The last inequality of (8.3.7) follows from (8.2.7).

Inequality (8.3.8) follows from (8.3.5)–(8.3.7) and (8.3.11). Comparing the leading coefficients of (8.3.8) gives (8.3.9). Relation (8.3.10) follows from (8.3.9). □

Theorem 8.3.2. *Let $d\mu$ be a measure supported in $[-1,1]$ and $m \in \mathbf{N}_2$. Then*

$$c(d\mu) \le \frac{1}{(m-2)!} \sum_{k=1}^{n} \frac{1}{\lambda_{k,m-2,n}(d\mu,m)P_n'(d\mu,m;x_{kn})^m} \le 1 \tag{8.3.12}$$

and

$$c(d\mu)\le\frac{1}{(m-2)!}\sum_{k=1}^{n}\frac{1}{\lambda_{k,m-2,n}(d\mu,m)P_n'(d\mu,m;x_{kn})^m(1-x_{kn}^2)}\le 2. \tag{8.3.13}$$

Proof. Comparing the leading coefficients in both the sides of (8.2.14) and using (8.3.7) gives (8.3.12).

To prove (8.3.13) we use (8.2.8) and (8.2.9) to obtain

$$\begin{aligned}&\sum_{k=1}^{n}\frac{\lambda_{kn}(d\mu_n)P_{n-1}(d\mu_n;x_{kn})^2}{1-x_{kn}^2}\\ =\;&\frac{\gamma_n(d\mu_n)^2}{(m-2)!\gamma_{n-1}(d\mu_n)^2}\sum_{k=1}^{n}\frac{1}{\lambda_{k,m-2,n}(d\mu,m)P_n'(d\mu,m;x_{kn})^m(1-x_{kn}^2)},\end{aligned}$$

that is

$$\begin{aligned}&\frac{1}{(m-2)!}\sum_{k=1}^{n}\frac{1}{\lambda_{k,m-2,n}(d\mu,m)P_n'(d\mu,m;x_{kn})^m(1-x_{kn}^2)}\\ =\;&\frac{\gamma_{n-1}(d\mu_n)^2}{\gamma_n(d\mu_n)^2}\sum_{k=1}^{n}\frac{\lambda_{kn}(d\mu_n)P_{n-1}(d\mu_n;x_{kn})^2}{1-x_{kn}^2}.\end{aligned} \tag{8.3.14}$$

By (8.3.2) we obtain

$$\frac{\gamma_{n-1}(d\mu_n)^2}{\gamma_n(d\mu_n)^2}\sum_{k=1}^{n}\frac{\lambda_{kn}(d\mu_n)P_{n-1}(d\mu_n;x_{kn})^2}{1-x_{kn}^2}\le 2 \tag{8.3.15}$$

and hence the right inequality of (8.3.13). To prove the left we use (8.3.12) to get

$$\begin{aligned}&\frac{1}{(m-2)!}\sum_{k=1}^{n}\frac{1}{\lambda_{k,m-2,n}(d\mu,m)P_n'(d\mu,m;x_{kn})^m(1-x_{kn}^2)}\\ \ge\;&\frac{1}{(m-2)!}\sum_{k=1}^{n}\frac{1}{\lambda_{k,m-2,n}(d\mu,m)P_n'(d\mu,m;x_{kn})^m}\ge c(d\mu).\quad\square\end{aligned}$$

As an immediate consequence of Theorem 8.3.2 we state

Corollary 8.3.2. *Let $d\mu$ be a measure supported in $[-1,1]$ and $m\in\mathbf{N}_2$. Then*

$$c(d\mu)\le\frac{1}{(m-2)!}\sum_{k=1}^{n}\frac{\int_{-1}^{1}[(x-x_{kn})\ell_{kn}(d\mu,m;x)]^m d\mu(x)}{\lambda_{k,m-2,n}(d\mu,m)(1-x_{kn}^2)}\le 2. \tag{8.3.16}$$

Proof. Multiplying (8.3.13) by the number $\int_{-1}^{1}P_n(d\mu,m;x)^m d\mu(x)=1$ we obtain

$$c(d\mu)\le\frac{1}{(m-2)!}\sum_{k=1}^{n}\frac{\int_{-1}^{1}P_n(d\mu,m;x)^m d\mu(x)}{\lambda_{k,m-2,n}(d\mu,m)P_n'(d\mu,m;x_{kn})^m(1-x_{kn}^2)}\le 2,$$

which is equivalent to (8.3.14) if we notice that

$$\frac{P_n(d\mu,m,x)}{P_n'(d\mu,m;x_{kn})}=(x-x_{kn})\ell_{kn}(d\mu,m;x).\quad\square$$

Lemma 8.3.3. *Let $-a=b=1$ and $m \geq 2$. Let the assumptions of Theorem 3.2.1 prevail. Then*

$$\lambda_{kn}(d\mu_n) \leq \left\{ \int_{x_{k-1,n}(d\mu,m)}^{x_{k+1,n}(d\mu,m)} d\mu(x) \right\}^{2/m}, \quad k=1,2,\ldots,n. \tag{8.3.17}$$

If relation (4.5.9) is true then

$$\lim_{n\to\infty} \max_{1\leq k\leq n} \lambda_{kn}(d\mu_n)=0. \tag{8.3.18}$$

Proof. By (1.2.7) and Hölder inequality for each k, $1\leq k\leq n$,

$$\begin{aligned}
\lambda_k(d\mu_n) &\leq \int_{x_{k-1}}^{x_{k+1}} d\mu_n(x) = \int_{x_{k-1}}^{x_{k+1}} |P_n(x)|^{m-2} d\mu(x) \\
&\leq \left\{ \int_{x_{k-1}}^{x_{k+1}} |P_n(x)|^m d\mu(x) \right\}^{(m-2)/m} \left\{ \int_{x_{k-1}}^{x_{k+1}} d\mu(x) \right\}^{2/m} \\
&\leq \left\{ \int_{x_{k-1}}^{x_{k+1}} d\mu(x) \right\}^{2/m}.
\end{aligned}$$

This proves (8.3.17). Relation (8.3.18) follows from (8.3.17) and (4.5.10). □

Theorem 8.3.3. *Let $d\mu$ be a measure supported in $[-1,1]$ and $m\in\mathbf{N}_2$. Let $\Delta\subset[-1,1]$ be a union of finitely many disjoint intervals. If $p\geq 1$ and*

$$\int_\Delta d\mu(x) > 0, \tag{8.3.19}$$

then

$$\liminf_{n\to\infty} \sum_{x_{kn}\in\Delta} \lambda_{kn}(d\mu_n)|P_{n-1}(d\mu_n;x_{kn})|^p > 0. \tag{8.3.20}$$

Moreover, (8.3.19) is necessary for validity of (8.3.20) provided $p<2$ and $\mu\in C[-1,1]$.

Proof. Since Δ is a union of finitely many disjoint intervals, inequality (8.3.19) implies that there exists an interval $\Omega\subset\Delta$ such that $\int_\Omega d\mu(x)>0$. So we may suppose without loss of generality that $\Delta=[c,d]$ and that if $\int_{c'}^{d'} d\mu(x)=\int_c^d d\mu(x)$ for $c\leq c'\leq d'\leq d$ then $c'=c$ and $d'=d$.

Now by Lemma 8.1.6 we can choose δ so that

$$\sigma(\Delta;\delta)>0 \tag{8.3.21}$$

and

$$\int_{c+\delta/3}^{d-\delta/3} d\mu(x) \geq \frac{1}{4}\int_\Delta d\mu(x).$$

Hence by Lemma 4.5.2 there exists a number N such that for every $n\geq N$, the polynomial $P_n(d\mu,m;x)$ has at least one zero in both $[c,c+\delta/3]$ and $[d-\delta/3,d]$, say, x_{i_n} and x_{j_n}, respectively. Then $c\leq x_{i_n}\leq c+\delta/3$, $d-\delta/3\leq x_{j_n}\leq d$ and

$$\int_{\Delta_n'} d\mu(x) \geq \frac{1}{4}\int_\Delta d\mu(x) \tag{8.3.22}$$

whenever $n \geq N$, where $\Delta'_n = [x_{i_n}, x_{j_n}]$. Applying (8.3.4) and (8.3.7) we have that for $n \geq N$

$$\begin{aligned}\sum_{x_k \in \Delta} \lambda_k(d\mu_n)|P_{n-1}(d\mu_n;x_k)| &\geq \sum_{k=i_n}^{j_n} \lambda_k(d\mu_n)|P_{n-1}(d\mu_n;x_k)| \\ &\geq \frac{\delta\sigma(\Delta'_n;\delta/3)}{12}\left\{\int_{\Delta'_n} d\mu(x)\right\}^{1/m}. \end{aligned} \tag{8.3.23}$$

By (8.1.39)

$$\begin{aligned}\sigma(\Delta'_n;\delta/3) &= \left(\int_{\Delta'_n} d\mu(x)\right)^{-1} \inf_{\substack{\Omega\in\mathbb{M}\\ \Omega\subset\Delta'_n\\ |\Omega|=\delta/3}} \int_{\Delta'_n\setminus\Omega} d\mu(x) \\ &= \left(\int_{\Delta'_n} d\mu(x)\right)^{-1} \inf_{\substack{\Omega\in\mathbb{M}\\ \Omega\subset\Delta'_n\\ |\Omega|=\delta/3}} \int_{\Delta\setminus([c,x_{i_n}]\cup[x_{j_n},d]\cup\Omega)} d\mu(x) \\ &\geq \left(\int_{\Delta} d\mu(x)\right)^{-1} \inf_{\substack{\Omega\in\mathbb{M}\\ \Omega\subset\Delta\\ |\Omega|=\delta}} \int_{\Delta\setminus\Omega} d\mu(x) \\ &= \sigma(\Delta;\delta)\end{aligned}$$

which, together with (8.3.22) and (8.3.23), yields

$$\sum_{x_k \in \Delta} \lambda_k(d\mu_n)|P_{n-1}(d\mu_n;x_k)| \geq \frac{\delta\sigma(\Delta;\delta)}{24}\left\{\int_{\Delta} d\mu(x)\right\}^{1/m}. \tag{8.3.24}$$

Hence by (8.3.21)

$$\liminf_{n\to\infty} \sum_{x_k \in \Delta} \lambda_k(d\mu_n)|P_{n-1}(d\mu_n;x_k)| > 0. \tag{8.3.25}$$

Using Hölder inequality, relation (8.3.20) follows from (8.3.25).

To prove the second part of the theorem suppose to the contrary that

$$\int_{\Delta} d\mu(x) = 0,$$

where Δ is a nontrivial interval. Applying Theorem 7.1.1 we see that for every n, the interval Δ contains no more than one zero of $P_n(d\mu, m; x)$. Therefore for $p < 2$

$$\begin{aligned}\lambda_k(d\mu_n)|P_{n-1}(d\mu_n;x_k)|^p &= \lambda_k(d\mu_n)^{1-\frac{p}{2}}[\lambda_k(d\mu_n)P_{n-1}(d\mu_n;x_k)^2]^{\frac{p}{2}} \\ &\leq \lambda_k(d\mu_n)^{1-\frac{p}{2}} \to 0, \quad n\to\infty.\end{aligned}$$

Here we use (8.3.18) and the inequality

$$\lambda_k(d\mu_n)P_{n-1}(d\mu_n;x_k)^2 \leq \sum_{k=1}^{n} \lambda_k(d\mu_n)P_{n-1}(d\mu_n;x_k)^2 = 1. \quad \square$$

8.4 Bounds and Inequalities for Power Orthogonal Polynomials

Theorem 8.4.1. *Let the assumptions of Theorem 8.3.3 prevail. If inequality (8.3.19) is true then the inequality*

$$|P_n(d\mu,m;x)| \leq c(d\mu,m,\Delta) \sum_{x_{kn}(d\mu,m)\in\Delta} |(x-x_{kn}(d\mu,m))\ell_{kn}(d\mu,m;x)| \tag{8.4.1}$$

holds for all $x \in [-1,1]$.

Proof. Rewrite (8.3.5) in the form

$$\begin{aligned}\frac{\gamma_{n-1}(d\mu_n)}{\gamma_n(d\mu_n)} &\sum_{x_{kn}(d\mu,m)\in\Delta} \lambda_{kn}(d\mu_n)|P_{n-1}(d\mu_n;x_{kn})|\,|P_n(d\mu_n;x)| \\ &= \sum_{x_{kn}(d\mu,m)\in\Delta} |(x-x_{kn})\ell_{kn}(d\mu,m;x)|,\end{aligned} \tag{8.4.2}$$

which, together with (8.3.19), (8.3.21), and (8.3.24), implies (8.4.1). □

Theorem 8.4.2. *Let* $d\mu$ *be a measure supported in* $[-1,1]$ *and* $m \in \mathbf{N}_2$. *Then*

$$\lambda_n(d\mu_n;x)P_n(d\mu,m;x)^2 \leq c(d\mu,m) \tag{8.4.3}$$

and

$$\lambda_{mn/2}(d\mu;x)P_n(d\mu,m;x)^m \leq c(d\mu,m). \tag{8.4.4}$$

Proof. By (8.4.1) and (8.2.15)

$$\begin{aligned}P_n(d\mu,m;x)^2 &\leq c\left[\sum_{k=1}^{n} |(x-x_{kn})\ell_{kn}(d\mu_n;x)|\right]^2 \\ &\leq c\left[\sum_{k=1}^{n} |\ell_{kn}(d\mu_n;x)|\right]^2 \\ &\leq c\left[\sum_{k=1}^{n} \lambda_{kn}(d\mu_n)\right]\left[\sum_{k=1}^{n} \frac{\ell_{kn}(d\mu_n;x)^2}{\lambda_{kn}(d\mu_n)}\right] \\ &\leq c\lambda_n(d\mu_n;x)^{-1},\end{aligned}$$

which is equivalent to (8.4.3). Further, inequality (8.4.4) follows from (8.4.3) and (8.2.17). □

Theorem 8.4.3. *Let* $d\mu$ *be a measure on* $\mathbf{R}$. *Let* $x_k = x_k(d\mu,m)$ *and* $2 \leq m < \infty$. *Then for any positive number* ε_n *there exists a set*

$$I_n = \bigcup_{k=1}^{n} (x_k - h_k, x_k + h_k') \tag{8.4.5}$$

with $h_k, h_k' > 0$ such that $|I_n| \le \varepsilon_n$ and the inequality

$$|P_n(d\mu, m; x)| \ge \frac{\varepsilon_n}{4} \left[\int_{\mathbf{R}} d\mu(x) \right]^{-1/m} \tag{8.4.6}$$

holds for all $x \in \mathbf{R} \setminus I_n$.

Proof. Use (8.4.2) with $\Delta = \mathbf{R}$, (8.1.5), (8.3.7), and (8.3.11). □

Theorem 8.4.4. *Let w be a weight in $[-1,1]$ and $w \in \mathbf{S}$. Then for $m \in \mathbf{N}_2$*

$$\left| \pi D(z)^2 [P_n(w,m;\zeta) z^n]^m \right| < \frac{1}{1-|z|^2}, \quad |z| < 1, \tag{8.4.7}$$

where $D(z)$ is defined in (1.2.17) with $d\nu(\theta) = w(\cos\theta)|\sin\theta|$ and $\zeta = \frac{1}{2}(z+z^{-1})$ is an arbitrary point of the cut plane, i.e., $\zeta \in \mathbf{C} \setminus [-1,1]$.

Proof. By use of (1.2.18) we have

$$\begin{aligned} 1 &= \int_{-1}^{1} P_n(w,m;x)^m w(x)dx \\ &= \frac{1}{2} \int_{-\pi}^{\pi} P_n(w,m;\cos\theta)^m w(\cos\theta)|\sin\theta| d\theta \\ &= \lim_{r \to 1-0} \frac{1}{2} \int_{-\pi}^{\pi} \left| P_n(w,m;\frac{1}{2}(z+z^{-1}))z^n \right|^m |D(z)|^2 d\theta, \quad z = re^{i\theta}. \end{aligned} \tag{8.4.8}$$

If $f(z) = \sum_{k=0}^{\infty} c_k z^k$ is regular in $|z| < 1$, using Cauchy inequality

$$\begin{aligned} |f(z)|^2 &\le \sum_{k=0}^{\infty} |c_k|^2 \sum_{k=0}^{\infty} |z^{2k}| \\ &\le \frac{1}{1-|z|^2} \lim_{r \to 1-0} \frac{1}{2\pi} \int_{-\pi}^{\pi} \left| f(re^{i\theta}) \right|^2 d\theta. \end{aligned} \tag{8.4.9}$$

Putting

$$f(z) = \pi^{1/2} D(z) \left[P_n(w,m;\zeta) z^n \right]^{m/2}$$

from (8.4.8) and (8.4.9) we obtain (8.4.7). □

Theorem 8.4.5. *Let w be a weight in $[-1,1]$ and $w \ge C > 0$. Then*

$$|P_n(w,m;x)| \le c \left| x + (x^2-1)^{1/2} \right|^n, \quad x \in \mathbf{R} \setminus [-1,1]. \tag{8.4.10}$$

Here $c = c(x,C,m)$ and $(x^2-1)^{1/2}$ is chosen so that $|x+(x^2-1)^{1/2}| > 1$.

Proof. By use of (1.2.17) and the formulas (10.2.11) and (10.2.13) in [137, p. 270] for the function

$$f(\theta) = w(\cos\theta)|\sin\theta| := f_1(\theta) f_2(\theta)$$

we have

$$|D(f;z)| = |D(f_1;z)D(f_2;z)| \ge |D(C;z)|\,|D(f_2;z)| = \left| \frac{C}{2}(1-z^2) \right|^{1/2},$$

which, coupled with (8.4.7), yields (8.4.10). □

Theorem 8.4.6. *Let $1 \leq m < \infty$ and let w be a weight in $\langle a,b\rangle$. If $b < +\infty$ and w is nondecreasing in the interval $\langle c,b]$, $c \geq a$, then the inequality*

$$w(x)^{1/m}|P_n(w,m;x)| \leq w(b)^{1/m}|P_n(w,m;b)| \tag{8.4.11}$$

holds for all $x \in \langle c,b]$.

Meanwhile, if $a > -\infty$ and w is nonincreasing in the interval $[a,c\rangle$, $c \leq b$, then the inequality

$$w(x)^{1/m}|P_n(w,m;x)| \leq w(a)^{1/m}|P_n(w,m;a)| \tag{8.4.12}$$

holds for all $x \in [a,c\rangle$.

Proof. We use simple symbols $P_n = P_n(w,m)$ and $x_k = x_k(w,m)$. It suffices to show (8.4.11). Our starting point is the identity

$$\begin{aligned} &|P_n(b)|^m w(b) - |P_n(x)|^m w(x) \\ = \;& m\int_x^b |P_n(t)|^{m-1}[\operatorname{sgn} P_n(t)]P_n'(t)w(t)dt + \int_x^b |P_n(t)|^m dw(t) \\ = \;& S_1(x) + S_2(x), \end{aligned} \tag{8.4.13}$$

which by partial integration follows from the formula

$$\frac{d}{dt}(|P_n(t)|^m) = m|P_n(t)|^{m-1}[\operatorname{sgn} P_n(t)]P_n'(t), \quad \text{a.e..} \tag{8.4.14}$$

It is enough to prove that the expression (8.4.13) is nonnegative for $x \in \langle c,b]$. Clearly, since w is nondecreasing in the interval $\langle c,b]$, we conclude $S_2(x) \geq 0$. On the other hand, it is easy to see that

$$[\operatorname{sgn} P_n(t)]P_n'(t) \begin{cases} > 0, & t > x_n, \\ < 0, & t < x_1. \end{cases}$$

Hence for $x \in [x_n,b]$ we have $S_1(x) \geq 0$ and for $x \in \langle a,x_1]$ we have

$$\begin{aligned} S_1(x) &= m\int_a^b |P_n(t)|^{m-1}[\operatorname{sgn} P_n(t)]P_n'(t)w(t)dt \\ &\quad -m\int_a^x |P_n(t)|^{m-1}[\operatorname{sgn} P_n(t)]P_n'(t)w(t)dt \\ &= -m\int_a^x |P_n(t)|^{m-1}[\operatorname{sgn} P_n(t)]P_n'(t)w(t)dt \geq 0. \end{aligned}$$

Then $S_1(x) \geq 0$ for $x \in \langle c,b] \cap \{\langle a,x_1] \cup [x_n,b]\}$.

For $x \in [x_k,x_{k+1}]$, $1 \leq k \leq n-1$, $n \geq 2$, we introduce the new weight

$$u(t) = w(t)\left|\prod_{i=1}^k (t-x_i)\right|^m.$$

We claim

$$P_{n-k}(u,m;t) = c\frac{P_n(u,m;t)}{\prod_{i=1}^k (t-x_i)}, \quad c > 0.$$

In fact, for $q \in \mathbf{P}_{n-k-1}$ we see

$$\begin{aligned}&\int_a^b |P_{n-k}(u,m;t)|^{m-1}[\operatorname{sgn} P_{n-k}(u,m;t)]q(t)u(t)dt\\ =\ & c^{m-1}\int_a^b |P_n(w,m;t)|^{m-1}[\operatorname{sgn} P_n(w,m;t)]\prod_{i=1}^{k}(t-x_i)q(t)w(t)dt = 0.\end{aligned}$$

This proves our claim. It is easy to see that u is increasing monotonically for $x \in \langle c,b] \cap [x_k,b]$. Then by the preceding argument, we conclude for $x \in \langle c,b] \cap [x_k,x_{k+1}]$

$$\begin{aligned}&|P_n(w,m;b)|^m w(b) - |P_n(w,m;x)|^m w(x)\\ =\ & c^{-m}\left[|P_{n-k}(u,m;b)|^m u(b) - |P_{n-k}(u,m;x)|^m u(x)\right] \geq 0. \quad \square\end{aligned}$$

Corollary 8.4.1. *Let $1 \leq m < \infty$. We have for $x \in [-1,1]$*

$$\left(\frac{1+x}{2}\right)^{\beta/m}|P_n(w^{(0,\beta)},m;x)| \leq |P_n(w^{(0,\beta)},m;1)|, \quad \beta \geq 0, \tag{8.4.15}$$

$$\left(\frac{1-x}{2}\right)^{\alpha/m}|P_n(w^{(\alpha,0)},m;x)| \leq |P_n(w^{(\alpha,0)},m;-1)|, \quad \alpha \geq 0, \tag{8.4.16}$$

$$|x|^{\alpha/m}|P_n(u^{(\alpha)},m;x)| \leq |P_n(u^{(\alpha)},m;1)|, \quad u^{(\alpha)}(x) = |x|^{\alpha}, \quad \alpha \geq 0, \tag{8.4.17}$$

and for $x \geq 0$

$$e^{-x/m}|P_n(\ell^{(1)},m;x)| \leq |P_n(\ell^{(1)},m;0)|. \tag{8.4.18}$$

Proof. Apply Theorem 8.4.6. $\square$

Theorem 8.4.7. *Let $d\mu$ and $d\nu$ be measures on $\mathbf{R}$. Let $m \in \mathbf{N}_2$ and*

$$d\mu \geq C d\nu. \tag{8.4.19}$$

Then for $P \in \mathbf{P}_n$

$$|P(x)|^m \leq C^{-1}\lambda_{mn/2+1}(d\nu;x)^{-1}\int_{\mathbf{R}}|P(t)|^m d\mu(t). \tag{8.4.20}$$

Furthermore, if $d\nu$ satisfies the condition

$$\lambda_{mn/2+1}(d\nu;x) \geq c(m)\lambda_n(d\nu;x), \tag{8.4.21}$$

then

$$|P(x)|^m \leq c(m)^{-1}C^{-1}\lambda_n(d\nu;x)^{-1}\int_{\mathbf{R}}|P(t)|^m d\mu(t). \tag{8.4.22}$$

Proof. By (1.2.4)

$$Q(x)^2 \leq \lambda_n(d\nu;x)^{-1}\int_{\mathbf{R}} Q(t)^2 d\nu(t), \quad Q \in \mathbf{P}_{n-1}. \tag{8.4.23}$$

This gives

$$|P(x)|^m \leq \lambda_{mn/2+1}(d\nu;x)^{-1}\int_{\mathbf{R}}|P(t)|^m d\nu(t),$$

which by (8.4.19) yields (8.4.20). Inequality (8.4.22) follows directly from (8.4.20) and (8.4.21). $\square$

Corollary 8.4.2. *Let the assumptions of Theorem 8.4.7 prevail. Then for* $P \in \mathbf{P}_n$

$$|P(x)|^m \le C^{-1}\left[\sum_{k=0}^{mn/2} P_k(d\nu;x)^2\right]\int_{\mathbf{R}} |P(x)|^m d\mu(x). \tag{8.4.24}$$

Proof. Apply (8.4.20) and (1.2.5). □

Theorem 8.4.8. *Let* $0 < m < +\infty$ *and let* $d\mu$ *be a measure in* $[-1,1]$ *satisfying*

$$\mu' \ge Cw^{(\alpha,\beta)}, \quad a.\ e.. \tag{8.4.25}$$

Then with a constant $c = c(\alpha,\beta,C,m) > 0$

$$\|P\| \le cn^{\gamma/m}\|P\|_{d\mu,m}, \quad P \in \mathbf{P}_n, \tag{8.4.26}$$

where $\gamma = \max\{2\alpha+2, 2\beta+2, 1\}$.

Proof. By (5.1.7)

$$|Q(x)|^m \le \lambda_n(d\mu,m;x)^{-1}\int_{\mathbf{R}} |Q(t)|^m d\mu(t), \quad Q \in \mathbf{P}_{n-1}. \tag{8.4.27}$$

This together with (5.3.3) and (8.4.25) gives

$$\begin{aligned} |P(x)|^m &\le cn\widehat{W}_n^{(\alpha,\beta)}(x)^{-1}\int_{-1}^{1} |P(t)|^m w^{(\alpha,\beta)}(t)dt \\ &\le cn\widehat{W}_n^{(\alpha,\beta)}(x)^{-1}\int_{-1}^{1} |P(t)|^m d\mu(t). \end{aligned} \tag{8.4.28}$$

Here

$$\widehat{W}_n^{(\alpha,\beta)}(x) = \left[(1-x)^{1/2}+\frac{1}{n}\right]^{2\alpha+1}\left[(1+x)^{1/2}+\frac{1}{n}\right]^{2\beta+1}. \tag{8.4.29}$$

We claim that

$$\widehat{W}_n^{(\alpha,\beta)}(x) \ge \frac{1}{6}n^{1-\gamma}. \tag{8.4.30}$$

In fact, for $x \ge 0$ we have

$$\left[(1-x)^{1/2}+\frac{1}{n}\right]^{2\alpha+1} \ge \begin{cases} n^{-(2\alpha+1)}, & 2\alpha+1 \ge 0, \\ 2^{2\alpha+1} \ge \frac{1}{2}, & 2\alpha+1 < 0, \end{cases}$$

$$\left[(1+x)^{1/2}+\frac{1}{n}\right]^{2\beta+1} \ge \begin{cases} 1, & 2\beta+1 \ge 0, \\ 3^{2\beta+1} \ge \frac{1}{3}, & 2\beta+1 < 0. \end{cases}$$

Hence for $x \ge 0$ we have

$$\widehat{W}_n^{(\alpha,\beta)}(x) \ge \frac{1}{6}n^{-\max\{2\alpha+1,0\}}.$$

Similarly for $x < 0$ we have

$$\widehat{W}_n^{(\alpha,\beta)}(x) \ge \frac{1}{6}n^{-\max\{2\beta+1,0\}}.$$

Thus inequality (8.4.30) follows. Using (8.4.30) and (8.4.28) we obtain (8.4.26). □

Theorem 8.4.9. *Let the assumptions of Theorem 8.4.8 prevail. Then for* $P \in \mathbf{P}_n$

$$|P(\cos\theta)| \le \begin{cases} c\theta^{-(2\alpha+1)/m} n^{1/m} \|P\|_{d\mu,m}, & Cn^{-1} \le \theta \le \pi/2, \\ cn^{(2\alpha+2)/m} \|P\|_{d\mu,m}, & 0 \le \theta \le Cn^{-1}. \end{cases} \tag{8.4.31}$$

Here C *is an arbitrary but fixed positive number and* $c = c(\alpha,\beta,C,m)$*. Similar bounds hold in the interval* $[\pi/2,\pi]$.

Proof. Use (8.4.28) and the estimation

$$\widehat{W}_n(\cos\theta) \sim \begin{cases} \theta^{2\alpha+1}, & Cn^{-1} \le \theta \le \pi/2, \\ n^{-2\alpha-1}, & 0 \le \theta \le Cn^{-1}. \quad \square \end{cases} \tag{8.4.32}$$

8.5 Bounds and Inequalities for integrals Involving Power Orthogonal Polynomials

Theorem 8.5.1. *Let* $d\mu$ *and* $d\nu$ *be measures on* $\mathbf{R}$*. Let* $2 \le m < \infty$, $0 < p < \infty$, $n \in \mathbf{N}_0$, $\Delta \in \mathbb{M}$*, and* $\delta < |\Delta|$*. Then*

$$\int_\Delta |P_n(d\mu,m;x)|^p d\nu(x) \ge \frac{\delta^p \sigma(d\nu,\Delta;\delta) \int_\Delta d\nu(x)}{4^p \left[\int_{\mathbf{R}} d\mu(x)\right]^{p/m}}. \tag{8.5.1}$$

Proof. Using (8.4.6) with $\varepsilon_n = \delta$ and (8.1.39) we obtain

$$\begin{aligned} \int_\Delta |P_n(d\mu,m;x)|^p d\nu(x) &\ge \int_{\Delta\setminus I_n} |P_n(d\mu,m;x)|^p d\nu(x) \\ &\ge \frac{\delta^p}{4^p}\left[\int_{\mathbf{R}} d\mu(x)\right]^{-p/m} \int_{\Delta\setminus I_n} d\nu(x) \\ &\ge \frac{\delta^p}{4^p}\left[\int_{\mathbf{R}} d\mu(x)\right]^{-p/m} \sigma(d\nu,\Delta;\delta) \int_\Delta d\nu(x). \quad \square \end{aligned}$$

Corollary 8.5.1. *Let* $d\mu$ *be a measure on* $\mathbf{R}$*. Let* $2 \le m < \infty$, $0 < p < \infty$*, and* $n \in \mathbf{N}_0$*. Then for any* $\Delta \in \mathbb{M}$

$$\int_\Delta |P_n(d\mu,m;x)|^p dx \ge \frac{|\Delta|^{p+1}}{2^{3p+1}}\left[\int_{\mathbf{R}} d\mu(x)\right]^{-p/m}. \tag{8.5.2}$$

Proof. Choosing $\delta = |\Delta|/2$ and $\nu(x) = x$ in (8.5.1) we obtain (8.5.2). $\square$

Theorem 8.5.2. *Let* $d\mu$ *and* $d\nu$ *be measures on* $\mathbf{R}$*. Let* $2 \le m < \infty$, $0 < p < \infty$*, and* $\Delta \in \mathbb{M}$*. Then*

$$\int_\Delta d\nu(x) > 0 \tag{8.5.3}$$

if and only if

$$\liminf_{n\to\infty} \int_\Delta |P_n(d\mu,m;x)|^p d\nu(x) > 0. \tag{8.5.4}$$

In particular,

$$\liminf_{n\to\infty} \int_\Delta |P_n(d\mu,m;x)|^p dx > 0 \tag{8.5.5}$$

if and only if $|\Delta| > 0$.

Proof. Relation (8.5.4) follows by Lemma 8.1.6 and (8.5.3) from (8.5.1); the converse is trivial. The second part of the theorem follows directly from the first. □

8.6 Notes

8.6.1. The technique of proving that $\Lambda_n(x)$ is "big" on a large set was developed by Erdős and Vértesi (cf. [23] or [135, Chapter III, pp. 71–123]). Main ideas of the proof of Theorem 8.1.1, including the origin of the crucial lemma (Lemma 8.1.1), may be found in the important paper of Vértesi [148].

8.6.2. Theorem 8.1.1 and Lemma 8.1.6 are due to Shi [23, 85, 86, 97]. They have many applications. The main application is in power orthogonal polynomials. In attempting to study convergence of orthogonal Fourier series or convergence of Lagrange interpolation at zeros of power orthogonal polynomials, one invariably encounters the need for bounds and inequalities of the power orthogonal polynomials on the interval of orthogonality. Historically, the problem of finding bounds and inequalities has lived under the shadow of the deeper asymptotics on the segment, because the latter are often the only way of obtaining the former. Of course, this way usually gives asymptotic estimates for certain "nice" measures only. The approach in this chapter may be applicable to finding bounds and inequalities of many important quantities in power orthogonal polynomials for general measures. Thus this makes it possible to extend many important results previously obtained. Further applications of these results will be discussed in Chapter 11.

8.6.3. Theorem 8.1.2 appears in [97]. There even the optimal constant in (8.1.5) is investigated.

8.6.4. The main results in Sections 8.2–8.4 are due to Shi [85, 86, 99, 117].

8.6.5. Turán [141] proved the following important inequality (called Tur án inequality).

Lemma 8.6.1. *Let $d\mu \in \mathbf{S}$ be a measure supported in $[-1,1]$ and let $-1 \le d-\delta < d+\delta \le 1$. Then the inequality*

$$\int_{d-\delta}^{d+\delta} P_n(d\mu;x)^2 d\mu(x) \ge c(d\mu,\delta) \tag{8.6.1}$$

holds for every $n \in \mathbf{N}$.

The inequality (8.6.1) was then extended by Máté, Nevai and Totik in [57, Theorem 13.3] to a measure satisfying $\mu'(x) > 0$, a. e. in $[-1,1]$.

Turán [141] also pointed out that $c(d\mu,\delta)$ here is an ineffective expression depending only upon $d\mu$ and δ, that is, it cannot be calculated explicitly, and it would be of special relevance to replace it by an effective one. The background of this requirement was explained in [141] and [142, pp. 70–73]. In a few words, the inequality (8.6.1) plays a crucial role in developing some results of Turán and Wiener, and to give the quantities in these results explicitly we need the explicit expression for $c(d\mu,\delta)$.

As a method to solve this problem, later Turán in [142, p. 71] further proposed the following

Problem 71. *Give an explicit estimate for $n_0(d\mu,\delta)$ such that, if $d\mu \in \mathbf{S}$, then*

$$\int_{d-\delta}^{d+\delta} P_n(d\mu;x)^2 d\mu(x) > \frac{1}{2\pi}\int_{d-\delta}^{d+\delta} \frac{dx}{\sqrt{1-x^2}} > \frac{\delta}{\pi}$$

holds for all $n > n_0(d\mu, \delta)$.

Since for $n \le n_0(d\mu, \delta)$ we can determine a quantity $c_0(d\mu, \delta)$ such that for all permitted d's

$$\int_{d-\delta}^{d+\delta} P_n(d\mu; x)^2 d\mu(x) \ge c_0(d\mu, \delta) > 0,$$

the quantity $c(d\mu, \delta)$ occurring in (8.6.1) can be chosen so that

$$c(d\mu, \delta) = \min\left\{ c_0(d\mu, \delta), \frac{\delta}{\pi} \right\},$$

which is an effective expression and, although, cannot be calculated easily.

Now if we put $\nu = \mu$, $m = p = 2$, and $\delta = |\Delta|/2$, then (8.5.1) gives the Turán inequality for general measures and provides a satisfactory solution to Problem 71 of Turán, i.e. an explicit expression of a lower bound of the quantity occurring in Tur ́an inequality. The inequality (8.5.1) for a general form with $m = 2$ was proved by Shi [85, Theorem 7] and [86, Theorem 5]. The present general inequality was given by Shi [99, 117].

Chapter 9

Asymptotics of General Polynomials

9.1 Nth Root Asymptotics

Given a triangular matrix X of nodes

$$x_{1n} < x_{2n} < \cdots < x_{nn}, \quad n = 1,2,\ldots \tag{9.1.1}$$

define the Lagrange interpolation polynomial of $f \in C[-1,1]$ by

$$\mathcal{L}_n(X,f) = \mathcal{L}_n(X,f;x) = \sum_{k=1}^{n} f(x_{kn})\ell_{kn}(x), \quad n = 1,2,\ldots, \tag{9.1.2}$$

where the fundamental polynomials $\ell_{kn}(x)$ are defined by

$$\ell_{kn}(x) = \frac{\omega_n(x)}{\omega_n'(x_{kn})(x - x_{kn})}, \quad k = 1,2,\ldots,n,$$

with $\omega_n(x) = \omega_n(X;x) = (x - x_{1n})(x - x_{2n})\cdots(x - x_{nn})$.

To state our results we need some definitions and notations. Write for $f \in C[-1,1]$ and $0 < p < \infty$

$$\|\mathcal{L}_n(X)\|_{d\mu,p} = \sup_{\|f\|\le 1} \|\mathcal{L}_n(X,f)\|_{d\mu,p}, \quad \|\mathcal{L}_n(X)\| = \left\| \sum_{k=1}^{n} |\ell_{kn}| \right\|, \tag{9.1.3}$$

$$\Lambda_n(X;x) = \sum_{k=1}^{n} |(x - x_{kn})\ell_{kn}(x)|, \tag{9.1.4}$$

$$\gamma_n(X) = \sum_{k=1}^{n} \frac{1}{|\omega_n'(x_{kn})|}, \tag{9.1.5}$$

$$\phi(z) = z + \sqrt{z^2 - 1}, \quad z \in \mathbf{C}. \tag{9.1.6}$$

Here and later we are to take those values of the nth and square roots, which are positive on the positive real axis. We write $X \subset [-1,1]$ if

$$x_{kn} \in [-1,1], \quad k = 1,2,\ldots,n, \quad n = 1,2,\ldots.$$

$N_n(X;c,d)$ stands for the cardinality of the set $\{k:\ x_{kn} \in [c,d]\}$, $[c,d] \subset [-1,1]$. If $x_{kn} = x_{kn}(d\mu,m)$ then we write $\mathcal{L}_n(X,f;x) = \mathcal{L}_n(d\mu,m,f;x)$, $N_n(X;c,d) = N_n(d\mu,m;c,d)$, etc. .

Definition 9.1.1. *We call $d\mu$ supported in $[-1,1]$ a regular measure, if*

$$\lim_{n\to\infty}\gamma_n(d\mu)^{1/n}=2. \tag{9.1.7}$$

The first main result of this section is the following

Theorem 9.1.1. *Let $d\mu$ be a measure supported in $[-1,1]$. Let $X\subset[-1,1]$ and $0<p<\infty$. Then the following statements are equivalent:*

(a) We have

$$\lim_{n\to\infty}\gamma_n(X)^{1/n}=2. \tag{9.1.8}$$

(b) We have

$$\lim_{n\to\infty}\|\Lambda_n(X)\|^{1/n}=1. \tag{9.1.9}$$

(c) We have

$$\lim_{n\to\infty}\|\mathcal{L}_n(X)\|^{1/n}=1. \tag{9.1.10}$$

Each of the above statements implies the following statements, which are equivalent to each other:

(d) The relation

$$\lim_{n\to\infty}|\omega_n(X;z)|^{1/n}=\frac{1}{2}|\phi(z)| \tag{9.1.11}$$

holds for every $z\in\mathbf{C}\setminus[-1,1]$.

(e) The relation

$$\lim_{n\to\infty}\omega_n(X;z)^{1/n}=\frac{1}{2}\phi(z) \tag{9.1.12}$$

holds for every $z\in\mathbf{C}\setminus[-1,1]$.

(f) We have

$$\lim_{n\to\infty}\|\omega_n(X)\|^{1/n}=\frac{1}{2}. \tag{9.1.13}$$

(g) The relation

$$\lim_{n\to\infty}\frac{N_n(X;c,d)}{n}=\frac{1}{\pi}\int_c^d\frac{dx}{\sqrt{1-x^2}} \tag{9.1.14}$$

holds for every interval $[c,d]\subset[-1,1]$.

(h) The limit

$$\lim_{n\to\infty}\frac{1}{n}\sum_{k=1}^{n}f(x_{kn})=\frac{1}{\pi}\int_{-1}^{1}f(x)\frac{dx}{\sqrt{1-x^2}} \tag{9.1.15}$$

holds for every $f\in C[-1,1]$.

The second main result of this section is the following

Theorem 9.1.2. *Let $d\mu$ be a regular measure supported in $[-1,1]$. Let $X\subset[-1,1]$ and $0<p<\infty$. Then Statement (a) of Theorem 9.1.1 is equivalent to each of the following statements:*

(i) We have

$$\lim_{n\to\infty}\|\Lambda_n(X)\|_{d\mu,p}^{1/n}=1. \tag{9.1.16}$$

(j) We have

$$\lim_{n\to\infty}\|\mathcal{L}_n(X)\|_{d\mu,p}^{1/n}=1. \tag{9.1.17}$$

Since the proofs of the theorems are long, we break them into some lemmas, which are of independent interest.

Definition 9.1.2. [19] *A matrix $X \subset [-1,1]$ is said to be an arcsine distribution if relation (9.1.14) holds for every interval $[c,d] \subset [-1,1]$.*

Lemma 9.1.1. *For every matrix X*

$$\liminf_{n\to\infty} \gamma_n(X)^{1/(n-1)}(x_{nn}-x_{1n}) \geq 4. \tag{9.1.18}$$

Proof. Let

$$x_k = \frac{1}{2}(x_1+x_n)+\frac{1}{2}\tau_k(x_n-x_1), \ \ k=1,\ldots,n, \tag{9.1.19}$$

and

$$z = \frac{1}{2}(x_1+x_n)+\frac{1}{2}\zeta(x_n-x_1). \tag{9.1.20}$$

Then $|\tau_k| \leq 1, \ k=1,\ldots,n$. Using the Lagrange interpolation formula based on the nodes $x_1,\ldots,x_n$ the $(n-1)$th Chebyshev polynomial of the first kind $T_{n-1}(\zeta)$ can be expressed as

$$T_{n-1}(\zeta) = \sum_{k=1}^{n} T_{n-1}(\tau_k)\ell_k(z) = \sum_{k=1}^{n} T_{n-1}(\tau_k)\frac{\omega_n(z)}{\omega_n'(x_k)(z-x_k)}. \tag{9.1.21}$$

Comparing the coefficients of z^{n-1} in (9.1.21) and noticing $|T_{n-1}(\tau_k)| \leq 1$ we get

$$2^{2n-3}(x_n-x_1)^{1-n} = \sum_{k=1}^{n} \frac{T_{n-1}(\tau_k)}{\omega_n'(x_k)} \leq \sum_{k=1}^{n} \frac{1}{|\omega_n'(x_k)|} = \gamma_n(X). \tag{9.1.22}$$

Thus (9.1.18) follows. □

Lemma 9.1.2. *Let $X \subset [-1,1]$. Then we have the inequality*

$$\gamma_n(X) \geq 2^{n-2}. \tag{9.1.23}$$

Equality holds only for $\omega_n(x) = 2^{2-n}(x^2-1)U_{n-2}(x)$, where $U_{n-2}(x)$ is the $(n-2)$th Chebyshev polynomial of the second kind.

Meanwhile, we have the inequality

$$\|\omega_n(X)\| \geq 2^{1-n}. \tag{9.1.24}$$

Equality holds only for $\omega_n(x) = 2^{1-n}T_n(x)$.

Proof. The inequality (9.1.23) follows directly from (9.1.22). For the polynomial $\omega_n(x) = 2^{2-n}(x^2-1)U_{n-2}(x)$ its zeros are

$$y_k = \cos\frac{n-k}{n-1}\pi, \quad k=1,2,\ldots,n.$$

The Lagrange interpolation of T_{n-1} on the zeros y_k's is

$$T_{n-1}(x) = \sum_{k=1}^{n} T_{n-1}(y_k)\frac{\omega_n(x)}{\omega_n'(y_k)(x-y_k)} = \sum_{k=1}^{n} \frac{1}{|\omega_n'(y_k)|}\cdot\frac{\omega_n(x)}{x-y_k}.$$

comparing the leading coefficients of the polynomials on both the sides of the above equation we obtain

$$\gamma_n(X) = \sum_{k=1}^{n} \frac{1}{|\omega_n'(y_k)|} = 2^{n-2}.$$

For the second half of the lemma we see

$$\|\omega_n(X)\| \geq \|2^{1-n}T_n\| = 2^{1-n}$$

and equality holds only for $\omega_n(x) = 2^{1-n}T_n(x)$. □

Lemma 9.1.3. *For every matrix X and every point $\zeta \in \mathbf{C} \setminus [-1,1]$*

$$\limsup_{n\to\infty} \frac{1}{n} \sum_{k=1}^{n} \ln \frac{1}{|\zeta - \tau_{kn}|} \leq \frac{1}{\pi} \int_{-1}^{1} \ln \frac{1}{|\zeta - x|} \frac{dx}{\sqrt{1-x^2}}$$
$$+ \ln \left\{ \limsup_{n\to\infty} \left[\frac{1}{4} \gamma_n(X)^{1/(n-1)} (x_{nn} - x_{1n}) \right] \right\}. \tag{9.1.25}$$

Proof. By (9.1.21) we have

$$|T_{n-1}(\zeta)| \leq \gamma_n(X)|\omega_n(z)| \max_{1\leq k\leq n} \frac{1}{|z - x_k|}. \tag{9.1.26}$$

Denoting by $\rho(\zeta)$ the Euclidean distance of ζ from $[-1,1]$ one has

$$|z - x_k| = \frac{1}{2}|(x_n - x_1)(\zeta - \tau_k)| \geq \frac{1}{2}(x_n - x_1)\rho(\zeta)$$

and hence

$$|T_{n-1}(\zeta)| \leq \gamma_n(X)|\omega_n(z)| \frac{2}{(x_n - x_1)\rho(\zeta)}. \tag{9.1.27}$$

Taking logarithms on both the sides in (9.1.27) and dividing by n yields

$$\frac{1}{n} \ln |T_{n-1}(\zeta)| \leq \frac{1}{n} \ln \gamma_n(X) + \frac{1}{n} \ln |\omega_n(z)| + \frac{1}{n} \ln \frac{2}{(x_n - x_1)\rho(\zeta)}.$$

On the other hand

$$\frac{1}{n} \ln |\omega_n(z)| = \frac{1}{n} \sum_{k=1}^{n} \ln |z - x_k| = \ln \frac{|x_1 - x_n|}{2} + \frac{1}{n} \sum_{k=1}^{n} \ln |\zeta - \tau_k|.$$

Thus

$$\frac{1}{n} \sum_{k=1}^{n} \ln \frac{1}{|\zeta - \tau_k|} \leq \frac{1}{n} \ln \frac{2^{n-2}}{|T_{n-1}(\zeta)|}$$
$$+ \frac{n-1}{n} \ln \left[\frac{1}{4} \gamma_n(X)^{1/(n-1)} (x_n - x_1) \right] + \frac{1}{n} \ln \frac{2}{\rho(\zeta)}. \tag{9.1.28}$$

If $t_{k,n-1}$'s stand for the zeros of T_{n-1} then the expression

$$\frac{\pi}{n-1} \ln \frac{2^{n-2}}{|T_{n-1}(\zeta)|} = \frac{\pi}{n-1} \sum_{k=1}^{n-1} \ln \frac{1}{|\zeta - t_{k,n-1}|}$$

is a Riemann sum of the integral

$$\int_0^{\pi} \ln \frac{1}{|\zeta - \cos\theta|} d\theta = \int_{-1}^{1} \ln \frac{1}{|\zeta - x|} \frac{dx}{\sqrt{1-x^2}}.$$

Thus

$$\lim_{n\to\infty} \frac{\pi}{n-1} \ln \frac{2^{n-2}}{|T_{n-1}(\zeta)|} = \int_{-1}^{1} \ln \frac{1}{|\zeta - x|} \frac{dx}{\sqrt{1-x^2}}. \tag{9.1.29}$$

Hence (9.1.28) implies (9.1.25). □

Lemma 9.1.4. *Let* $X \subset [-1,1]$. *Then the condition*

$$\lim_{n\to\infty} \gamma_n(X)^{1/(n-1)}(x_{nn} - x_{1n}) = 4 \tag{9.1.30}$$

implies that X *is arcsine.*

Proof. It follows from (9.1.25) and (9.1.30) that

$$\limsup_{n\to\infty} \frac{1}{n} \sum_{k=1}^{n} \ln \frac{1}{|\zeta - \tau_{kn}|} \le \frac{1}{\pi} \int_{-1}^{1} \ln \frac{1}{|\zeta - x|} \frac{dx}{\sqrt{1-x^2}}. \tag{9.1.31}$$

Let $Q(x) = \prod_{k=1}^{n}(x - \zeta_k)$ be arbitrary, where $\zeta_k \in \mathbf{C} \setminus [-1,1]$, $k = 1,2,\ldots,n$. Substituting $\zeta = \zeta_k$ into (9.1.31) and adding up, we obtain

$$\limsup_{n\to\infty} \frac{1}{n} \sum_{k=1}^{n} \ln \frac{1}{|Q(\tau_{kn})|} \le \frac{1}{\pi} \int_{-1}^{1} \ln \frac{1}{|Q(x)|} \frac{dx}{\sqrt{1-x^2}}. \tag{9.1.32}$$

If f is a bounded upper semicontinuous function in $[-1,1]$, then there exists a sequence of polynomials $\{Q_j\}$ such that

$$Q_j(x) > Q_{j-1}(x) > \cdots > Q_1(x) > c > 0 \tag{9.1.33}$$

and

$$\lim_{j\to\infty} \ln \frac{1}{Q_j(x)} = f(x). \tag{9.1.34}$$

By (9.1.32) for each $j \in \mathbf{N}$

$$\begin{aligned} \limsup_{n\to\infty} \frac{1}{n} \sum_{k=1}^{n} f(\tau_{kn}) &\le \limsup_{n\to\infty} \frac{1}{n} \sum_{k=1}^{n} \ln \frac{1}{|Q_j(\tau_{kn})|} \\ &\le \frac{1}{\pi} \int_{-1}^{1} \ln \frac{1}{|Q_j(x)|} \frac{dx}{\sqrt{1-x^2}}. \end{aligned}$$

As $j \to \infty$ by dominated convergence it follows from (9.1.33) and (9.1.34) that

$$\limsup_{n\to\infty} \frac{1}{n} \sum_{k=1}^{n} f(\tau_{kn}) \le \frac{1}{\pi} \int_{-1}^{1} f(x) \frac{dx}{\sqrt{1-x^2}}. \tag{9.1.35}$$

Now let $c \in [-1,1]$ be arbitrary. Inserting in (9.1.35) for f the characteristic function of the interval $[c,1]$, we see

$$\limsup_{n\to\infty} \frac{1}{n} \sum_{\tau_{kn}\geq c} 1 \leq \frac{1}{\pi}\int_c^1 \frac{dx}{\sqrt{1-x^2}}. \tag{9.1.36}$$

Similarly, if we choose f as the characteristic function of the interval $[-1,c]$ then we conclude

$$\limsup_{n\to\infty} \frac{1}{n} \sum_{\tau_{kn}\leq c} 1 \leq \frac{1}{\pi}\int_{-1}^c \frac{dx}{\sqrt{1-x^2}}. \tag{9.1.37}$$

So

$$\begin{aligned}\liminf_{n\to\infty} \frac{1}{n} \sum_{\tau_{kn}\geq c} 1 &\geq 1-\limsup_{n\to\infty} \frac{1}{n} \sum_{\tau_{kn}\leq c} 1 \geq 1-\frac{1}{\pi}\int_{-1}^c \frac{dx}{\sqrt{1-x^2}} \\ &= \frac{1}{\pi}\int_c^1 \frac{dx}{\sqrt{1-x^2}} = \frac{1}{\pi}\arccos c,\end{aligned}$$

which together with (9.1.36) gives

$$\lim_{n\to\infty} \frac{1}{n} \sum_{\tau_{kn}\geq c} 1 = \frac{1}{\pi}\arccos c.$$

This proves that X is an arcsine distribution. □

Lemma 9.1.5. *Let $X \subset [-1,1]$. Then Statement (a) of Theorem 9.1.1 implies Statement (g) of Theorem 9.1.1.*

Proof. Statement (a) yields

$$\limsup_{n\to\infty} \gamma_n(X)^{1/(n-1)}(x_{nn}-x_{1n}) \leq 4,$$

which, coupled with (9.1.18), implies (9.1.30) and $-\lim_{n\to\infty} x_{1n} = \lim_{n\to\infty} x_{nn} = 1$. Thus X is arcsine on $[-1,1]$. Namely Statement (g) is satisfied. □

Lemma 9.1.6. *Let $X \subset [-1,1]$. Then Statements (d), (e), (f), (g), and (h) of Theorem 9.1.1 are equivalent to each other.*

Proof. (d)$\Longrightarrow$(e) This follows from the two facts: the relations (9.1.11) and (9.1.12) are the same for $z>1$; both the sides of (9.1.12) are one-valued and regular functions of z on $\mathbf{C}\setminus[-1,1]$.

(e)$\Longrightarrow$(f) Choose the curve $C_\delta : \phi(z) = 1+\delta$, which surrounds the interval $[-1,1]$ for every $\delta>0$. Applying the maximum principle to $\omega_n(z)$ inside C_δ and then letting $\delta \to 0$, we get

$$\limsup_{n\to\infty} \|\omega_n\|^{1/n} \leq \frac{1}{2},$$

which, together with (9.1.24), gives (9.1.13).

(f)$\Longrightarrow$(g) This can be found in [143, Lemma 6.1]. We omit the details.

(g)$\Longrightarrow$(h) Statement (g) means that relation (9.1.15) holds for the characteristic function f of every interval $[c,d] \subset [-1,1]$. Then Statement (h) follows.

(h)$\Longrightarrow$(d) Choosing $f(t) = \ln|z-t|$, which is continuous for every $z \in \mathbf{C} \setminus [-1,1]$, we obtain

$$\lim_{n\to\infty} |\omega_n(z)|^{1/n} = \exp\left\{\frac{1}{\pi}\int_{-1}^{1} \ln|z-x| \frac{dx}{\sqrt{1-x^2}}\right\}.$$

By (9.1.29)

$$\begin{aligned}\frac{1}{\pi}\int_{-1}^{1} \ln|z-x|\frac{dx}{\sqrt{1-x^2}} &= \lim_{n\to\infty} \ln|2^{1-n}T_n(z)|^{1/n} \\ &= \ln\left[\frac{1}{2}\left|z+\sqrt{z^2-1}\right|\right] = \ln\left[\frac{1}{2}|\phi(z)|\right]. \qquad (9.1.38)\end{aligned}$$

Hence relation (9.1.11) follows. □

Lemma 9.1.7. *Let a matrix $X \subset [-1,1]$ have an arcsine distribution. Then the relation*

$$\lim_{n\to\infty} \|\omega_n\|_\Delta^{1/n} = \frac{1}{2} \qquad (9.1.39)$$

holds for every interval $\Delta \subset [-1,1]$.

Proof. Assume that $\Delta = [c,d] \subset [-1,1]$. Using (9.1.11) and then applying (1.5.9) to the polynomial $Q_n(z) = \omega_n(\frac{1}{2}(c+d) + \frac{1}{2}(d-c)z)$ and $z = i\varepsilon$, we have

$$\begin{aligned}\frac{1}{2}\left|\phi\left(\frac{1}{2}(c+d)+\frac{1}{2}(d-c)\varepsilon i\right)\right| &= \lim_{n\to\infty} |Q_n(i\varepsilon)|^{1/n} \\ &\le \lim_{n\to\infty} |T_n(i\varepsilon)|^{1/n} \liminf_{n\to\infty} \|Q_n\|^{1/n} = |\phi(i\varepsilon)| \liminf_{n\to\infty} \|\omega_n\|_\Delta^{1/n}.\end{aligned}$$

Since ϕ is continuous and $|\phi(z)| = 1$ for $z \in [-1,1]$, as $\varepsilon \to 0$ the above inequality yields

$$\liminf_{n\to\infty} \|\omega_n\|_\Delta^{1/n} \ge \frac{1}{2},$$

which, coupled with (9.1.13), gives (9.1.39). □

Lemma 9.1.8. *Let $X \subset [-1,1]$. Then Statements (a), (b), and (c) of Theorem 9.1.1 are equivalent to each other.*

Proof. (a)$\Longrightarrow$(b) It is easy to see that

$$\Lambda_n(X;x) = \gamma_n(X)|\omega_n(X;x)|. \qquad (9.1.40)$$

Since by Lemmas 9.1.5 and 9.1.6 Statement (a) implies Statement (f), Statement (b) follows.

(b)$\Longrightarrow$(c) Let n be fixed and let $\sum_{k=1}^{n} |\ell_k(\xi)| = \|\mathcal{L}_n(X)\|$ hold for some $\xi \in [-1,1]$. If $|\xi - x_k| > 1/(2n^2)$, $k = 1,\ldots,n$, then

$$\|\mathcal{L}_n(X)\| = \sum_{k=1}^{n} \frac{|(\xi-x_k)\ell_k(\xi)|}{|\xi-x_k|} < 2n^2\|\Lambda_n(X)\|. \qquad (9.1.41)$$

If $|\xi - x_j| \le 1/(2n^2)$ holds for some j, $1 \le j \le n$, then by Markov inequality

$$\begin{aligned}\|\mathcal{L}_n(X)\| &= \sum_{k=1}^{n}[\mathrm{sgn}\ \ell_k(\xi)]\ell_k(\xi) = \mathrm{sgn}\ \ell_j(\xi) + \int_{x_j}^{\xi}\left\{\sum_{k=1}^{n}[\mathrm{sgn}\ \ell_k(\xi)]\ell_k(x)\right\}' dx \\ &\le 1 + n^2|\xi - x_j|\ \|\mathcal{L}_n(X)\| \le 1 + \frac{1}{2}\|\mathcal{L}_n(X)\|.\end{aligned}$$

Namely $\|\mathcal{L}_n(X)\| \le 2$. But for every $X \subset [-1,1]$ by (9.1.23) and (9.1.24)

$$\|\Lambda_n(X)\| \ge \frac{1}{2}. \tag{9.1.42}$$

Thus inequality (9.1.41) is also true in this case. On the other hand, clearly

$$\mathcal{L}_n(X;x) \ge \frac{1}{2}\Lambda_n(X;x), \tag{9.1.43}$$

which, coupled with (9.1.41), proves that Statement (b) implies Statement (c).

(c)$\Longrightarrow$(a) By (9.1.24)

$$\|\mathcal{L}_n(X)\| \ge 2^{-n}\sum_{k=1}^{n}\frac{1}{|\omega_n'(x_k)|} = 2^{-n}\gamma_n(X),$$

from which it follows by (9.1.10) that $\limsup_{n\to\infty}\gamma_n(X)^{1/n} \le 2$. Then by (9.1.23) we obtain (9.1.8). □

Lemma 9.1.9. *Let $d\mu$ be a measure supported in $[-1,1]$ and let $2 \le p < \infty$. Then*

$$\lim_{n\to\infty}\gamma_n(X(d\mu,p))^{1/n} = \lim_{n\to\infty}\gamma_n(d\mu,p)^{1/n}, \tag{9.1.44}$$

provided one of the two limits exists, where

$$X(d\mu,p) = \{x_{kn}(d\mu,p) :\ k = 1,2,\ldots,n,\ n = 1,2,\ldots\}.$$

Proof. By definition

$$\gamma_n(X(d\mu,p)) = \sum_{k=1}^{n}\frac{1}{|\omega_n'(d\mu,p;x_{kn}(d\mu,p))|} = \gamma_n(d\mu,p)\sum_{k=1}^{n}\frac{1}{|P_n'(d\mu,p;x_{kn}(d\mu,p))|}. \tag{9.1.45}$$

By (8.3.10) the relation (9.1.44) follows, provided one of the two limits in (9.1.44) exists. □

Lemma 9.1.10. *Let $d\mu$ be a measure supported in $[-1,1]$. Then the inequality*

$$\limsup_{n\to\infty}\left\{\sup_{Q_n\in\mathbf{P}_n\setminus\{0\}}\frac{\|Q_n\|}{\|Q_n\|_{d\mu,p}}\right\}^{1/n} \le 1 \tag{9.1.46}$$

is the same for every p, $0 < p < \infty$.

Moreover, the inequality (9.1.46) is equivalent to that the limit

$$\lim_{n\to\infty}\left\{\frac{\|Q_n\|}{\|Q_n\|_{d\mu,p}}\right\}^{1/n} = 1 \tag{9.1.47}$$

holds for every sequence $\{Q_n \in \mathbf{P}_n \setminus \{0\} :\ n = 1,2,\ldots\}$.

Proof. Let $0 < p < q < \infty$. By Hölder inequality

$$\begin{aligned}
\|Q_n\|_{d\mu,p} &= \left\{\int_{-1}^{1} |Q_n(x)|^p d\mu(x)\right\}^{1/p} \\
&\le \left\{\left[\int_{-1}^{1} |Q_n(x)^q d\mu(x)\right]^{p/q}\left[\int_{-1}^{1} d\mu(x)\right]^{(q-p)/q}\right\}^{1/p} \\
&\le \|Q_n\|_{d\mu,q}\left[\int_{-1}^{1} d\mu(x)\right]^{(q-p)/(pq)},
\end{aligned}$$

from which inequality (9.1.46) implies

$$\limsup_{n\to\infty}\left\{\sup_{Q_n\in\mathbf{P}_n\setminus\{0\}}\frac{\|Q_n\|}{\|Q_n\|_{d\mu,q}}\right\}^{1/n} \le 1. \tag{9.1.48}$$

Conversely, we have

$$\|Q_n\|_{d\mu,q}^q = \int_{-1}^{1} |Q_n(x)|^{q-p}|Q_n(x)|^p d\mu(x) \le \|Q_n\|^{q-p}\|Q_n\|_{d\mu,p}^p$$

and hence

$$\frac{\|Q_n\|}{\|Q_n\|_{d\mu,p}} \le \left(\frac{\|Q_n\|}{\|Q_n\|_{d\mu,q}}\right)^{q/p},$$

from which inequality (9.1.48) implies (9.1.46).

To prove the second half of the lemma we observe that

$$\|Q_n\|_{d\mu,p} \le \|Q_n\|\left\{\int_{-1}^{1} d\mu(x)\right\}^{1/p}$$

and hence

$$\liminf_{n\to\infty}\left\{\frac{\|Q_n\|}{\|Q_n\|_{d\mu,p}}\right\}^{1/n} \ge 1. \tag{9.1.49}$$

This proves the equivalence of (9.1.46) and (9.1.47). □

Lemma 9.1.11. *Let $d\mu$ be a regular measure supported in $[-1,1]$. Then inequalities (9.1.46) and (9.1.47) hold.*

Moreover, we have

$$\lim_{n\to\infty}\|Q_n\|^{1/n} = \lim_{n\to\infty}\|Q_n\|_{d\mu,p}^{1/n}, \tag{9.1.50}$$

provided one of the two limits exists.

Proof. By Lemma 9.1.10 it suffices to show that equation (9.1.47) with $p = 2$ holds for every sequence $\{Q_n \in \mathbf{P}_n \setminus \{0\} : n = 1,2,\ldots\}$. To this end we write

$$Q_n(x) = \sum_{k=0}^{n} \alpha_{kn} P_k(d\mu; x).$$

Then

$$\|Q_n\|_{d\mu,2} = \left\{ \sum_{k=0}^{n} \alpha_{kn}^2 \right\}^{1/2}$$

and hence

$$\frac{\|Q_n\|}{\|Q_n\|_{d\mu,2}} = \frac{\|\sum_{k=0}^{n} \alpha_{kn} P_k(d\mu)\|}{\{\sum_{k=0}^{n} \alpha_{kn}^2\}^{1/2}}$$

$$\leq \sum_{k=0}^{n} \|P_k(d\mu)\| \leq (n+1) \sup_{0\leq k\leq n} \|P_k(d\mu)\|.$$

Since $d\mu$ is regular,

$$\limsup_{n\to\infty} \left\{ \frac{\|Q_n\|}{\|Q_n\|_{d\mu,2}} \right\}^{1/n} \leq \limsup_{n\to\infty} \left\{ (n+1) \sup_{0\leq k\leq n} \|P_k(d\mu)\| \right\}^{1/n} = 1,$$

which, coupled with (9.1.49), yields (9.1.47) with $p = 2$.

The second half of the lemma is a direct consequence of the first. □

Proof of Theorem 9.1.1. (a)$\Longleftrightarrow$(b)$\Longleftrightarrow$(c) Apply Lemma 9.1.8.

(a)$\Longrightarrow$(g) Apply Lemma 9.1.5.

(d)$\Longleftrightarrow$(e)$\Longleftrightarrow$(f) $\Longleftrightarrow$(g)$\Longleftrightarrow$(h) Apply Lemma 9.1.6. □

Proof of Theorem 9.1.2. By Theorem 9.1.1 it is enough to show (b)$\Longleftrightarrow$(i) and (c)$\Longleftrightarrow$(j), which may be derived by applying Lemma 9.1.11 to $Q_n = \Lambda_n(X)$ and $Q_n = \mathcal{L}_n(X)$, respectively. □

Remark 9.1.1. In general, Statement (h) of Theorem 9.1.1 does not imply Statement (a) of Theorem 9.1.1. For example, let

$$x_{kn} = \begin{cases} t_{k,n-1}, & 1 \leq k \leq n-1, \\ t_{n-1,n-1} - 2^{-n}, & k = n, \end{cases}$$

where $t_{k,n-1}$'s are the zeros of $T_{n-1}(x)$. Then

$$\|\omega_n(X)\| = \|2^{2-n}(x - x_{nn})T_{n-1}(x)\| \leq 2^{3-n},$$

which, coupled with (9.1.24), implies Statement (h). On the other hand, by Markov inequality

$$|\omega_n'(x_{n-1,n})| = 2^{2-2n}|T_{n-1}'(t_{n-1,n-1})| \leq 2^{2-2n}(n-1)^2$$

and hence

$$\liminf_{n\to\infty} \gamma_n(X)^{1/n} \geq |\omega_n'(x_{n-1,n})|^{-1/n} \geq 4.$$

9.2 Ratio Asymptotics (I)

Given two triangular matrices of nodes ($n \geq 2$)

$$-1 \leq x_{1n} < x_{2n} < \cdots < x_{nn} \leq 1 \tag{9.2.1}$$

and

$$-1 \leq y_{1,n-1} < y_{2,n-1} < \cdots < y_{n-1,n-1} \leq 1, \tag{9.2.2}$$

put

$$\omega_n(x) = \prod_{k=1}^{n}(x - x_{kn}), \quad \Omega_{n-1}(x) = \prod_{k=1}^{n-1}(x - y_{k,n-1}).$$

The main aim of this section is to give conditions such that the ratio asymptotics

$$\lim_{n\to\infty} \frac{\omega_n(z)}{\Omega_{n-1}(z)} = \frac{1}{2}\phi(z) \tag{9.2.3}$$

holds for every $z \in \mathbf{C} \setminus \langle -1, 1\rangle$, where ϕ is defined in (9.1.6) and the symbol $\langle -1, 1\rangle$ here stands for one of the four symbols $[-1, 1]$, $(-1, 1]$, $[-1, 1)$, and $(-1, 1)$.

To state our results we need some notations. Denote by $L[-1, 1]$ the set of complex valued and Riemann integrable functions on $[-1, 1]$. Write $f_1(x) = \ln(1 - x)$ and $f_2(x) = \ln(1 + x)$. Then the first main result in this section is as follows.

Theorem 9.2.1. *Let the assumptions (9.2.1) and (9.2.2) prevail and let*

$$-1 \leq x_{1n} \leq y_{1,n-1} \leq x_{2n} \leq y_{2,n-1} \leq \cdots \leq x_{n-1,n} \leq y_{n-1,n-1} \leq x_{nn} \leq 1. \tag{9.2.4}$$

Then the following statements are equivalent:

(a) The relation (9.2.3) holds for every $z \in \mathbf{C} \setminus [-1, 1]$.

(b) The relation

$$\lim_{n\to\infty} \sum_{k=1}^{n} f(x_{kn}) \frac{\Omega_{n-1}(x_{kn})}{\omega_n'(x_{kn})} = \frac{2}{\pi} \int_{-1}^{1} f(x)\sqrt{1 - x^2}\,dx \tag{9.2.5}$$

holds for every $f \in L[-1, 1]$.

(c) The relation

$$\lim_{n\to\infty} \left[\sum_{k=1}^{n} f(x_{kn}) - \sum_{k=1}^{n-1} f(y_{k,n-1})\right] = \frac{1}{\pi} \int_{-1}^{1} f(x) \frac{dx}{\sqrt{1 - x^2}} \tag{9.2.6}$$

holds for every $f \in C^1[-1, 1]$.

We also give the other three results concerning the ratio asymptotics (9.2.3) including the endpoints of the interval $[-1, 1]$.

Theorem 9.2.2. *Let the assumptions (9.2.1) and (9.2.2) prevail and let*

$$-1 < x_{1n} \leq y_{1,n-1} \leq x_{2n} \leq y_{2,n-1} \leq \cdots \leq x_{n-1,n} \leq y_{n-1,n-1} \leq x_{nn} < 1. \tag{9.2.7}$$

Then the following statements are equivalent:

(a) The relation (9.2.3) holds for every $z \in \mathbf{C} \setminus (-1,1)$.

(b) The relation

$$\lim_{n\to\infty} \sum_{k=1}^{n} f(x_{kn}) \frac{\Omega_{n-1}(x_{kn})}{\omega_n'(x_{kn})(1-x_{kn}^2)} = \frac{2}{\pi} \int_{-1}^{1} f(x) \frac{dx}{\sqrt{1-x^2}} \tag{9.2.8}$$

holds for every $f \in L[-1,1]$.

(c) The relation (9.2.6) holds for every $f \in C^1[-1,1] \cup \{f_1, f_2\}$.

Theorem 9.2.3. *Let the assumptions (9.2.1) and (9.2.2) prevail and let*

$$-1 \le x_{1n} \le y_{1,n-1} \le x_{2n} \le y_{2,n-1} \le \cdots \le x_{n-1,n} \le y_{n-1,n-1} \le x_{nn} < 1. \tag{9.2.9}$$

Then the following statements are equivalent:

(a) The relation (9.2.3) holds for every $z \in \mathbf{C} \setminus [-1,1)$.

(b) The relation

$$\lim_{n\to\infty} \sum_{k=1}^{n} f(x_{kn}) \frac{\Omega_{n-1}(x_{kn})}{\omega_n'(x_{kn})(1-x_{kn})} = \frac{2}{\pi} \int_{-1}^{1} f(x) \sqrt{\frac{1+x}{1-x}} dx \tag{9.2.10}$$

holds for every $f \in L[-1,1]$.

(c) The relation (9.2.6) holds for every $f \in C^1[-1,1] \cup \{f_1\}$.

Theorem 9.2.4. *Let the assumptions (9.2.1) and (9.2.2) prevail and let*

$$-1 < x_{1n} \le y_{1,n-1} \le x_{2n} \le y_{2,n-1} \le \cdots \le x_{n-1,n} \le y_{n-1,n-1} \le x_{nn} \le 1. \tag{9.2.11}$$

Then the following statements are equivalent:

(a) The relation (9.2.3) holds for every $z \in \mathbf{C} \setminus (-1,1]$.

(b) The relation

$$\lim_{n\to\infty} \sum_{k=1}^{n} f(x_{kn}) \frac{\Omega_{n-1}(x_{kn})}{\omega_n'(x_{kn})(1+x_{kn})} = \frac{2}{\pi} \int_{-1}^{1} f(x) \sqrt{\frac{1-x}{1+x}} dx \tag{9.2.12}$$

holds for every $f \in L[-1,1]$.

(c) The relation (9.2.6) holds for every $f \in C^1[-1,1] \cup \{f_2\}$.

To prove these theorems we need some auxiliary lemmas.

Lemma 9.2.1. [81, p. 45] *We have*

$$\frac{1}{\pi} \int_{-1}^{1} \frac{\ln(z-x)}{\sqrt{1-x^2}} dx = \ln\left[\frac{1}{2}\phi(z)\right], \quad z \in \mathbf{C} \setminus (-1,1). \tag{9.2.13}$$

Lemma 9.2.2. [65, p. 62] *We have*

$$\frac{1}{\pi} \int_{-1}^{1} \frac{\sqrt{1-x^2}}{z-x} dx = \phi(z)^{-1}, \quad z \in \mathbf{C} \setminus (-1,1). \tag{9.2.14}$$

Lemma 9.2.3. *Let the assumptions (9.2.1) and (9.2.2) prevail. If the relation (9.2.3) is true for every $z \in \mathbf{C} \setminus [-1,1]$ and if*

$$\sup_n \sum_{k=1}^{n} \left| \frac{\Omega_{n-1}(x_{kn})}{\omega_n'(x_{kn})} \right| = C < +\infty, \tag{9.2.15}$$

then the relation (9.2.5) holds for every $f \in L[-1,1]$.

Conversely, if the relation (9.2.5) is true for every $f \in L[-1,1]$, then the relation (9.2.3) holds for every $z \in \mathbf{C} \setminus [-1,1]$.

Proof. To prove the first conclusion by Banach-Steinhaus theorem it suffices to show that relation (9.2.5) holds for every polynomial, or equivalently, for every monomial x^m, $m = 0,1,\ldots$. To this end using (9.2.14) and

$$\frac{1}{z-x} = \sum_{i=0}^{\infty} z^{-i-1} x^i, \quad |z| > 1$$

we obtain the identity

$$\phi(z)^{-1} = \frac{1}{\pi} \sum_{i=0}^{\infty} z^{-i-1} \int_{-1}^{1} x^i \sqrt{1-x^2}\,dx, \quad |z| > 1. \tag{9.2.16}$$

Further, we need the Lagrange interpolation formula

$$\Omega_{n-1}(z) = \sum_{k=1}^{n} \frac{\Omega_{n-1}(x_{kn})\omega_n(z)}{\omega_n'(x_{kn})(z-x_{kn})},$$

or equivalently,

$$\frac{\Omega_{n-1}(z)}{\omega_n(z)} = \sum_{k=1}^{n} \frac{\Omega_{n-1}(x_{kn})}{\omega_n'(x_{kn})(z-x_{kn})}. \tag{9.2.17}$$

By (9.2.3) we have

$$\begin{aligned} 2\phi(z)^{-1} &= \lim_{n\to\infty} \frac{\Omega_{n-1}(z)}{\omega_n(z)} = \lim_{n\to\infty} \sum_{k=1}^{n} \frac{\Omega_{n-1}(x_{kn})}{\omega_n'(x_{kn})(z-x_{kn})} \\ &= \lim_{n\to\infty} \sum_{k=1}^{n} \frac{\Omega_{n-1}(x_{kn})}{\omega_n'(x_{kn})} \sum_{i=0}^{\infty} z^{-i-1} x_{kn}^i, \end{aligned}$$

which, coupled with (9.2.16), yields

$$\lim_{n\to\infty} \sum_{k=1}^{n} \sum_{i=0}^{\infty} z^{-i-1} \frac{x_{kn}^i \Omega_{n-1}(x_{kn})}{\omega_n'(x_{kn})} = \frac{2}{\pi} \sum_{i=0}^{\infty} z^{-i-1} \int_{-1}^{1} x^i \sqrt{1-x^2}\,dx. \tag{9.2.18}$$

Clearly, by (3.3.3) we have

$$\sum_{k=1}^{n} \frac{\Omega_{n-1}(x_{kn})}{\omega_n'(x_{kn})} = 1 = \frac{2}{\pi} \int_{-1}^{1} \sqrt{1-x^2}\,dx,$$

which shows that relation (9.2.5) holds for the function $f=1$. Now suppose, as an induction hypothesis, that relation (9.2.5) holds for every $f(x)=x^i$, $i\le m$. Then it follows from the hypothesis and (9.2.18) that

$$\lim_{n\to\infty}\sum_{k=1}^{n}\sum_{i=m+1}^{\infty} z^{-i-1}\frac{x_{kn}^i\Omega_{n-1}(x_{kn})}{\omega_n'(x_{kn})}=\frac{2}{\pi}\sum_{i=m+1}^{\infty} z^{-i-1}\int_{-1}^{1}x^i\sqrt{1-x^2}dx$$

and hence, multiplying the factor z^{m+2} on both the sides,

$$\lim_{n\to\infty}\sum_{k=1}^{n}\sum_{i=m+1}^{\infty} z^{m+1-i}\frac{x_{kn}^i\Omega_{n-1}(x_{kn})}{\omega_n'(x_{kn})}=\frac{2}{\pi}\sum_{i=m+1}^{\infty} z^{m+1-i}\int_{-1}^{1}x^i\sqrt{1-x^2}dx. \tag{9.2.19}$$

According to (9.2.15) we get the estimations

$$\begin{aligned}&\left|\sum_{k=1}^{n}\sum_{i=m+2}^{\infty} z^{m+1-i}\frac{x_{kn}^i\Omega_{n-1}(x_{kn})}{\omega_n'(x_{kn})}\right|=\left|\sum_{i=m+2}^{\infty} z^{m+1-i}\sum_{k=1}^{n}\frac{x_{kn}^i\Omega_{n-1}(x_{kn})}{\omega_n'(x_{kn})}\right|\\ \le\ &\sum_{i=m+2}^{\infty}|z|^{m+1-i}\sum_{k=1}^{n}\left|\frac{\Omega_{n-1}(x_{kn})}{\omega_n'(x_{kn})}\right|\le C\sum_{i=m+2}^{\infty}|z|^{m+1-i}=\frac{C}{|z|-1}\end{aligned}$$

and

$$\left|\frac{2}{\pi}\sum_{i=m+2}^{\infty} z^{m+1-i}\int_{-1}^{1}x^i\sqrt{1-x^2}dx\right|\le\frac{2}{\pi}\sum_{i=m+2}^{\infty}|z|^{m+1-i}\int_{-1}^{1}\sqrt{1-x^2}dx=\frac{1}{|z|-1}.$$

Since $|z|$ may be large enough, from (9.2.19) we conclude that relation (9.2.5) holds for $f(x)=x^{m+1}$. By induction this proves that relation (9.2.5) holds for every monomial.

To prove the second conclusion of the lemma inserting $f(x)=1/(z-x)$, $z\in\mathbf{C}\setminus[-1,1]$, into (9.2.5) it follows from (9.2.17) and (9.2.14) that

$$\lim_{n\to\infty}\frac{\Omega_{n-1}(z)}{\omega_n(z)}=\lim_{n\to\infty}\sum_{k=1}^{n}\frac{\Omega_{n-1}(x_{kn})}{\omega_n'(x_{kn})(z-x_{kn})}=\frac{2}{\pi}\int_{-1}^{1}\frac{\sqrt{1-x^2}}{z-x}dx=2\phi(z)^{-1}.\quad\square$$

Remark 9.2.1. The assumption (9.2.15) is necessary for the first conclusion of Lemma 9.2.3 in general. In fact, by Banach-Steinhaus theorem it is enough to show that Statement (a) of Theorem 9.2.1 does not imply (9.2.15). This is the case from the following example.

Let

$$\begin{cases}\omega_n(x)\ =2^{2-n}(x-x_{nn})T_{n-1}(x),\\ \Omega_{n-1}(x)=\omega_{n-1}(x),\end{cases}\tag{9.2.20}$$

where T_{n-1} stands for the $(n-1)$th Chebyshev polynomial of the first kind, $t_{k,n-1}$'s its zeros, and

$$x_{nn}=t_{n-1,n-1}+n^{-n}. \tag{9.2.21}$$

Then for $z\in\mathbf{C}\setminus[-1,1]$

$$\begin{aligned}&\lim_{n\to\infty}\frac{\omega_n(z)}{\omega_{n-1}(z)}=\lim_{n\to\infty}\frac{2^{2-n}(z-x_{nn})T_{n-1}(z)}{2^{3-n}(z-x_{n-1,n-1})T_{n-2}(z)}\\ =\ &\lim_{n\to\infty}\frac{T_{n-1}(z)}{2T_{n-2}(z)}=\frac{1}{2}\phi(z).\end{aligned}$$

Since $x_{nn} - t_{k,n-2} \geq x_{nn} - t_{k+1,n-1}$ for $k = 1, \ldots, n-3$, we have

$$\begin{aligned}
\left|\frac{\omega_{n-1}(x_{nn})}{\omega_n'(x_{nn})}\right| &= \left|\frac{2^{3-n}(x_{nn} - x_{n-1,n-1})T_{n-2}(x_{nn})}{2^{2-n}T_{n-1}(x_{nn})}\right| \\
&\geq \frac{(x_{nn} - x_{n-1,n-1})^2}{x_{nn} - t_{n-1,n-1}} \\
&\geq n^n[t_{n-1,n-1} - t_{n-2,n-2} - (n-1)^{-(n-1)}]^2 \\
&\geq cn^{n-2},
\end{aligned}$$

where $c > 0$ is a certain constant. Thus relation (9.2.15) is violated.

Lemma 9.2.4. *Let the assumptions (9.2.1) and (9.2.2) prevail. If the relation (9.2.3) is true for every $z \in \mathbf{C} \setminus (-1,1)$ and if*

$$\sup_n \sum_{k=1}^{n} \left|\frac{\Omega_{n-1}(x_{kn})}{\omega_n'(x_{kn})(1-x_{kn}^2)}\right| < +\infty, \tag{9.2.22}$$

then the relation (9.2.8) holds for every $f \in L[-1,1]$.

Conversely, if the relation (9.2.8) is true for every $f \in L[-1,1]$, then the relation (9.2.3) holds for every $z \in \mathbf{C} \setminus (-1,1)$.

Proof. Let us prove the first conclusion of the lemma. Since relation (9.2.22) implies (9.2.15), by Lemma 9.2.3 relation (9.2.5) holds for every $f \in L[-1,1]$. Inserting $x = 1$ or $x = -1$ into (9.2.17) give

$$\frac{\Omega_{n-1}(1)}{\omega_n(1)} = \sum_{k=1}^{n} \frac{\Omega_{n-1}(x_{kn})}{\omega_n'(x_{kn})(1-x_{kn})} \tag{9.2.23}$$

or

$$\frac{\Omega_{n-1}(-1)}{\omega_n(-1)} = -\sum_{k=1}^{n} \frac{\Omega_{n-1}(x_{kn})}{\omega_n'(x_{kn})(1+x_{kn})}, \tag{9.2.24}$$

respectively. Hence by (9.2.3)

$$\lim_{n\to\infty} \sum_{k=1}^{n} \frac{\Omega_{n-1}(x_{kn})}{\omega_n'(x_{kn})(1-x_{kn})} = 2 \tag{9.2.25}$$

and

$$\lim_{n\to\infty} \sum_{k=1}^{n} \frac{\Omega_{n-1}(x_{kn})}{\omega_n'(x_{kn})(1+x_{kn})} = 2. \tag{9.2.26}$$

It follows from (9.2.25) and (9.2.26) that

$$\lim_{n\to\infty} \sum_{k=1}^{n} \frac{\Omega_{n-1}(x_{kn})}{\omega_n'(x_{kn})(1-x_{kn}^2)} = 2, \tag{9.2.27}$$

which means that relation (9.2.8) holds for $f = 1$.

Further, since

$$\sum_{k=1}^{n} \frac{x_{kn}\Omega_{n-1}(x_{kn})}{\omega_n'(x_{kn})(1-x_{kn}^2)} = \sum_{k=1}^{n} \frac{\Omega_{n-1}(x_{kn})}{\omega_n'(x_{kn})(1-x_{kn}^2)} - \sum_{k=1}^{n} \frac{\Omega_{n-1}(x_{kn})}{\omega_n'(x_{kn})(1+x_{kn})},$$

by means of (9.2.26) and (9.2.27) we conclude

$$\lim_{n\to\infty}\sum_{k=1}^{n}\frac{x_{kn}\Omega_{n-1}(x_{kn})}{\omega_n'(x_{kn})(1-x_{kn}^2)}=0,$$

which means that relation (9.2.8) holds for $f(x)=x$. Now suppose, as an induction hypothesis, that relation (9.2.8) holds for every $f(x)=x^i$, $i\le m$ $(m\ge 1)$. Using the recurrence relation

$$\sum_{k=1}^{n}\frac{x_{kn}^{m+2}\Omega_{n-1}(x_{kn})}{\omega_n'(x_{kn})(1-x_{kn}^2)}=\sum_{k=1}^{n}\frac{x_{kn}^{m}\Omega_{n-1}(x_{kn})}{\omega_n'(x_{kn})(1-x_{kn}^2)}-\sum_{k=1}^{n}\frac{x_{kn}^{m}\Omega_{n-1}(x_{kn})}{\omega_n'(x_{kn})} \tag{9.2.28}$$

and the hypothesis it follows from (9.2.5) that

$$\begin{aligned}&\lim_{n\to\infty}\sum_{k=1}^{n}\frac{x_{kn}^{m+2}\Omega_{n-1}(x_{kn})}{\omega_n'(x_{kn})(1-x_{kn}^2)}\\ =\;&\frac{2}{\pi}\int_{-1}^{1}\frac{x^m}{\sqrt{1-x^2}}dx-\frac{2}{\pi}\int_{-1}^{1}x^m\sqrt{1-x^2}dx\\ =\;&\frac{2}{\pi}\int_{-1}^{1}\frac{x^{m+2}}{\sqrt{1-x^2}}dx,\end{aligned}$$

which means that relation (9.2.8) holds for $f(x)=x^{m+2}$. By induction this proves that relation (9.2.8) holds for every monomial and hence for every polynomial. By Banach-Steinhaus theorem it follows from (9.2.22) that relation (9.2.8) holds for every $f\in L[-1,1]$.

Let us prove the second conclusion of the lemma. If relation (9.2.8) holds for every $f\in L[-1,1]$, then relation (9.2.5) also holds for every $f\in L[-1,1]$. According to Lemma 9.2.3 relation (9.2.3) holds for every $z\in\mathbf{C}\setminus[-1,1]$.

Meanwhile, relation (9.2.8) with $f(x)=1+x$ becomes

$$\lim_{n\to\infty}\sum_{k=1}^{n}\frac{\Omega_{n-1}(x_{kn})}{\omega_n'(x_{kn})(1-x_{kn})}=\frac{2}{\pi}\int_{-1}^{1}\sqrt{\frac{1+x}{1-x}}dx=2,$$

which, together with (9.2.23), gives

$$\lim_{n\to\infty}\frac{\omega_n(1)}{\Omega_{n-1}(1)}=\frac{1}{2}.$$

This proves (9.2.3) for $z=1$. Similarly, we can prove (9.2.3) for $z=-1$. □

By the same arguments as that of Lemma 9.2.4 we can get the following two lemmas, omitting the details.

Lemma 9.2.5. *Let the assumptions (9.2.1) and (9.2.2) prevail. If the relation (9.2.3) is true for every* $z\in\mathbf{C}\setminus[-1,1)$ *and if*

$$\sup_n\sum_{k=1}^{n}\left|\frac{\Omega_{n-1}(x_{kn})}{\omega_n'(x_{kn})(1-x_{kn})}\right|<+\infty,$$

then the relation (9.2.10) holds for every $f\in L[-1,1]$.

Conversely, if the relation (9.2.10) is true for every $f\in L[-1,1]$, *then the relation (9.2.3) holds for every* $z\in\mathbf{C}\setminus[-1,1)$.

Lemma 9.2.6. *Let the assumptions (9.2.1) and (9.2.2) prevail. If the relation (9.2.3) is true for every* $z \in \mathbf{C} \setminus (-1,1]$ *and if*

$$\sup_n \sum_{k=1}^{n} \left| \frac{\Omega_{n-1}(x_{kn})}{\omega_n'(x_{kn})(1+x_{kn})} \right| < +\infty,$$

then the relation (9.2.12) holds for every $f \in L[-1,1]$.

Conversely, if the relation (9.2.12) is true for every $f \in L[-1,1]$, *then the relation (9.2.3) holds for every* $z \in \mathbf{C} \setminus (-1,1]$.

Lemma 9.2.7. *Let the assumptions (9.2.1) and (9.2.2) prevail. If the relation (9.2.3) is true for every* $z \in \mathbf{C} \setminus [-1,1]$ *and if*

$$\sup_n \sum_{k=1}^{n-1} |x_{kn} - y_{k,n-1}| = C < +\infty, \tag{9.2.29}$$

then the relation (9.2.6) holds for every $f \in C^1[-1,1]$.

Conversely, if the relation (9.2.6) is true for every $f \in C^1[-1,1]$, *then the relation (9.2.3) holds for every* $z \in \mathbf{C} \setminus [-1,1]$.

Proof. To prove the first conclusion of the lemma we use (9.2.3) and (9.2.13) to obtain

$$\lim_{n\to\infty} \left[\sum_{k=1}^{n} \ln(z - x_{kn}) - \sum_{k=1}^{n-1} \ln(z - y_{k,n-1}) \right] = \ln \frac{\phi(z)}{2} = \frac{1}{\pi} \int_{-1}^{1} \frac{\ln(z-x)}{\sqrt{1-x^2}} dx.$$

Let $P(x) = \prod_{k=1}^{M} (x - z_k)$, $z_k \in \mathbf{C} \setminus [-1,1]$. Then the above relation yields

$$\lim_{n\to\infty} \left[\sum_{k=1}^{n} \ln P(x_{kn}) - \sum_{k=1}^{n-1} \ln P(y_{k,n-1}) \right] = \frac{1}{\pi} \int_{-1}^{1} \frac{\ln P(x)}{\sqrt{1-x^2}} dx. \tag{9.2.30}$$

Now let us show that relation (9.2.6) holds for every $f(x) = x^m$, $m = 0, 1, \ldots$. Obviously, relation (9.2.6) holds for $f = 1$. Assume that $m \geq 1$ is fixed. Let $N \in \mathbf{N}_1$ and let

$$P_1(x) = \sum_{k=0}^{N} \frac{x^{km}}{k!}, \quad P_2(x) = \sum_{k=N+1}^{\infty} \frac{x^{km}}{k!}.$$

Clearly, $P_1(x) > 0$, $P_2(x) > 0$, $x \in [-1,1]$, and $e^{x^m} = P_1(x) + P_2(x)$. Given an arbitrary positive number ε, choose $N \in \mathbf{N}_1$ so large that

$$\left\| \frac{P_2}{P_1} \right\| \leq \varepsilon, \quad \left\| \frac{m}{N! P_1} \right\| \leq \varepsilon.$$

For this fixed number N write

$$g(x) = \ln \left[1 + \frac{P_2(x)}{P_1(x)} \right].$$

It is easy to see that

$$g(x) = \ln \frac{P_1(x) + P_2(x)}{P_1(x)} = \ln \frac{e^{x^m}}{P_1(x)} = x^m - \ln P_1(x) \tag{9.2.31}$$

and

$$\|g\| \le \left\|\frac{P_2}{P_1}\right\| \le \varepsilon. \tag{9.2.32}$$

Meanwhile

$$\begin{aligned} g'(x) &= [x^m - \ln P_1(x)]' = mx^{m-1} - \frac{P_1'(x)}{P_1(x)} \\ &= mx^{m-1} - \frac{\sum_{k=1}^N mx^{km-1}/(k-1)!}{P_1(x)} \\ &= mx^{m-1} - \frac{mx^{m-1}[P_1(x) - x^{mN}/N!]}{P_1(x)} \\ &= \frac{mx^{m(N+1)-1}}{N!P_1(x)}. \end{aligned}$$

Hence

$$\|g'\| = \left\|\frac{mx^{m(N+1)-1}}{N!P_1(x)}\right\| \le \left\|\frac{m}{N!P_1}\right\| \le \varepsilon. \tag{9.2.33}$$

Thus by the mean value theorem for derivative it follows from (9.2.29), (9.2.32), and (9.2.33) that

$$\begin{aligned} &\left|\sum_{k=1}^n g(x_{kn}) - \sum_{k=1}^{n-1} g(y_{k,n-1}) - \frac{1}{\pi}\int_{-1}^1 \frac{g(x)}{\sqrt{1-x^2}}dx\right| \\ = &\left|\sum_{k=1}^{n-1} [g(x_{kn}) - g(y_{kn-1})] + g(x_{nn}) - \frac{1}{\pi}\int_{-1}^1 \frac{g(x)}{\sqrt{1-x^2}}dx\right| \\ \le &\|g'\| \sum_{k=1}^{n-1} |x_{kn} - y_{k,n-1}| + 2\|g\| \\ \le &(C+2)\varepsilon. \end{aligned} \tag{9.2.34}$$

On the other hand, by (9.2.30) there is a number n_0 such that for $n > n_0$

$$\left|\sum_{k=1}^n \ln P_1(x_{kn}) - \sum_{k=1}^{n-1} \ln P_1(y_{k,n-1}) - \frac{1}{\pi}\int_{-1}^1 \frac{\ln P_1(x)}{\sqrt{1-x^2}}dx\right| < \varepsilon. \tag{9.2.35}$$

Thus using (9.2.31), (9.2.34), and (9.2.35) we have that for $n > n_0$

$$\left|\sum_{k=1}^n x_{kn}^m - \sum_{k=1}^{n-1} y_{k,n-1}^m - \frac{1}{\pi}\int_{-1}^1 \frac{x^m}{\sqrt{1-x^2}}dx\right| < (C+3)\varepsilon,$$

which proves that the relation (9.2.6) holds for $f(x) = x^m$.

Denote

$$\mathcal{F}_n(f) = \sum_{k=1}^n f(x_{kn}) - \sum_{k=1}^{n-1} f(y_{k,n-1}).$$

Then by the mean value theorem for derivative it follows from (9.2.29) that

$$\begin{aligned}
\|\mathcal{F}_n\| &:= \sup_{\|f\|\le 1,\|f'\|\le 1} |\mathcal{F}_n(f)| \\
&= \sup_{\|f\|\le 1,\|f'\|\le 1} \left|\sum_{k=1}^{n-1}[f(x_{kn})-f(y_{k,n-1})]+f(x_{nn})\right| \\
&= \sup_{\|f\|\le 1,\|f'\|\le 1} \left|\sum_{k=1}^{n-1} f'(\xi_k)(x_{kn}-y_{k,n-1})+f(x_{nn})\right| \\
&\le \sum_{k=1}^{n-1}|x_{kn}-y_{k,n-1}|+1 \\
&\le C+1.
\end{aligned}$$

Then by Banach-Steinhaus theorem relation (9.2.6) holds for every $f\in C^1[-1,1]$.

Let us prove the second conclusion of the lemma. Inserting $f(x)=\ln(z-x)$ into (9.2.6) we get

$$\lim_{n\to\infty}\ln\left[\frac{\omega_n(z)}{\Omega_{n-1}(z)}\right]=\frac{1}{\pi}\int_{-1}^{1}\frac{\ln(z-x)}{\sqrt{1-x^2}}dx.$$

It remains to apply (9.2.13). □

Proof of Theorem 9.2.1. Since relations (9.2.1), (9.2.2), and (9.2.4) imply

$$\frac{\Omega_{n-1}(x_{kn})}{\omega_n'(x_{kn})}\ge 0,\quad k=1,2,\ldots,n,$$

we have by (3.3.3)

$$\sum_{k=1}^{n}\left|\frac{\Omega_{n-1}(x_{kn})}{\omega_n'(x_{kn})}\right|=\sum_{k=1}^{n}\frac{\Omega_{n-1}(x_{kn})}{\omega_n'(x_{kn})}=1. \tag{9.2.36}$$

Meanwhile relation (9.2.4) also yields

$$\sum_{k=1}^{n-1}|x_{kn}-y_{k,n-1}|\le 2. \tag{9.2.37}$$

Then both (9.2.15) and (9.2.29) hold. Hence Theorem 9.2.1 follows from Lemmas 9.2.3 and 9.2.7. □

Proof of Theorem 9.2.2. (a)$\Longrightarrow$(b) In the proof of Lemma 9.2.4 Statement (a) alone implies (9.2.27). By (9.2.1), (9.2.2), (9.2.7), and (9.2.27)

$$\lim_{n\to\infty}\sum_{k=1}^{n}\left|\frac{\Omega_{n-1}(x_{kn})}{\omega_n'(x_{kn})(1-x_{kn}^2)}\right|=\lim_{n\to\infty}\sum_{k=1}^{n}\frac{\Omega_{n-1}(x_{kn})}{\omega_n'(x_{kn})(1-x_{kn}^2)}=2.$$

We conclude that relation (9.2.22) is true. It remains to apply Lemma 9.2.4.

(b)$\Longrightarrow$(a) Apply Lemma 9.2.4.

(a)$\Longleftrightarrow$(c) By Theorem 9.2.1 it is enough to show that relation (9.2.3) holds for $x = 1$ (or $x = -1$) if and only if relation (9.2.6) holds for $f = f_1$ (or $f = f_2$). But this is indeed the case by (9.2.13). □

Proof of Theorem 9.2.3. Use the same argument as that of Theorem 9.2.2 applying Lemma 9.2.5 instead of Lemma 9.2.4. □

Proof of Theorem 9.2.4. Use the same argument as that of Theorem 9.2.2 applying Lemma 9.2.6 instead of Lemma 9.2.4. □

Our investigation of the ratio asymptotics (9.2.3) for general polynomials is motivated by the ratio asymptotics for orthogonal polynomials. Actually we have ([65, Theorems 3.2.3 and 3.2.4, pp. 17–19; Theorems 4.1.12 and 4.1.13, pp. 32–34]; [66, Theorem 10]; [116, Theorem 1]), in which

$$\omega_n(z) = \Omega_n(z) = \frac{1}{\gamma_n(d\mu)} P_n(d\mu; z).$$

Theorem 9.2.5. *Let $d\mu$ be a measure supported in $[-1,1]$ and let $x_{kn} = x_{kn}(d\mu)$. Then the following statements are equivalent:*

(a) We have

$$\lim_{n\to\infty} \frac{\gamma_{n-1}(d\mu)}{\gamma_n(d\mu)} = \frac{1}{2}, \quad \lim_{n\to\infty} \int_{-1}^{1} x P_n(d\mu; x)^2 d\mu(x) = 0. \tag{9.2.38}$$

(b) The relation (9.2.3) holds for every $z \in \mathbf{C} \setminus (-1,1)$.

(c) The relation

$$\lim_{n\to\infty} \sum_{k=1}^{n} f(x_{kn}) \lambda_{kn}(d\mu) P_{n-1}(d\mu; x_{kn})^2 = \frac{2}{\pi} \int_{-1}^{1} f(x) \sqrt{1-x^2}\, dx \tag{9.2.39}$$

holds for every $f \in L[-1,1]$.

(d) The relation

$$\lim_{n\to\infty} \sum_{k=1}^{n} f(x_{kn}) \lambda_{kn}(d\mu) \frac{P_{n-1}(d\mu; x_{kn})^2}{1-x_{kn}^2} = \frac{2}{\pi} \int_{-1}^{1} f(x) \frac{dx}{\sqrt{1-x^2}} \tag{9.2.40}$$

holds for every $f \in L[-1,1]$.

(e) The relation

$$\lim_{n\to\infty} \sum_{k=1}^{n} f(x_{kn}) \lambda_{kn}(d\mu) \frac{P_{n-1}(d\mu; x_{kn})^2}{1-x_{kn}} = \frac{2}{\pi} \int_{-1}^{1} f(x) \sqrt{\frac{1+x}{1-x}}\, dx \tag{9.2.41}$$

holds for every $f \in L[-1,1]$.

(f) The relation

$$\lim_{n\to\infty} \sum_{k=1}^{n} f(x_{kn}) \lambda_{kn}(d\mu) \frac{P_{n-1}(d\mu; x_{kn})^2}{1+x_{kn}} = \frac{2}{\pi} \int_{-1}^{1} f(x) \sqrt{\frac{1-x}{1+x}}\, dx \tag{9.2.42}$$

holds for every $f \in L[-1,1]$.

(g) The relation (9.2.6) holds for every $f \in C^1[-1,1]$.

We point out that there is a close relationship between Theorems 9.2.1–9.2.4 and Theorem 9.2.5. Using (8.2.12) we obtain

$$\lambda_{kn}(d\mu)P_{n-1}(d\mu;x_{kn})^2 = \frac{\gamma_n(d\mu)P_{n-1}(d\mu;x_{kn})}{\gamma_{n-1}(d\mu)P_n'(d\mu;x_{kn})} = \frac{\omega_{n-1}(d\mu;x_{kn})}{\omega_n'(d\mu;x_{kn})} \tag{9.2.43}$$

and hence relations (9.2.39)–(9.2.42) are equivalent to (9.2.5), (9.2.8), (9.2.10), and (9.2.12), respectively. Thus Theorems 9.2.1–9.2.4 for the ratio asymptotics of general polynomials extend and improve Theorem 9.2.5 for that of orthogonal polynomials.

The second motivation of investigation of the ratio asymptotics for general polynomials is related to study of the asymptotics for polynomials with disturbed nodes. We restrict ourselves to the result for polynomials with disturbed nodes corresponding to Theorem 9.2.1 in details, the other results will be left to the reader.

Given other two triangular matrices of nodes ($n \ge 2$)

$$-1 \le \xi_{1n} < \xi_{2n} < \cdots < \xi_{nn} \le 1 \tag{9.2.44}$$

and

$$-1 \le \eta_{1,n-1} < \eta_{2,n-1} < \cdots < \eta_{n-1,n-1} \le 1, \tag{9.2.45}$$

put

$$\pi_n(x) = \prod_{k=1}^{n}(x-\xi_{kn}), \quad \Pi_{n-1}(x) = \prod_{k=1}^{n-1}(x-\eta_{k,n-1}).$$

The last main result in this section is the following

Theorem 9.2.6. *Let the assumptions (9.2.1), (9.2.2), (9.2.4), (9.2.44), and (9.2.45) prevail and let*

$$-1 \le \xi_{1n} \le \eta_{1,n-1} \le \xi_{2n} \le \eta_{2,n-1} \le \cdots \le \xi_{n-1,n} \le \eta_{n-1,n-1} \le \xi_{nn} \le 1. \tag{9.2.46}$$

If one of Statements (a)–(c) of Theorem 9.2.1 is true and if

$$\lim_{n\to\infty}\left[\sum_{k=1}^{n}|\xi_{kn}-x_{kn}| + \sum_{k=1}^{n-1}|\eta_{k,n-1}-y_{k,n-1}|\right] = 0, \tag{9.2.47}$$

then the following statements hold:

(a) The relation

$$\lim_{n\to\infty}\frac{\pi_n(z)}{\Pi_{n-1}(z)} = \frac{1}{2}\phi(z) \tag{9.2.48}$$

holds for every $z \in \mathbf{C}\setminus[-1,1]$.

(b) The relation

$$\lim_{n\to\infty}\sum_{k=1}^{n} f(\xi_{kn})\frac{\Pi_{n-1}(\xi_{kn})}{\pi_n'(\xi_{kn})} = \frac{2}{\pi}\int_{-1}^{1} f(x)\sqrt{1-x^2}dx \tag{9.2.49}$$

holds for every $f \in L[-1,1]$.

(c) The relation

$$\lim_{n\to\infty}\left[\sum_{k=1}^{n} f(\xi_{kn}) - \sum_{k=1}^{n-1} f(\eta_{k,n-1})\right] = \frac{1}{\pi}\int_{-1}^{1} f(x)\frac{dx}{\sqrt{1-x^2}} \tag{9.2.50}$$

holds for every $f \in C^1[-1,1]$.

Proof. According to Theorem 9.2.1 it is sufficient to show that Statement (c) of Theorem 9.2.1 implies Statement (c) here. To this end we write

$$\begin{aligned}
&\left|\sum_{k=1}^{n} f(\xi_{kn}) - \sum_{k=1}^{n-1} f(\eta_{k,n-1}) - \frac{1}{\pi}\int_{-1}^{1} f(x)\frac{dx}{\sqrt{1-x^2}}\right| \\
= &\left|\sum_{k=1}^{n}[f(\xi_{kn}) - f(x_{kn})] + \sum_{k=1}^{n-1}[f(y_{k,n-1}) - f(\eta_{k,n-1})]\right. \\
&\left.+\left[\sum_{k=1}^{n} f(x_{kn}) - \sum_{k=1}^{n-1} f(y_{k,n-1}) - \frac{1}{\pi}\int_{-1}^{1} f(x)\frac{dx}{\sqrt{1-x^2}}\right]\right|.
\end{aligned}$$

Applying the mean value theorem for derivative the above relation is equal to

$$\begin{aligned}
&\left|\sum_{k=1}^{n} f'(q_{kn})(\xi_{kn} - x_{kn}) + \sum_{k=1}^{n-1} f'(r_{k,n-1})(y_{k,n-1} - \eta_{k,n-1})\right. \\
&\left.+\left[\sum_{k=1}^{n} f(x_{kn}) - \sum_{k=1}^{n-1} f(y_{k,n-1}) - \frac{1}{\pi}\int_{-1}^{1} f(x)\frac{dx}{\sqrt{1-x^2}}\right]\right|
\end{aligned}$$

and hence does not exceed

$$\begin{aligned}
&\|f'\|\left[\sum_{k=1}^{n}|\xi_{kn} - x_{kn}| + \sum_{k=1}^{n-1}|y_{k,n-1} - \eta_{k,n-1}|\right] \\
&+\left|\sum_{k=1}^{n} f(x_{kn}) - \sum_{k=1}^{n-1} f(y_{k,n-1}) - \frac{1}{\pi}\int_{-1}^{1} f(x)\frac{dx}{\sqrt{1-x^2}}\right|.
\end{aligned}$$

Then relation (9.2.49) follows from (9.2.47) and (9.2.6). □

Remark 9.2.2. Although Statement (b) of Theorem 9.2.1 is equivalent to Statement (b) of Theorem 9.2.2 in the case of orthogonal polynomials, this equivalence does not remain true for general polynomials, even if the assumption (9.2.7) is valid.

For example, let ω_n and Ω_{n-1} be given in (9.2.20) but here

$$x_{nn} = 1 - n^{-n} \tag{9.2.51}$$

instead of (9.2.21). Clearly the assumption (9.2.7) is valid and hence the relation (9.2.36) holds, which implies (9.2.15). By the same argument as that of Remark 9.2.1 we conclude that Statement (a) of Theorem 9.2.1 is true and hence Statement (b) of Theorem 9.2.1 holds. But

$$\begin{aligned}
\frac{\omega_{n-1}(x_{nn})}{\omega_n'(x_{nn})(1-x_{nn}^2)} &= \frac{2^{3-n}(x_{nn} - x_{n-1,n-1})T_{n-2}(x_{nn})}{2^{2-n}T_{n-1}(x_{nn})(1-x_{nn}^2)} \\
&\geq \frac{x_{nn} - x_{n-1,n-1}}{2(1-x_{nn})} = \frac{(n-1)^{-(n-1)} - n^{-n}}{2n^{-n}} \\
&= \frac{1}{2}\left[n\left(\frac{n}{n-1}\right)^{n-1} - 1\right] \geq \frac{1}{2}(n-1),
\end{aligned}$$

which implies that relation (9.2.8) is not true for $f = 1$.

9.3 Ratio Asymptotics (II)

Given a sequence of monic polynomials

$$\omega_n(x) = x^n + \mu_{1n}x^{n-1} + \cdots + \mu_{n-1,n}x + \mu_{nn}, \quad n = 0, 1, \ldots, \tag{9.3.1}$$

the polynomial $x\omega_{n-1}(x)$ may be written as

$$x\omega_{n-1}(x) = \omega_n(x) + \sum_{i=1}^{n} a_{i,1,n}\omega_{n-i}(x). \tag{9.3.2}$$

A sequence of monic polynomials $\{\omega_n\}_{n=0}^{\infty}$ is said to have an *asymptotic recurrence relation* if the limit

$$a_{i1} := \lim_{n\longrightarrow\infty} a_{i,1,n} \tag{9.3.3}$$

exists for every $i \in \mathbf{N}$. The main aim of this section is to investigate asymptotics for such a sequence of polynomials.

To state our results we need some notations. In general for each $k \in \mathbf{N}_0$ the polynomial $x^k\omega_{n-1}(x)$ may be written as

$$x^k\omega_{n-1}(x) = R_{k-1,n}(x)\omega_n(x) + \sum_{i=1}^{n} a_{i,k,n}\omega_{n-i}(x), \tag{9.3.4}$$

where $R_{k-1,n} \in \mathbf{P}_{k-1}$ and $R_{0n}(x) = 1$. Here we agree $R_{-1,n}(x) = 0,\ a_{0,1,n} = 1,\ a_{i,0,n} = 1$ for $i = 1$, and $a_{i,0,n} = 0$ otherwise. The first main result of this section is the following

Theorem 9.3.1. *The following statements are equivalent:*

(a) The limit (9.3.3) exists for every $i \in \mathbf{N}_0$.

(b) The limit

$$a_{1k} := \lim_{n\longrightarrow\infty} a_{1,k,n} \tag{9.3.5}$$

exists for every $k \in \mathbf{N}_0$.

Moreover, if either Statement (a) or Statement (b) is true, then the recurrence relation

$$a_{1k} = \sum_{i=1}^{k} c_{i,k-1}a_{i1}, \tag{9.3.6}$$

holds for every $k \in \mathbf{N}$, where $c_{0,0} = 0,\ c_{1,0} = 1$, and

$$c_{ij} = \sum_{p=0}^{j+1-i} c_{i-1+p,j-1}a_{p1}, \quad 1 \le i \le j+1. \tag{9.3.7}$$

Particularly interesting is the case when the conditions (9.2.1) are satisfied and when the equality

$$a_{1k} = \int_{-1}^{1} x^k d\mu(x) \tag{9.3.8}$$

holds for every $k \in \mathbf{N}_0$. In this case we can obtain a further conclusion, which is the second main result of this section.

Theorem 9.3.2. *Let $d\mu$ be a measure supported in $[-1,1]$ and let assumption (9.2.1) prevail. Then the following statements are equivalent:*

(a) The condition

$$\sup_{n\geq 1}\sum_{i=1}^{n}\left|\frac{\omega_{n-1}(x_{in})}{\omega_n'(x_{in})}\right| = C < +\infty \tag{9.3.9}$$

is true and relation (9.3.8) holds for every $k \in \mathbf{N}_0$.

(b) Condition (9.3.9) is true and the a_{k1}'s are given by the recurrence relation

$$a_{k1} = a_{1,k} - \sum_{i=1}^{k-1} c_{i,k-1}a_{i1}. \tag{9.3.10}$$

(c) Condition (9.3.9) is true and the relation

$$\lim_{n\to\infty}\sum_{i=1}^{n}\frac{x_{in}^k\omega_{n-1}(x_{in})}{\omega_n'(x_{in})} = \int_{-1}^{1} x^k d\mu(x) \tag{9.3.11}$$

holds for every $k \in \mathbf{N}_0$.

(d) The relation

$$\lim_{n\to\infty}\sum_{i=1}^{n} f(x_{in})\frac{\omega_{n-1}(x_{in})}{\omega_n'(x_{in})} = \int_{-1}^{1} f(x)d\mu(x) \tag{9.3.12}$$

holds for every $f \in C[-1,1]$.

(e) Condition (9.3.9) is true and the relation

$$\lim_{n\to\infty}\frac{\omega_{n-1}(z)}{\omega_n(z)} = \int_{-1}^{1}\frac{d\mu(x)}{z-x} \tag{9.3.13}$$

holds for every $z \in \mathbf{C}\setminus[-1,1]$.

An important case is the case when assumption (9.2.1) is true and

$$-1 \leq x_{1n} \leq x_{1,n-1} \leq x_{2n} \leq x_{2,n-1} \leq \cdots \leq x_{n-1,n} \leq x_{n-1,n-1} \leq x_{nn} \leq 1. \tag{9.3.14}$$

In this case it is clear that

$$\frac{\omega_{n-1}(x_{in})}{\omega_n'(x_{in})} \geq 0, \quad i = 1,2,\ldots,n. \tag{9.3.15}$$

Hence by (3.3.3) it follows from (9.3.15) that

$$\sum_{i=1}^{n}\left|\frac{\omega_{n-1}(x_{in})}{\omega_n'(x_{in})}\right| = \sum_{i=1}^{n}\frac{\omega_{n-1}(x_{in})}{\omega_n'(x_{in})} = 1.$$

That is, condition (9.3.9) must be valid. Then we can formulate the corresponding corollary of Theorems 9.3.2 as follows.

Corollary 9.3.1. *Let $d\mu$ be a measure supported in $[-1,1]$ and let assumptions (9.2.1) and (9.3.14) prevail. Then the following statements are equivalent:*

(a) Relation (9.3.8) holds for every $k \in \mathbf{N}_0$.

(b) The a_{k1}'s are given by (9.3.10).

(c) Relation (9.3.11) holds for every $k \in \mathbf{N}_0$.

(d) Relation (9.3.12) holds for every $f \in C[-1,1]$.

(e) Relation (9.3.13) holds for every $z \in \mathbf{C}\setminus[-1,1]$.

Our investigation of the asymptotics for general polynomials having an asymptotic recurrence relation is motivated by the ratio asymptotics for orthogonal polynomials. In fact, most interesting is the case when the asymptotic recurrence relation

$$\lim_{n\to\infty} a_{i,1,n} = \begin{cases} \frac{1}{4}, & i=2, \\ 0, & i\neq 2, \end{cases} \tag{9.3.16}$$

holds. In this case we have Theorem 9.2.5.

To prove these theorems we need some auxiliary lemmas.

Lemma 9.3.1. *We have the recurrence relations for* $R_{k-1,n}(x)$ *and* $a_{i,k,n}$ $(k\geq 1,\ 1\leq i\leq n)$*:*

$$R_{k-1,n}(x) = xR_{k-2,n}(x) + a_{1,k-1,n} = \sum_{i=0}^{k-1} a_{1,i,n}x^{k-1-i} \tag{9.3.17}$$

and

$$a_{i,k,n} = \sum_{p=1}^{i+1} a_{i+1-p,1,n+1-p}a_{p,k-1,n}. \tag{9.3.18}$$

Proof. We use induction with respect to k. Recalling that $R_{0n}(x)=1$, $R_{-1,n}(x)=0$, and $a_{1,0,n}=1$, it is nothing to prove for $k=1$. Suppose, as an induction hypothesis, that relations (9.3.17) and (9.3.18) hold for some index $k=r\geq 1$. Replacing k by r and i by p in (9.3.4), respectively, and then multiplying the obtained formula with x, we have

$$x^{r+1}\omega_{n-1}(x) = xR_{r-1,n}(x)\omega_n(x) + \sum_{p=1}^{n} a_{p,r,n}x\omega_{n-p}(x).$$

By (9.3.2) we have

$$\begin{aligned} & x^{r+1}\omega_{n-1}(x) \\ = {} & xR_{r-1,n}(x)\omega_n(x) + \sum_{p=1}^{n} a_{p,r,n} \sum_{q=0}^{n+1-p} a_{q,1,n+1-p}\omega_{n+1-p-q}(x) \\ = {} & xR_{r-1,n}(x)\omega_n(x) + \sum_{i=0}^{n}\left[\sum_{p+q-1=i} a_{q,1,n+1-p}a_{p,r,n}\right]\omega_{n-i}(x) \\ = {} & R_{rn}(x)\omega_n(x) + \sum_{i=1}^{n} a_{i,r+1,n}\omega_{n-i}(x), \end{aligned}$$

where

$$R_{rn}(x) = xR_{r-1,n}(x) + a_{1,r,n} = \sum_{i=0}^{r} a_{1,i,n}x^{r-i}$$

and

$$a_{i,r+1,n} = \sum_{p+q-1=i} a_{q,1,n+1-p}a_{p,r,n} = \sum_{p=1}^{i+1} a_{i+1-p,1,n+1-p}a_{p,r,n}.$$

These formulas show that relations (9.3.17) and (9.3.18) hold for $k=r+1$. By induction this proves the lemma. □

Lemma 9.3.2. *For $k \geq 2$ and $1 \leq j < k$ we have the relation*

$$a_{1,k,n} = \sum_{i=1}^{j+1} c_{i,j,n} a_{i,k-j,n}, \tag{9.3.19}$$

where $c_{0,0,n} = 0$, $c_{1,0,n} = 1$, and

$$c_{i,j,n} = \sum_{p=0}^{j+1-i} c_{i-1+p,j-1,n} a_{p,1,n+1-i}, \quad 1 \leq i \leq j+1. \tag{9.3.20}$$

Proof. We use induction with respect to j. Relation (9.3.18) with $i = 1$ reads

$$a_{1,k,n} = a_{1,1,n} a_{1,k-1,n} + a_{2,k-1,n}, \tag{9.3.21}$$

which shows that relations (9.3.19) and (9.3.20) hold for $j = 1$. Suppose now, as an induction hypothesis, that relations (9.3.19) and (9.3.20) hold for some index $j = r \geq 1$, that is,

$$a_{1,k,n} = \sum_{i=1}^{r+1} c_{i,r,n} a_{i,k-r,n} \tag{9.3.22}$$

with

$$c_{i,r,n} = \sum_{p=0}^{r+1-i} c_{i-1+p,r-1,n} a_{p,1,n+1-i}, \quad 1 \leq i \leq r+1.$$

Using (9.3.18) relation (9.3.22) may be written as

$$a_{1,k,n} = \sum_{i=1}^{r+1} c_{i,r,n} \sum_{p=1}^{i+1} a_{i+1-p,1,n+1-p} a_{p,k-r-1,n}.$$

Interchanging the indices i and p, we obtain

$$\begin{aligned} a_{1,k,n} &= \sum_{p=1}^{r+1} c_{p,r,n} \sum_{i=1}^{p+1} a_{p+1-i,1,n+1-i} a_{i,k-r-1,n} \\ &= \sum_{i=1}^{r+2} \left[\sum_{p=i-1}^{r+1} c_{p,r,n} a_{p+1-i,1,n+1-i} \right] a_{i,k-r-1,n} \\ &= \sum_{i=1}^{r+2} c_{i,r+1,n} a_{i,k-r-1,n}, \end{aligned}$$

where

$$c_{i,r+1,n} = \sum_{p=i-1}^{r+1} c_{p,r,n} a_{p+1-i,1,n+1-i} = \sum_{p=0}^{r+2-i} c_{i-1+p,r,n} a_{p,1,n+1-i}.$$

This shows that relations (9.3.19) and (9.3.20) hold for $j = r+1$. By induction this proves the lemma. □

Particularly interesting is the case when $j = k-1$. We formulate the result in this case as follows.

Lemma 9.3.3. *We have the relation*

$$a_{1,k,n} = \sum_{i=1}^{k} c_{i,k-1,n} a_{i,1,n}, \quad k = 1,2,\ldots. \tag{9.3.23}$$

Lemma 9.3.4. *Let p be an arbitrary but fixed positive integer. If limit (9.3.3) exists for every $i \le p$ then limit (9.3.5) exists for every $k \le p$ and may be uniquely determined by (9.3.6) and (9.3.7) from a_{k1}, $1 \le k \le p$.*

Conversely, if limit (9.3.5) exists for every $k \le p$ then limit (9.3.3) exists for every $i \le p$ and may be uniquely determined by (9.3.10) and (9.3.7) from a_{1k}, $1 \le k \le p$.

Proof. Let limit (9.3.3) exist for every $i \le p$. According to (9.3.23) and (9.3.20) we conclude by induction that relations (9.3.6) and (9.3.7) hold.

Conversely, assume that limit (9.3.5) exists for every $k \le p$. It is nothing to prove that limit (9.3.3) exists for $i = 1$, because in this case limits (9.3.3) and (9.3.5) are same. Let $p > 1$ and suppose, as an induction hypothesis, that limit (9.3.3) exists for every $i \le r$ with $r < p$. Then by (9.3.20) the limits

$$c_{ir} = \lim_{n\to\infty} c_{i,r,n}, \quad 1 \le i \le r, \tag{9.3.24}$$

exist. Relation (9.3.23) with $k = r+1$ reads

$$a_{1,r+1,n} = \sum_{i=1}^{r+1} c_{i,r,n} a_{i,1,n},$$

from which we obtain

$$a_{r+1,1,n} = a_{1,r+1,n} - \sum_{i=1}^{r} c_{i,r,n} a_{i,1,n}.$$

Remembering the assumption and the induction hypothesis, it follows from (9.3.24) that the limit

$$a_{r+1,1} = \lim_{n\to\infty} a_{r+1,1,n}$$

exists and

$$a_{r+1,1} = a_{1,r+1} - \sum_{i=1}^{r} c_{i,r} a_{i,1}.$$

This shows that limit (9.3.3) exists for $i = r+1$. By induction this proves that limit (9.3.3) exists for every $i \le p$ and that relations (9.3.10) and (9.3.7) hold. □

Proof of Theorem 9.3.1. We need only to apply Lemma 9.3.4. □

Proof of Theorem 9.3.2. (a)$\Longleftrightarrow$(b) Apply Theorem 9.3.1.

(a)$\Longleftrightarrow$(c) Using Lemma 9.3.1 it follows from (9.3.4) that

$$a_{1,k,n} = \sum_{i=1}^{n} \frac{x_{i,n}^{k} \omega_{n-1}(x_{in})}{\omega_n'(x_{in})}.$$

Hence

$$a_{1k} = \lim_{n\to\infty} a_{1,k,n} = \lim_{n\to\infty} \sum_{i=1}^{n} \frac{x_{in}^{k} \omega_{n-1}(x_{in})}{\omega_n'(x_{in})}$$

provided one of the two limits exists.

(c)$\Longleftrightarrow$(d) Apply Banach-Steinhaus theorem.

(d)$\Longrightarrow$(e) First we observe that by Banach-Steinhaus theorem condition (9.3.9) is valid. To prove the second conclusion we apply Statement (d) to the function $f(x) = 1/(z-x)$, $z \in \mathbf{C} \setminus [-1,1]$. We see that the limit

$$\lim_{n\to\infty} \sum_{i=1}^{n} \frac{\omega_{n-1}(x_{in})}{\omega_n'(x_{in})(z-x_{in})} = \int_{-1}^{1} \frac{d\mu(x)}{z-x} \tag{9.3.25}$$

holds. Relation (9.2.17) with $\Omega_{n-1} = \omega_{n-1}$ becomes

$$\frac{\omega_{n-1}(z)}{\omega_n(z)} = \sum_{i=1}^{n} \frac{\omega_{n-1}(x_{in})}{\omega_n'(x_{in})(z-x_{in})}. \tag{9.3.26}$$

By means of (9.3.26) relation (9.3.25) shows that relation (9.3.13) holds for $z \in \mathbf{C} \setminus [-1,1]$.

(e)$\Longrightarrow$(c) Relations (9.3.13) and (9.3.26) yield (9.3.25). Since the functions

$$f_n(z) = \sum_{i=1}^{n} \frac{\omega_{n-1}(x_{in})}{\omega_n'(x_{in})(z-x_{in})}$$

are analytic and uniformly bounded for $|z| \geq R > 1$, the coefficients of the Laurent expansion of $f_n(z)$ tend to those of the function

$$\int_{-1}^{1} \frac{d\mu(x)}{z-x} = \sum_{k=0}^{\infty} z^{-k-1} \int_{-1}^{1} x^k d\mu(x). \quad \square$$

9.4 Power Asymptotics

Lemma 9.4.1. *Let $d\nu$ be a measure supported in $[-\pi,\pi]$ and $d\nu \in \mathbf{S}$. Let $0 < p < \infty$. If $\rho_n(z) = z^n + a_{n-1}z^{n-1} + \cdots + a_1 z + a_0$ has no zeros in $|z| > 1$ and satisfies*

$$\frac{1}{2\pi} \int_{-\pi}^{\pi} |d_n^{-1/p} \rho_n(e^{i\theta})|^p d\nu(\theta) \leq C, \tag{9.4.1}$$

where $d_n = d_n(d\nu)$ is defined in (1.2.19), then

$$\left| \frac{d_n^{-1/p} \rho_n(z)}{z^n} \right| \leq c(d\nu, C, R, p), \quad |z| \geq R > 1. \tag{9.4.2}$$

Proof. Let us consider the function

$$D(z)\psi_n^*(z)^{p/2} - 1 = [D(0)d_n^{-1/2} - 1] + d_{1n}z + d_{2n}z^2 + \cdots, \tag{9.4.3}$$

where $D(z)$ is defined in (1.2.17),

$$\psi_n(z) = d_n^{-1/p} \rho_n(z), \quad \psi_n^*(z) = z^n \overline{\psi}_n(z^{-1}),$$

and $\psi_n^*(0) = d_n^{-1/p}$. We see that the function (9.4.3) is regular for $|z| < 1$. If $r < 1$, we have

$$\begin{aligned}
&\frac{1}{2\pi}\int_{-\pi}^{\pi}\left|D(re^{i\theta})\psi_n^*(re^{i\theta})^{p/2} - 1\right|^2 d\theta \\
= &\frac{1}{2\pi}\int_{-\pi}^{\pi}\left|D(re^{i\theta})\right|^2\left|\psi_n^*(re^{i\theta})\right|^p d\theta \\
&1 - 2\Re\left\{\frac{1}{2\pi}\int_{-\pi}^{\pi} D(re^{i\theta})\psi_n^*(re^{i\theta})^{p/2} d\theta\right\}.
\end{aligned}$$

The last term is equal to

$$-2\Re\left\{D(0)\psi_n^*(0)^{p/2}\right\} = -2D(0)d_n^{-1/2}.$$

Thus by (1.2.18)

$$\begin{aligned}
&\lim_{r\to 1-0}\frac{1}{2\pi}\int_{-\pi}^{\pi}\left|D(re^{i\theta})\psi_n^*(re^{i\theta})^{p/2} - 1\right|^2 d\theta \\
= &\frac{1}{2\pi}\int_{-\pi}^{\pi}\left|\psi_n^*(e^{i\theta})\right|^p d\mu(\theta) + 1 - 2D(0)d_n^{-1/2} \\
= &\frac{1}{2\pi}\int_{-\pi}^{\pi}\left|\psi_n(e^{i\theta})\right|^p d\mu(\theta) + 1 - 2D(0)d_n^{-1/2} \\
\leq\ &C + 1 - 2D(0)d_n^{-1/2}.
\end{aligned}$$

That is,

$$[D(0)d_n^{-1/2} - 1]^2 + |d_{1n}|^2 + |d_{2n}|^2 + \cdots \leq C + 1 - 2D(0)d_n^{-1/2}.$$

Using Chauchy inequality

$$\begin{aligned}
&|d_{1n}z + d_{2n}z^2 + \cdots|^2 \leq (|d_{1n}|^2 + |d_{2n}|^2 + \cdots)\frac{|z|^2}{1-|z|^2} \\
\leq\ &\frac{|z|^2}{1-|z|^2}[C + 1 - 2D(0)d_n^{-1/2}].
\end{aligned} \tag{9.4.4}$$

By (1.2.20)

$$\lim_{n\to\infty} d_n(d\nu) = G(d\nu) = D(0)^2. \tag{9.4.5}$$

So we have

$$\limsup_{n\to\infty}\left|\sum_{k=1}^{\infty} d_{kn}z^k\right|^2 \leq (C-1)\frac{|z|^2}{1-|z|^2}$$

and hence by (9.4.3) and (9.4.5)

$$\limsup_{n\to\infty}\left|D(z)\psi_n^*(z)^{p/2} - 1\right| \leq (C-1)\frac{|z|^2}{1-|z|^2}.$$

Thus

$$\limsup_{n\to\infty}\left|\bar{D}(z^{-1})[z^{-n}\psi_n(z)]^{p/2} - 1\right| \leq \frac{C-1}{|z|^2-1}, \quad |z| \geq R > 1,$$

or

$$\left|\bar{D}(z^{-1})[z^{-n}d_n^{-1/p}\rho_n(z)]^{p/2}\right| \leq c(C,R), \quad |z| \geq R > 1.$$

This gives (9.4.2). □

Lemma 9.4.2. *Let $d\nu$ be a measure supported in $[-\pi,\pi]$ and $d\nu \in \mathbf{S}$. Let $0 < p < \infty$. If $\rho_n(z) = z^n + a_{n-1}z^{n-1} + \cdots + a_1 z + a_0$ has no zeros in $|z| > 1$ and satisfies*

$$\int_{-\pi}^{\pi} |\rho_n(e^{i\theta})|^p d\nu(\theta) \le C, \tag{9.4.6}$$

then

$$\left|\frac{\rho_n(z)}{z^n}\right| \le c(d\nu, C, R, p), \quad |z| \ge R > 1. \tag{9.4.7}$$

Proof. By (9.4.5) inequality (9.4.6) implies (9.4.1) and inequality (9.4.2) implies (9.4.7). Then we apply Lemma 9.4.1. □

Theorem 9.4.1. *Let $d\mu$ be a measure supported in $[-1,1]$ and $d\mu \in \mathbf{S}$. Let $X \subset [-1,1]$ and $0 < p < \infty$. Then*

$$\int_{-1}^{1} |2^n \omega_n(X;x)|^p d\mu(x) \ge \frac{\pi}{2} G(d\mu). \tag{9.4.8}$$

Moreover, if

$$\int_{-1}^{1} |2^n \omega_n(X;x)|^p d\mu(x) \le C, \tag{9.4.9}$$

then

$$\left|\frac{2^n \omega_n(X;z)}{\phi(z)^n}\right| \le c, \quad z \in \Omega, \tag{9.4.10}$$

whenever $\Omega \subset \mathbf{C} \setminus [-1,1]$ is compact.

Proof. By introducing $Q_n(x) = 2^n \omega_n(X;x)$ and

$$x = \frac{1}{2}\left(z + z^{-1}\right)$$

we get

$$\begin{aligned} Q_n(x) &= \sum_{k=0}^{n} b_k x^k = \sum_{k=0}^{n} 2^{-k} b_k (z + z^{-1})^k \\ &= z^{-n} \sum_{k=0}^{2n} a_k z^k := z^{-n} \rho_{2n}(z). \end{aligned} \tag{9.4.11}$$

It is easy to check that

$$a_k = a_{2n-k}, \quad k = 0, 1, \ldots, n, \quad a_0 = a_{2n} = 1$$

and

$$|Q_n(x)| = |\rho_{2n}(z)|, \quad z = e^{i\theta}.$$

Meanwhile we see that $\rho_{2n}(z) \ne 0$ for $|z| \ne 1$.

By (1.2.21) and (1.2.22)

$$\begin{aligned} &\int_{-\pi}^{\pi} |\rho_{2n}(e^{i\theta})|^p d\nu(\theta) = \int_{-\pi}^{\pi} |Q_n(\cos\theta)|^p |\sin\theta| d\mu(\cos\theta) \\ &= 2\int_{-1}^{1} |Q_n(x)|^p d\mu(x). \end{aligned} \tag{9.4.12}$$

Let $d\nu$ be defined in (1.2.21). By (1.2.19) we have $d_n(d\nu) \geq d_{n+1}(d\nu)$, which by (1.2.20), (1.2.15), and (1.2.24) gives

$$d_{2n}(d\nu) \geq G(d\nu) \geq \frac{1}{2}G(d\mu). \tag{9.4.13}$$

Then inequality (9.4.8) follows from (9.4.12) and (9.4.13). Applying Lemma 9.4.2 and using (9.4.9) and (9.4.12) yield

$$\left|\frac{\rho_{2n}(z)}{z^{2n}}\right| \leq c, \quad |z| \geq R > 1.$$

By virtue of (9.4.11) we have

$$\left|\frac{Q_n(x)}{z^n}\right| \leq c, \quad |z| \geq R > 1,$$

from which (9.4.10) follows at once. □

Theorem 9.4.2. *Let $X \subset [-1,1]$. Then the following statements are equivalent:*
(a) The inequalities

$$c_1 \leq \left|\frac{2^n \omega_n(X;z)}{\phi(z)^n}\right| \leq c_2, \quad z \in \Omega \tag{9.4.14}$$

hold for every compact set Ω in $\mathbf{C} \setminus [-1,1]$.
(b) The inequalities

$$c_1 \leq \left|\frac{\Lambda_n(X;z)}{\phi(z)^n}\right| \leq c_2, \quad z \in \Omega \tag{9.4.15}$$

hold for every compact set Ω in $\mathbf{C} \setminus [-1,1]$.
(c) We have

$$\left|\sum_{k=1}^{n} f(x_{kn}) - \frac{n}{\pi}\int_{-1}^{1} f(x)\frac{dx}{\sqrt{1-x^2}}\right| \leq c\|f\|_\Delta, \tag{9.4.16}$$

whenever f is analytic in an open set $\Delta \supset [-1,1]$, here

$$\|f\|_\Delta = \sup_{z\in\Delta}|f(z)|.$$

Proof. (a)$\Longrightarrow$(c) Let $\Delta \supset [-1,1]$ be an open set. Choose a closed curve Γ in Δ surrounding $[-1,1]$ so that dist$(\Gamma,[-1,1]) > 0$ and put $\Omega = \Gamma$. Then, for an arbitrary sequence $\{\varepsilon_n\}$ satisfying $\varepsilon_n \to 0$ as $n \to \infty$, it follows from (9.4.14) that the limit

$$\lim_{n\to\infty} \frac{\varepsilon_n 2^n \omega_n(z)}{\phi(z)^n} = 0$$

holds uniformly on Ω. Thus by differentiation we conclude that the limit

$$\lim_{n\to\infty} \frac{\varepsilon_n 2^n \omega_n(z)}{\phi(z)^n}\left\{\frac{\omega_n'(z)}{\omega_n(z)} - \frac{n}{\sqrt{z^2-1}}\right\} = 0$$

holds uniformly on Ω. Since $\{\varepsilon_n\}$ is arbitrary, by (9.4.14) the function

$$R_n(z) := \frac{\omega'_n(z)}{\omega_n(z)} - \frac{n}{\sqrt{z^2-1}}$$

satisfies $|R_n(z)| \leq c,\ z \in \Omega$. Then for every f analytic in Δ

$$\frac{1}{2\pi i}\int_\Gamma f(z)\left\{\frac{\omega'_n(z)}{\omega_n(z)} - \frac{n}{\sqrt{z^2-1}}\right\}dz = \frac{1}{2\pi i}\int_\Gamma f(z)R_n(z)dz.$$

Applying the residue theorem one has

$$\frac{1}{2\pi i}\int_\Gamma f(z)\frac{\omega'_n(z)}{\omega_n(z)}dz = \sum_{k=1}^n f(x_{kn}).$$

Meanwhile

$$\frac{n}{2\pi i}\int_\Gamma \frac{f(z)}{\sqrt{z^2-1}}dz = \frac{n}{\pi}\int_{-1}^1 f(x)\frac{dx}{\sqrt{1-x^2}}$$

and

$$\left|\frac{1}{2\pi i}\int_\Gamma f(z)R_n(z)dz\right| \leq c\|f\|_\Delta.$$

Then (9.4.16) follows at once.

(c)$\Longrightarrow$(a) Let $\Omega \subset \mathbf{C}\setminus[-1,1]$ be an arbitrary compact set. Choose an open set Δ so that $\Delta \supset [-1,1]$ and dist$(\Delta,\Omega) > 0$, which is possible, since Ω is compact and $\Omega \cap [-1,1] = \phi$. Then $f(x) = \ln(z-x)$ with $z \in \Omega$ is analytic in Δ. For this function inequality (9.4.16) gives

$$\left|\ln\omega_n(z) - \frac{n}{\pi}\int_{-1}^1 \frac{\ln(z-x)dx}{\sqrt{1-x^2}}\right| \leq c\|f\|_\Delta \leq c, \quad z \in \Omega,$$

which by (9.2.14) yields

$$\left|\ln\frac{2^n\omega_n(X,z)}{\phi(z)^n}\right| \leq c, \quad z \in \Omega.$$

This gives (9.4.14) provided we put $c_1 = e^{-c_0}$ and $c_2 = e^{c_0}$.

Similarly, if we replace 2^n by $\gamma_n(X)$ then by the same argument as above we can show the equivalence of Statements (b) and (c). □

Theorem 9.4.3. *Let $X \subset [-1,1]$. Then inequality*

$$\left|\frac{\Lambda_n(X;z)}{\phi(z)^n}\right| \geq c, \quad z \in \Omega, \tag{9.4.17}$$

holds for every compact set Ω in $\mathbf{C}\setminus[-1,1]$.

Meanwhile, if

$$\frac{\gamma_n(X)}{2^n} \leq c, \tag{9.4.18}$$

then inequality

$$\left|\frac{2^n\omega_n(X;z)}{\phi(z)^n}\right| \geq c, \quad z \in \Omega, \tag{9.4.19}$$

holds for every compact set Ω in $\mathbf{C}\setminus[-1,1]$.

Proof. Using the Lagrange interpolation formula based on the nodes $x_1,\ldots,x_n$, the $(n-1)$th Chebyshev polynomial of the first kind $T_{n-1}(z)$, $z\in\Omega$, can be expressed as

$$T_{n-1}(z)=\sum_{k=1}^{n}T_{n-1}(x_k)\ell_k(z)=\sum_{k=1}^{n}T_{n-1}(x_k)\frac{\omega_n(z)}{\omega_n'(x_k)(z-x_k)}.$$

Then

$$|T_{n-1}(z)|\le|\Lambda_n(z)|\max_{1\le k\le n}\frac{1}{|z-x_k|}\le|\Lambda_n(z)|d(z), \tag{9.4.20}$$

where $d(z)=1/\rho(z)$ and $\rho(z)$ is the Euclidean distance of z from $[-1,1]$. Taking logarithms on both the sides in (9.4.20) yields

$$\ln|T_{n-1}(z)|\le\ln\Lambda_n(z)+\ln d(z)\le\ln|\Lambda_n(z)|+c,$$

or in another form,

$$\left|\frac{\Lambda_n(z)}{T_{n-1}(z)}\right|\ge c,\quad z\in\Omega.$$

From the formula [50, p. 239]

$$\lim_{n\to\infty}\frac{T_n(z)}{\phi(z)^n}=\frac{1}{2},\quad z\in\mathbf{C}\setminus[-1,1], \tag{9.4.21}$$

inequality (9.4.17) follows at once.

Meanwhile, by (9.1.40) inequality (9.4.19) is an immediate consequence of (9.4.17) and (9.4.18). □

Theorem 9.4.4. *Let $X\subset[-1,1]$ and let $\Omega\subset\mathbf{C}\setminus[-1,1]$ be compact. Then the inequality*

$$\left|\frac{\Lambda_n(X;z)}{\phi(z)^n}\right|\le c,\quad z\in\Omega, \tag{9.4.22}$$

holds if and only if both (9.4.10) and (9.4.18) hold.

Proof. Inequality (9.4.22) by (9.4.17) is equivalent to (9.4.15) and hence by Theorem 9.4.2 to (9.4.14), which by (9.1.40) and (9.1.23) is equivalent to inequalities (9.4.10) and (9.4.18). □

9.5 Notes

9.5.1. Lemmas 9.1.1, 9.1.3, and 9.1.4 for $X=X(d\mu)$ are due to Erdős and Freud [19]. These results may be easily extended to an arbitrary matrix X. Lemma 9.1.9 is also due to Erdős and Freud [19]. Lemma 9.1.10 was given by Saff and Totik [80] and Lemma 9.1.11 by Li, Saff, and Sha [48]. Other results in Section 9.1 may be found in [94].

9.5.2. The main results in Section 9.2 are due to Shi [120].

9.5.3. The work of Section 9.3 follows the same lines as in Nevai's work about orthogonal polynomials [65].

9.5.4. The proof of Lemma 9.4.1 follows the line of the proof of Theorem 12.1.1 in [137, p. 295]. The proof of Theorem 9.4.2 follows the line of [144].

Chapter 10

Asymptotics of Power Orthogonal Polynomials

10.1 Nth Root Asymptotics

The aim of this section is to investigate the nth root asymptotics of the m orthogonal polynomials $P_n(d\mu, m; x)$ on $[-1, 1]$ as $n \to \infty$.

Theorem 10.1.1. *Let a measure $d\mu$ supported in $[-1, 1]$ satisfy that $|\{x \in [-1, 1] : \mu'(x) > 0\}| > 0$ and let $1 \le m < \infty$. Then the following statements are equivalent.*

(a) We have

$$\lim_{n \to \infty} \gamma_n(d\mu, m)^{1/n} = 2. \tag{10.1.1}$$

(b) We have

$$\lim_{n \to \infty} \gamma_n(d\mu)^{1/n} = 2. \tag{10.1.2}$$

(c) The relation

$$\lim_{n \to \infty} |P_n(d\mu, m; z)|^{1/n} = |\phi(z)| \tag{10.1.3}$$

holds for every $z \in \mathbf{C} \setminus [-1, 1]$.

(d) The relation

$$\lim_{n \to \infty} P_n(d\mu, m; z)^{1/n} = \phi(z) \tag{10.1.4}$$

holds for every $z \in \mathbf{C} \setminus [-1, 1]$.

(e) We have

$$\lim_{n \to \infty} \|P_n(d\mu, m)\|^{1/n} = 1. \tag{10.1.5}$$

(f) The relation

$$\lim_{n \to \infty} |\omega_n(d\mu, m; z)|^{1/n} = \frac{1}{2} |\phi(z)| \tag{10.1.6}$$

holds for every $z \in \mathbf{C} \setminus [-1, 1]$.

(g) The relation

$$\lim_{n \to \infty} \omega_n(d\mu, m; z)^{1/n} = \frac{1}{2} \phi(z) \tag{10.1.7}$$

holds for every $z \in \mathbf{C} \setminus [-1, 1]$.

(h) We have

$$\lim_{n\to\infty} \|\omega_n(d\mu,m)\|^{1/n} = \frac{1}{2}. \tag{10.1.8}$$

(i) The relation

$$\lim_{n\to\infty} \frac{N_n(d\mu,m;c,d)}{n} = \frac{1}{\pi}\int_c^d \frac{dx}{\sqrt{1-x^2}} \tag{10.1.9}$$

holds for every interval $[c,d] \subset [-1,1]$.

(j) The relation

$$\lim_{n\to\infty} \frac{1}{n}\sum_{k=1}^{n} f(x_{kn}(d\mu,m)) = \frac{1}{\pi}\int_c^d f(x)\frac{dx}{\sqrt{1-x^2}} \tag{10.1.10}$$

holds for every $f \in C[-1,1]$.

(k) We have

$$\lim_{n\to\infty} \left\{ \frac{\|Q_n\|}{\|Q_n\|_{d\mu,m}} \right\}^{1/n} = 1 \tag{10.1.11}$$

holds for every sequence $\{Q_n\}$, *where* $Q_n \in \mathbf{P}_n \setminus \{0\}$, $n \in \mathbf{N}$.

Remark 10.1.1. The equivalence of Statements (a) and (b) shows that each statement of the theorem, except Statement (b), is "m-independent". For example, relation (10.1.5) for each m, $1 \le m < \infty$, is equivalent to

$$\lim_{n\to\infty} \|P_n(d\mu)\|^{1/n} = 1.$$

To prove this theorem we need two auxiliary lemmas.

Lemma 10.1.1. [51, Lemma 3.2] *For every polynomial* $Q_n \in \mathbf{P}_n$, *every interval* $\Delta \subset [-1,1]$, *and every number* ε, $0 < \varepsilon < 1$, *there exists a measurable subset* Δ_ε *of* Δ *with* $|\Delta_\varepsilon| \ge \lambda(\varepsilon)|\Delta|$, *where* $\lambda(\varepsilon) = \frac{1}{4}\varepsilon^2 - \frac{1}{16}\varepsilon^4$, *such that, for every* $x \in \Delta_\varepsilon$, *we have*

$$|Q_n(x)| > (1-\varepsilon)^n \|Q_n\|_\Delta. \tag{10.1.12}$$

Lemma 10.1.2. *Let* $d\mu$ *be a measure supported in* $[-1,1]$. *Let* $n \ge p \ge 1$ *and* $0 < m < \infty$. *Then*

$$\frac{\gamma_n(d\mu,m)}{\gamma_{n-p}(d\mu,m)} \ge 2^{p-1}. \tag{10.1.13}$$

In particular

$$\frac{\gamma_n(d\mu,m)}{2^n} \ge \frac{1}{2}\left[\int_{-1}^{1} d\mu(x)\right]^{-1/m}. \tag{10.1.14}$$

Moreover, we have

$$\liminf_{n\to\infty} \gamma_n(d\mu,m)^{1/n} \ge 2. \tag{10.1.15}$$

Proof. By definition

$$\begin{aligned}
\gamma_n(d\mu,m)^{-1} &= \left\{\int_{-1}^{1} |\omega_n(d\mu,m;x)|^m d\mu(x)\right\}^{1/m} \\
&\leq \left\{\int_{-1}^{1} |2^{1-p}T_p(x)\omega_{n-p}(d\mu,m;x)|^m d\mu(x)\right\}^{1/m} \\
&\leq 2^{1-p}\left\{\int_{-1}^{1} |\omega_{n-p}(d\mu,m;x)|^m d\mu(x)\right\}^{1/m} \\
&= 2^{1-p}\gamma_{n-p}(d\mu,m)^{-1},
\end{aligned}$$

which is equivalent to (10.1.13).

The inequality (10.1.13) with $p=n$ reads

$$\frac{\gamma_n(d\mu,m)}{\gamma_0(d\mu,m)} \geq 2^{n-1},$$

which yields

$$\frac{\gamma_n(d\mu,m)}{2^n} \geq \frac{1}{2}\gamma_0(d\mu,m) = \frac{1}{2}\left[\int_{-1}^{1} d\mu(x)\right]^{-1/m}.$$

This proves (10.1.14). Inequality (10.1.15) is a direct consequence of (10.1.14). □

Proof of Theorem 10.1.1. We will give the proof along the following implications (a)⟹(b)⟹(k)⟹(e) ⟹(h)⟹ (g)⟹(f)⟹(j)⟹(i) ⟹(a) and (e)⟹(c)⟹(d) ⟹(e).

(a)⟹(b) We distinguish the cases when $m \geq 2$ and $1 \leq m < 2$.

Case 1. $m \geq 2$. Applying characterization theorems of L^p $(1 \leq p < \infty)$ best approximation [128, Theorem 1.7, p. 46, Theorem 1.11, p. 56] we observe that for the measure $d\nu_n(x) = |\omega_n(d\mu,m;x)|^{m-1}d\mu(x)$ we have

$$\omega_n(d\mu,m) = \omega_n(d\nu_n,1)$$

(see Section 8.2). Using this equality we have

$$\begin{aligned}
\gamma_n(d\mu,m)^{-m} &= \int_{-1}^{1} |\omega_n(d\mu,m;x)|^m d\mu(x) \\
&= \int_{-1}^{1} |\omega_n(d\mu,m;x)| d\nu_n(x) \\
&\leq \int_{-1}^{1} |\omega_n(d\mu;x)|\cdot|\omega_n(d\mu,m;x)|^{m-1}d\mu(x).
\end{aligned}$$

Applying Schwarz inequality we get

$$\begin{aligned}
\gamma_n(d\mu,m)^{-m} &\leq \left[\int_{-1}^{1} |\omega_n(d\mu;x)|^2 d\mu(x)\right]^{1/2}\left[\int_{-1}^{1} |\omega_n(d\mu,m;x)|^{2(m-1)}d\mu(x)\right]^{1/2} \\
&\leq \gamma_n(d\mu)^{-1}\|\omega_n(d\mu,m)\|^{m-1}\left[\int_{-1}^{1} d\mu(x)\right]^{1/2}. \qquad (10.1.16)
\end{aligned}$$

On the other hand, in this case by Theorem 9.1.1 and (9.1.44) relation (10.1.1) implies (10.1.8), which, coupled with (10.1.16), yields

$$\begin{aligned}\limsup_{n\to\infty}\gamma_n(d\mu)^{1/n} &\le \limsup_{n\to\infty}\left\{\gamma_n(d\mu,m)^m\|\omega_n(d\mu,m)\|^{m-1}\left[\int_{-1}^1 d\mu(x)\right]^{1/2}\right\}^{1/n}\\ &= 2^m 2^{-(m-1)} = 2.\end{aligned}$$

Hence by (10.1.15) we get (10.1.2).

Case 2. $1 \le m < 2$. By definition and by Hölder inequality

$$\begin{aligned}\gamma_n(d\mu,m)^{-m} &= \int_{-1}^1 |\omega_n(d\mu,m;x)|^m d\mu(x)\\ &\le \int_{-1}^1 |\omega_n(d\mu;x)|^m d\mu(x)\\ &\le \left[\int_{-1}^1 |\omega_n(d\mu;x)|^2 d\mu(x)\right]^{m/2}\left[\int_{-1}^1 d\mu(x)\right]^{(2-m)/2}\\ &= \gamma_n(d\mu)^{-m}\left[\int_{-1}^1 d\mu(x)\right]^{(2-m)/2}.\end{aligned}$$

Thus

$$\gamma_n(d\mu) \le \gamma_n(d\mu,m)\left[\int_{-1}^1 d\mu(x)\right]^{(2-m)/(2m)}, \quad m<2, \tag{10.1.17}$$

and hence

$$\limsup_{n\to\infty}\gamma_n(d\mu)^{1/n} \le 2.$$

Again using (10.1.15) we get (10.1.2).

(b)$\Longrightarrow$(k) Apply Lemma 9.1.11.

(k)$\Longrightarrow$(e) Inserting $Q_n = P_n(d\mu,m)$ into (10.1.11) we obtain (10.1.5).

(e)$\Longrightarrow$(h) By (10.1.15) and (10.1.5)

$$\limsup_{n\to\infty}\|\omega_n(d\mu,m)\|^{1/n} = \limsup_{n\to\infty}\|\gamma_n(d\mu,m)^{-1}P_n(d\mu,m)\|^{1/n} \le \frac{1}{2}$$

which, together with (9.1.24), gives (10.1.8).

(h)$\Longrightarrow$ (g)$\Longrightarrow$(f)$\Longrightarrow$(j)$\Longrightarrow$(i) Apply Theorem 9.1.1.

(i)$\Longrightarrow$(a) We use the ideas of Erdős and Freud [19]. Choose a small constant $c>0$ so that the subset

$$E_c = E_c(d\mu) = \{x\in[-1,1]:\ \mu'(x)\ge c\}$$

has a positive measure. Given an arbitrary number ε, $0<\varepsilon<1$, for each δ,

$$0<\delta<\frac{1}{2}\lambda(\varepsilon), \tag{10.1.18}$$

where $\lambda(\varepsilon)$ is given in Lemma 10.1.1, there exists an interval $I_\delta \subset [-1,1]$ such that

$$|I_\delta \cap E_c| \ge (1-\delta)|I_\delta|. \tag{10.1.19}$$

By means of Lemma 9.1.7 we conclude that

$$\lim_{n\to\infty} \|\omega_n(d\mu,m)\|_{I_\delta}^{1/n} = \frac{1}{2},$$

or equivalently, for n large enough

$$\|\omega_n(d\mu,m)\|_{I_\delta} \geq (1-\varepsilon)^n 2^{-n}.$$

Now applying Lemma 10.1.1 we can choose a subset Δ_δ of I_δ so that

$$|\Delta_\delta| \geq \lambda(\varepsilon)|I_\delta| \tag{10.1.20}$$

and

$$|\omega_n(d\mu,m;x)| \geq (1-\varepsilon)^n \|\omega_n(d\mu,m)\|_{I_\delta} \geq (1-\varepsilon)^{2n} 2^{-n}, \quad x \in \Delta_\delta. \tag{10.1.21}$$

By definition we see that

$$\Delta_\delta \setminus E_c \subset I_\delta \setminus E_c = I_\delta \setminus (I_\delta \cap E_c),$$

which by (10.1.19) and (10.1.18) implies

$$\begin{aligned} |\Delta_\delta \setminus E_c| &\leq |I_\delta \setminus (I_\delta \cap E_c)| = |I_\delta| - |I_\delta \cap E_c| \\ &\leq |I_\delta| - (1-\delta)|I_\delta| = \delta |I_\delta| \leq \frac{1}{2}\lambda(\varepsilon)|I_\delta|. \end{aligned}$$

Hence for the subset $E_\varepsilon = \Delta_\delta \cap E_c$ by (10.1.20) we have

$$\begin{aligned} |E_\varepsilon| &= |\Delta_\delta \cap E_c| = |\Delta_\delta \setminus (\Delta_\delta \setminus E_c)| = |\Delta_\delta| - |\Delta_\delta \setminus E_c| \\ &\geq \lambda(\varepsilon)|I_\delta| - \frac{1}{2}\lambda(\varepsilon)|I_\delta| = \frac{1}{2}\lambda(\varepsilon)|I_\delta|. \end{aligned} \tag{10.1.22}$$

We also see that

$$\mu'(x) \geq c, \quad x \in E_\varepsilon. \tag{10.1.23}$$

Thus for n large enough by the definition of $\gamma_n(d\mu,m)$ it follows from (10.1.21)–(10.1.23) that

$$\begin{aligned} \gamma_n(d\mu,m)^{-m} &= \int_{-1}^{1} |\omega_n(d\mu,m;x)|^m d\mu(x) \\ &\geq \int_{-1}^{1} |\omega_n(d\mu,m;x)|^m \mu'(x)dx \geq \int_{E_\varepsilon} |\omega_n(d\mu,m;x)|^m \mu'(x)dx \\ &\geq c|E_\varepsilon|(1-\varepsilon)^{2mn}2^{-mn} \geq \frac{c}{2}\lambda(\varepsilon)|I_\delta|(1-\varepsilon)^{2mn}2^{-mn}. \end{aligned}$$

Therefore

$$\limsup_{n\to\infty} \gamma_n(d\mu,m)^{1/n} \leq 2(1-\varepsilon)^{-2}.$$

Since $\varepsilon > 0$ is arbitrary, we obtain

$$\limsup_{n\to\infty} \gamma_n(d\mu,m)^{1/n} \leq 2,$$

which, coupled with (10.1.15), gives Statement (a).

(e)$\Longrightarrow$(c) We have shown that Statement (e) implies Statements (a) and (f), which gives Statement (c).

(c)$\Longrightarrow$(d)$\Longrightarrow$(e) By the same arguments as that of Lemma 9.1.6 we can prove these implications. □

10.2 Ratio Asymptotics

Theorem 10.2.1. *Let $d\mu$ be a measure supported in $[-1,1]$ and let $0 < m < \infty$. Then*

$$\limsup_{n\to\infty} \frac{\gamma_n(d\mu,m)}{\gamma_{n-1}(d\mu,m)} \geq 2. \tag{10.2.1}$$

Proof. Suppose to the contrary that

$$\limsup_{n\to\infty} \frac{\gamma_n(d\mu,m)}{\gamma_{n-1}(d\mu,m)} < 2.$$

Then for n large enough, say, $n \geq n_0$,

$$\frac{\gamma_n(d\mu,m)}{\gamma_{n-1}(d\mu,m)} \leq c < 2,$$

which implies that

$$\gamma_n(d\mu,m) \leq c\gamma_{n-1}(d\mu,m) \leq c^{n-n_0}\gamma_{n_0}(d\mu,m), \quad n \geq n_0.$$

So

$$\limsup_{n\to\infty} \gamma_n(d\mu,m)^{1/n} \leq c < 2,$$

a contradiction to (10.1.15). This contradiction proves (10.2.1). □

The expansions (9.3.2) and (9.3.4) with $\omega_n = \omega_n(d\mu,m)$ read

$$x\omega_{n-1}(d\mu,m;x) = \omega_n(d\mu,m;x) + \sum_{i=1}^{n} a_{i,1,n}\omega_{n-i}(d\mu,m;x) \tag{10.2.2}$$

and

$$x^k\omega_{n-1}(d\mu,m;x) = R_{k-1,n}(x)\omega_n(d\mu,m;x) + \sum_{i=1}^{n} a_{i,k,n}\omega_{n-i}(d\mu,m;x), \tag{10.2.3}$$

respectively, where $R_{k-1,n} \in \mathbf{P}_{k-1}$ and $R_{0n}(x) = 1$. Here we also agree $R_{-1,n}(x) = 0$, $a_{0,1,n} = 1$, $a_{i,0,n} = 1$ for $i = 1$, and $a_{i,0,n} = 0$ otherwise. Then Corollary 9.3.1 with replacing $d\mu$ by $d\nu$ becomes

Theorem 10.2.2. *Let $d\mu$ and $d\nu$ be measures supported in $[-1,1]$ and $1 \leq m < \infty$. Then the following statements are equivalent ($x_{in} = x_{in}(d\mu,m)$).*

(a) The relation

$$\lim_{n\to\infty} a_{1,k,n} = \int_{-1}^{1} x^k d\nu(x) \tag{10.2.4}$$

holds for every $k \in \mathbf{N}_0$.

(b) The a_{k1}'s are given by the recurrence relation (9.3.10).

(c) The relation

$$\lim_{n\to\infty} \sum_{i=1}^{n} \frac{x_{in}^k\omega_{n-1}(d\mu,m;x_{in})}{\omega_n'(d\mu,m;x_{in})} = \int_{-1}^{1} x^k d\nu(x) \tag{10.2.5}$$

holds for every $k \in \mathbf{N}_0$.

(d) The relation

$$\lim_{n\to\infty} \sum_{i=1}^{n} f(x_{in}) \frac{\omega_{n-1}(d\mu, m; x_{in})}{\omega_n'(d\mu, m; x_{in})} = \int_{-1}^{1} f(x) d\nu(x) \tag{10.2.6}$$

holds for every $f \in C[-1,1]$.

(e) The relation

$$\lim_{n\to\infty} \frac{\omega_{n-1}(d\mu, m; z)}{\omega_n(d\mu, m; z)} = \int_{-1}^{1} \frac{d\nu(x)}{z-x} \tag{10.2.7}$$

holds for every $z \in \mathbf{C} \setminus [-1,1]$.

More interesting is the case when the asymptotic recurrence relation (9.3.16) holds. In this case Theorems 10.2.2 and 9.2.1 give

Theorem 10.2.3. *Let* $d\mu$ *be a measure supported in* $[-1,1]$ *and* $1 \le m < \infty$. *Then the following statements are equivalent* ($x_{in} = x_{in}(d\mu, m)$).

(a) The relation (9.3.16) holds.

(b) The relation

$$\lim_{n\to\infty} a_{1,k,n} = \frac{2}{\pi} \int_{-1}^{1} x^k \sqrt{1-x^2}\, dx \tag{10.2.8}$$

holds for every $k \in \mathbf{N}_0$.

(c) The relation

$$\lim_{n\to\infty} \sum_{i=1}^{n} \frac{x_{in}^k \omega_{n-1}(d\mu, m; x_{in})}{\omega_n'(d\mu, m; x_{in})} = \frac{2}{\pi} \int_{-1}^{1} x^k \sqrt{1-x^2}\, dx \tag{10.2.9}$$

holds for every $k \in \mathbf{N}_0$.

(d) The relation

$$\lim_{n\to\infty} \sum_{i=1}^{n} f(x_{in}) \frac{\omega_{n-1}(d\mu, m; x_{in})}{\omega_n'(d\mu, m; x_{in})} = \frac{2}{\pi} \int_{-1}^{1} f(x) \sqrt{1-x^2}\, dx \tag{10.2.10}$$

holds for every $f \in C[-1,1]$.

(e) The relation

$$\lim_{n\to\infty} \frac{\omega_n(d\mu, m; z)}{\omega_{n-1}(d\mu, m; z)} = \frac{1}{2} \phi(z). \tag{10.2.11}$$

holds for every $z \in \mathbf{C} \setminus [-1,1]$.

(f) The relation

$$\lim_{n\to\infty} \left[\sum_{i=1}^{n} f(x_{in}) - \sum_{i=1}^{n-1} f(x_{i,n-1}) \right] = \frac{1}{\pi} \int_{-1}^{1} f(x) \frac{dx}{\sqrt{1-x^2}} \tag{10.2.12}$$

holds for every $f \in C^1[-1,1]$.

Proof. By Theorem 10.2.2 the measure $d\nu$ is uniquely determined by the a_{1k}'s or equivalently by the a_{k1}'s. It is easy to check that relation (9.3.16) holds for the measure $d\nu(x) = \frac{2}{\pi}\sqrt{1-x^2}\,dx$. It remains to apply Theorems 10.2.2 and 9.2.1. □

10.3 Power Asymptotics

Let for a weight w supported in $[-1,1]$ and $0<m<\infty$

$$\sigma_m = 2\left\{\frac{\Gamma(\frac{1}{2})\Gamma(\frac{m+1}{2})}{2\Gamma(\frac{m}{2}+1)}\right\}^{1/m}, \tag{10.3.1}$$

$$E_n(w,m) = \|\omega_n(w,m)\|_{w,m} = \gamma_n(w,m)^{-1}, \tag{10.3.2}$$

and

$$\bar{u}_m(\theta) = [|\sin\theta| w(\cos\theta)]^{1/m}. \tag{10.3.3}$$

The first main result in this section is based on an explicit formula of Bernstein for $\gamma_n(w,m)$ and $\omega_n(w,m)$ for a special weight w (see, say, [52, Lemma 2.1]).

Lemma 10.3.1. *Given $q\in\mathbf{N}$ let $S\in\mathbf{P}_{2q}$ satisfy that $S(x)>0,\ x\in(-1,1)$, possibly with simple zeros at ±1, and let*

$$R(x) = \left[\frac{1-x^2}{S(x)}\right]^{1/2} \tag{10.3.4}$$

and for $0<m<\infty$

$$R_m(x) = (1-x^2)^{-1/2}R(x)^m. \tag{10.3.5}$$

Let $n\geq q$ and with $x=\cos\theta,\ z=e^{i\theta},\ \theta\in[0,\pi]$,

$$\tau_n(x) = 2^{-n-1/m}G(R_m)^{1/m}[z^{-n}D(R(\cos\theta);z)^{-2}+z^nD(R(\cos\theta);z^{-1})^{-2}], \tag{10.3.6}$$

where the functions G and D are defined in (1.2.24) and (1.2.17), respectively. Then

$$\frac{\gamma_n(R_m,m)}{2^n} \geq \sigma_m^{-1}G(R_m)^{-1/m}, \quad 0<m<1, \tag{10.3.7}$$

$$\frac{\gamma_n(R_m,m)}{2^n} = \sigma_m^{-1}G(R_m)^{-1/m}, \quad 1\leq m<\infty, \tag{10.3.8}$$

$$\|\tau_n\|_{R_m,m} = 2^{-n}\sigma_m G(R_m)^{1/m}, \quad 0<m<\infty, \tag{10.3.9}$$

$$\omega_n(R_m,m) = \tau_n, \quad 1\leq m<\infty, \tag{10.3.10}$$

$$|\tau_n(x)R(x)| \leq 2^{-n+1-1/m}G(R_m)^{1/m}, \quad x\in[-1,1], \quad 1\leq m<\infty, \tag{10.3.11}$$

and for $z\in\mathbf{C}\setminus[-1,1]$ and $1\leq m<\infty$

$$\left|\frac{\tau_n(z)}{2^{-n-1/m}G(R_m)^{1/m}\phi(z)^nD(R(\cos\theta);\phi(z)^{-1})^{-2}}-1\right| \leq |\phi(z)|^{2q-2n-2}. \tag{10.3.12}$$

Theorem 10.3.1. *Let w be a weight supported in $[-1,1]$ and $0<m<\infty$. Then*

$$\liminf_{n\to\infty}\frac{\gamma_n(w,m)}{2^n} \geq \sigma_m^{-1}G(w)^{-1/m}. \tag{10.3.13}$$

Here we agree $G(w)=0$ if the integral in (1.2.24) diverges to $-\infty$.

Proof. Let R and R_m be given in (10.3.4) and (10.3.5), respectively, satisfying the conditions of Lemma 10.3.1. Let $w_s(x) = \max\{w(x), s\}$, $x \in [-1,1]$, $s \in (0,1)$. Let a measurable set $\Delta \subset [-1,1]$ satisfy $\|w\|_\Delta < \infty$ and let $E = [-1,1] \setminus \Delta$. By definition we have for $n \geq q$

$$\begin{aligned} \gamma_n(w,m)^{-m} &= E_n(w,m)^m \leq E_n(w_s,m)^m \\ &\leq \|\tau_n\|_{w_s,m}^m = \|\tau_n\|_{L_{w_s}^m(\Delta)}^m + \|\tau_n\|_{L_{w_s}^m(E)}^m. \end{aligned}$$

By (10.3.9)

$$\|\tau_n\|_{L_{w_s}^m(\Delta)}^m \leq \|\tau_n\|_{L_{R_m}^m(\Delta)}^m \|R_m^{-1} w_\varepsilon\|_\Delta \leq (2^{-n}\sigma_m)^m G(R_m) \|R_m^{-1} w_\varepsilon\|_\Delta$$

and by (10.3.11)

$$\|\tau_n\|_{L_{w_s}^m(E)}^m \leq (2^{-n+1-1/m})^m G(R_m) \|R^{-1}\|_{L_{w_\varepsilon}^m(E)}^m.$$

Using these estimations we obtain

$$\gamma_n(w,m)^{-m} \leq (2^{-n}\sigma_m)^m G(R_m) \left[\|R_m^{-1} w_\varepsilon\|_\Delta + 2^{m-1} \sigma_m^{-m} \|R^{-1}\|_{L_{w_\varepsilon}^m(E)}^m \right]. \tag{10.3.14}$$

Assume that $g \in C[-1,1]$ and $g(x) > 0$, $x \in [-1,1]$. Then there exists a sequence $\{r_n\}$, $r_n \in \mathbf{P}_n$, $r_n(x) > 0$, $x \in [-1,1]$, $n \in \mathbf{N}$, such that $\lim_{n\to\infty} \|r_n - g\| = 0$. Choose $S(x) = (1-x^2) r_n(x)^2$. Then relations (10.3.4) and (10.3.5) become

$$R(x) = r_n(x)^{-1}. \tag{10.3.15}$$

and

$$R_m(x) = (1-x^2)^{-1/2} r_n(x)^{-m}, \tag{10.3.16}$$

respectively. It follows from (10.3.14)–(10.3.16) that

$$\begin{aligned} &\limsup_{n\to\infty} \left[\gamma_n(w,m) 2^{-n}\right]^{-m} \leq \sigma_m^m G((1-x^2)^{-1/2} g(x)^{-m}) \\ &\times \left[\|(1-x^2)^{1/2} g(x)^m w_s(x)\|_\Delta + 2^{m-1} \sigma_m^{-m} \|g\|_{L_{w_s}^m(E)}^m \right]. \end{aligned} \tag{10.3.17}$$

We can further claim that inequality (10.3.17) holds for any measurable function g which is bounded above and below by positive constants. In fact, for such a function g, first for example we choose for each $N \in \mathbf{N}$

$$g_N(x) = \left\{ \int_{[x-1/N, x+1/N] \cap [-1,1]} dt \right\}^{-1} \int_{[x-1/N, x+1/N] \cap [-1,1]} g(t)\,dt$$

which is bounded above and below by positive constants and satisfies that the limit

$$\lim_{N\to\infty} g_N(x) = g(x) \tag{10.3.18}$$

holds almost everywhere in $[-1,1]$.

Next, we choose a measurable set $\Delta' \subset \Delta$ such that $|\Delta \setminus \Delta'|$ is small enough and the limit (10.3.18) holds uniformly on Δ'. Write $E' = [-1,1] \setminus \Delta'$. Since both g_N and g are bounded above and below by positive constants independent of n, inequality (10.3.17) gives for each N

$$\limsup_{n\to\infty}\left[\gamma_n(w,m)2^{-n}\right]^{-m} \le \sigma_m^m G((1-x^2)^{-1/2}g_N(x)^{-m}) \times\left[\|(1-x^2)^{1/2}g_N(x)^m w_s(x)\|_{\Delta'}+2^{m-1}\sigma_m^{-m}\|g_N\|^m_{L^m_{w_s}(E')}\right]. \quad (10.3.19)$$

Hence as $N\to\infty$ we obtain

$$\limsup_{n\to\infty}\left[\gamma_n(w,m)2^{-n}\right]^{-m} \le \sigma_m^m G((1-x^2)^{-1/2}g(x)^{-m}) \times\left[\|(1-x^2)^{1/2}g(x)^m w_s(x)\|_{\Delta'}+2^{m-1}\sigma_m^{-m}\|g\|^m_{L^m_{w_s}(E')}\right].$$

Since $|\Delta\setminus\Delta'|$ may be as small as possible, by Dominated convergence theorem inequality (10.3.17) holds for any measurable function g which is bounded above and below by positive constants. This proves our claim.

We can choose the interval Δ to omit small intervals containing the end points ± 1 so that the function $(1-x^2)^{1/2}w_s(x)$ is bounded above and below by positive constants on Δ. Thus the same is valid for the function

$$g(x)=\begin{cases}(1-x^2)^{-1/(2m)}w_s(x)^{-1/m}, & x\in\Delta,\\ 1, & x\in E.\end{cases}$$

Denote by χ_Δ and χ_E the characteristic functions of Δ and E, respectively. By (10.3.17)

$$\begin{aligned}&\limsup_{n\to\infty}\left[\gamma_n(w,m)2^{-n}\right]^{-m}\\ \le\ &\sigma_m^m G\left(w_s(x)\chi_\Delta(x)+(1-x^2)^{-1/2}\chi_E(x)\right)\left[1+2^{m-1}\sigma_m^{-m}\|1\|^m_{L^m_{w_s}(E)}\right].\end{aligned}$$

If Δ is chosen so that $|E|$ is small enough, then the above inequality gives

$$\limsup_{n\to\infty}[\gamma_n(w,m)2^{-n}]^{-m}\le\sigma_m^m G(w_s).$$

Finally, it is easy to see that

$$\ln w_s^{-1}=(\ln w_s^{-1})_+-(\ln w_s)_+.$$

Noticing that $(\ln w_s(\cos\theta))_+$ is bounded above by the integrable function $(\ln(1+w_s(\cos\theta)))_+$ for $s<1$, by Monotone convergence theorem

$$\lim_{s\to 0+}\int_0^\pi(\ln w_s(\cos\theta))_+\,d\theta=\int_0^\pi(\ln w(\cos\theta))_+\,d\theta$$

and

$$\lim_{s\to 0+}\int_0^\pi\left(\ln w_s(\cos\theta)^{-1}\right)_+d\theta=\int_0^\pi\left(\ln w(\cos\theta)^{-1}\right)_+d\theta.$$

Therefore inequality (10.3.13) follows. □

Lemma 10.3.2. *Let a weight w supported in $[-1,1]$ satisfy $w^{-1} \in L^q[-1,1]$ for every $q < \infty$ and let $1 < m < \infty$. Then*

$$\limsup_{n\to\infty} \frac{\gamma_n(w,m)}{2^n} \le \sigma_m^{-1} G(w)^{-1/m}. \tag{10.3.20}$$

Moreover, inequality (10.3.20) remains true for $m = 1$ provided $w \in C[-1,1]$ and $w(x) > 0$, $x \in [-1,1]$.

Proof. First let $1 < m < \infty$. Choose $q > s > 1$ so that $s \le m$ and $1/q + 1/s = 1$. By Hölder inequality it is easy to check that if $g^{-1} \in L^{mq/s}$ and $gh \in L^m$, then

$$\|gh\|_{L^m} \ge \|h\|_{L^{m/s}} \|g^{-1}\|_{L^{mq/s}}^{-1} \tag{10.3.21}$$

(see [52, Formula (2.17)]). Let $R_{m/s}$ be given in (10.3.5) satisfying the conditions of Lemma 10.3.1. Inequality (10.3.21) with $h = \omega_n(w,m) R_{m/s}^{s/m}$ and $g = R_{m/s}^{-s/m} w^{1/m}$ by (10.3.8) yields

$$\begin{aligned}\gamma_n(w,m)^{-1} &\ge \gamma_n(R_{m/s}, m/s)^{-1} \|R_{m/s}^{s/m} w^{-1/m}\|_{L^{mq/s}}^{-1}\\ &= 2^{-n}\sigma_{m/s} G(R_{m/s})^{1/m} \|R_{m/s}^{s/m} w^{-1/m}\|_{L^{mq/s}}^{-1}.\end{aligned}$$

We choose $S(x) = r(x)^2 > 0$, $x \in [-1,1]$, and r to uniformly approximate a function $g \in C[-1,1]$ with $g(x) > 0$, $x \in [-1,1]$. Then

$$\begin{aligned}&\liminf_{n\to\infty} \gamma_n(w,m)^{-1} 2^n\\ &\ge \sigma_{m/s} G\left[(1-x^2)^{(m-s)/(2s)} g(x)^{-m/s}\right]^{1/m}\\ &\quad \times \|(1-x^2)^{(m-s)/(2m)} g(x)^{-1} w(x)^{-1/m}\|_{L^{mq/s}}^{-1}.\end{aligned}$$

Further we choose g to approximate the function $(1-x^2)^{(m-s)/(2m)} w(x)^{-1/m}$. Then the above inequality gives

$$\liminf_{n\to\infty} \gamma_n(w,m)^{-1} 2^n \ge \sigma_{m/s} G[w^{1/s}]^{1/m} \|1\|_{L^{mq/s}}^{-1}.$$

But

$$\lim_{q\to\infty} \|1\|_{L^{mq/s}} = \lim_{q\to\infty} 2^{s/(mq)} = 1$$

and $\lim_{q\to\infty} s = 1$. Thus $\lim_{q\to\infty} \sigma_{m/s} = \sigma_m$ and inequality (10.3.20) follows.

Next, it is simple to prove the lemma for the case when $m = 1$. In fact, in this case $w \in C[-1,1]$ and $w(x) > 0$, $x \in [-1,1]$. So we can choose a sequence $\{r_n\}$ of polynomials such that $\lim_{n\to\infty} \|r_n w - 1\| = 0$. Then we choose an suitable function R_1 and obtain (10.3.20). □

Lemma 10.3.3. *Let w be a weight supported in $[-1,1]$ and $2 \le m < \infty$. Assume that $Q_n(z) = a_n z^n + a_{n-1} z^{n-1} + \cdots + a_1 z + a_0$. Then for $z \in \mathbf{C} \setminus [-1,1]$*

$$\begin{aligned}&\left|\frac{Q_n(z)}{P_n(w,m;z)} - 1\right|\\ &\le \left|\frac{a_n}{\gamma_n(w,m)} - 1\right| + \rho(z)^{-1}\left[\|Q_n\|_{w,m}^2 - \frac{a_n^2}{\gamma_n(w,m)^2}\right]^{1/2},\end{aligned} \tag{10.3.22}$$

where $\rho(z)$ denotes the Euclidean distance of z from $[-1,1]$.

Proof. The polynomial

$$Q(x) = Q_n(x) - \frac{a_n}{\gamma_n(w,m)} P_n(w,m;x) \tag{10.3.23}$$

is of degree at most $n-1$ and may be written as

$$Q(x) = \sum_{k=1}^{n} Q(x_{kn}) \ell_{kn}(w_n;x)$$

where $w_n = |P_n(w,m)|^{m-2} w$ and $x_{kn} = x_{kn}(w,m)$. By (10.3.23), (8.2.11), and (8.3.7) from the above formula it follows that

$$\begin{aligned} \left| \frac{Q_n(z)}{P_n(w,m;z)} - \frac{a_n}{\gamma_n(w,m)} \right| &= \left| \frac{Q(z)}{P_n(w,m;z)} \right| \\ &\le \sum_{k=1}^{n} \lambda_{kn}(w_n) |P_{n-1}(w_n;x_{kn}) Q(x_{kn})| \cdot |z - x_{kn}|^{-1}. \end{aligned}$$

Applying Schwarz inequality we get

$$\begin{aligned} & \left| \frac{Q_n(z)}{P_n(w,m;z)} - \frac{a_n}{\gamma_n(w,m)} \right| \\ \le \;& \rho(z)^{-1} \left\{ \sum_{k=1}^{n} \lambda_{kn}(w_n) P_{n-1}(w_n;x_{kn})^2 \right\}^{1/2} \left\{ \sum_{k=1}^{n} \lambda_{kn}(w_n) Q(x_{kn})^2 \right\}^{1/2} \\ = \;& \rho(z)^{-1} \left\{ \sum_{k=1}^{n} \lambda_{kn}(w_n) Q(x_{kn})^2 \right\}^{1/2} \\ = \;& \rho(z)^{-1} \left\{ \int_{-1}^{1} Q_n(x)^2 w_n(x) dx \right\}^{1/2}. \end{aligned} \tag{10.3.24}$$

Further, by (10.3.23) and Hölder inequality we have

$$\begin{aligned} & \int_{-1}^{1} Q(x)^2 w_n(x) dx \\ = \;& \int_{-1}^{1} Q_n(x)^2 w_n(x) dx - \frac{2a_n}{\gamma_n(w,m)} \int_{-1}^{1} Q_n(x) P_n(w,m;x) w_n(x) dx + \frac{a_n^2}{\gamma_n(w,m)^2} \\ = \;& \int_{-1}^{1} Q_n(x)^2 |P_n(w,m;x)|^{m-2} w(x) dx - \frac{a_n^2}{\gamma_n(w,m)^2} \\ \le \;& \left\{ \int_{-1}^{1} |Q_n(x)|^m w(x) dx \right\}^{2/m} \left\{ \int_{-1}^{1} |P_n(w,m;x)|^m w(x) dx \right\}^{1-2/m} - \frac{a_n^2}{\gamma_n(w,m)^2} \\ = \;& \|Q_n\|_{w,m}^2 - \frac{a_n^2}{\gamma_n(w,m)^2}, \end{aligned}$$

which by (10.3.24) gives (10.3.22). □

Denote

$$\Psi(x) = \left[\left(\frac{x^m + 1}{2} \right)^{1/(m-1)} - 1 \right]^{(m-1)/m}, \quad x \ge 1, \quad m > 1. \tag{10.3.25}$$

Lemma 10.3.4. *Let $1 < m < 2 < q$ satisfy $1/m + 1/q = 1$. Let w and u be weights supported in $[-1,1]$ which satisfy $uw^{-1} \in L^q[-1,1]$. Let Q_n and $\rho(z)$ be defined in Lemma 10.3.3. Then for $z \in \mathbf{C} \setminus [-1,1]$*

$$\begin{aligned} & \left| \frac{Q_n(z)}{P_n(w,m;z)} - 1 \right| \\ \leq & \left| \frac{a_n}{\gamma_n(w,m)} - 1 \right| + 2\rho(z)^{-1} \left| \frac{a_n P_n(u;z)}{\gamma_n(w,m) P_n(w,m;z)} \right| \Psi\left(\frac{\gamma_n(w,m)}{a_n} \|Q_n\|_{w,m} \right) \\ & \times \left\{ \|P_{n-1}(u) u w^{-1/m}\|_{L^q} + \rho(z)^{-1} \|P_n(u) u w^{-1/m}\|_{L^q} \right\}. \end{aligned} \tag{10.3.26}$$

Proof. We need Clarkson inequality [52] for $1 < m < 2$

$$\|f+g\|_{L^m}^{m/(m-1)} + \|f-g\|_{L^m}^{m/(m-1)} \leq 2\{\|f\|_{L^m}^m + \|g\|_{L^m}^m\}^{1/(m-1)}.$$

Put $f = a_n^{-1} Q_n w^{1/m}$ and $g = \omega_n(w,m) w^{1/m}$. We have

$$\|f+g\|_{L^m} = 2\|(f+g)/2\|_{L^m} \geq 2E_n(w,m)$$

and by Clarkson inequality

$$\begin{aligned} & \|a_n^{-1} Q_n - \omega_n(w,m)\|_{w,m}^{m/(m-1)} \\ = & \|(a_n^{-1} Q_n - \omega_n(w,m)) w^{1/m}\|_{L^m}^{m/(m-1)} \\ \leq & 2\left\{\|a_n^{-1} Q_n\|_{w,m}^m + E_n(w,m)^m\right\}^{1/(m-1)} - \{2E_n(w,m)\}^{m/(m-1)}. \end{aligned}$$

Using the symbol (10.3.25) the above inequality may be written as

$$\|a_n^{-1} Q_n - \omega_n(w,m)\|_{w,m} \leq 2E_n(w,m) \Psi\left(\frac{\gamma_n(w,m)}{a_n} \|Q_n\|_{w,m} \right). \tag{10.3.27}$$

Relation (1.2.11) with $d\mu(t) = u(t)dt$ reads

$$K_n(u;x,t) = \frac{\gamma_{n-1}(u)}{\gamma_n(u)} \cdot \frac{P_{n-1}(u;t) P_n(u;x) - P_{n-1}(u;x) P_n(u;t)}{x-t}.$$

Then for $z \in \mathbf{C} \setminus [-1,1]$ and $t \in [-1,1]$

$$|K_n(u;z,t)| \leq \rho(z)^{-1} [|P_{n-1}(u;t) P_n(u;z)| + |P_{n-1}(u;z) P_n(u;t)|]. \tag{10.3.28}$$

Since $a_n^{-1} Q_n - \omega_n(w,m) \in \mathbf{P}_{n-1}$, by (1.2.9)

$$a_n^{-1} Q_n(z) - \omega_n(w,m;z) = \int_{-1}^{1} [a_n^{-1} Q_n(t) - \omega_n(w,m;t)] K_n(u;z,t) u(t) dt.$$

This formula by Hölder inequality and by (10.3.28) yields

$$\begin{aligned} & |a_n^{-1} Q_n(z) - \omega_n(w,m;z)| \leq \|a_n^{-1} Q_n - \omega_n(w,m)\|_{w,m} \\ & \times \rho(z)^{-1} \left[|P_n(u;z)| \; \|P_{n-1}(u) u w^{-1/m}\|_{L^q} \right. \\ & \left. + |P_{n-1}(u;z)| \; \|P_n(u) u w^{-1/m}\|_{L^q} \right]. \end{aligned} \tag{10.3.29}$$

Now by the same argument as that of Lemma 10.3.3 from the identity

$$P_{n-1}(u;z) = \frac{\gamma_{n-1}(u)}{\gamma_n(u)} \sum_{k=1}^{n} \lambda_{kn}(u) P_{n-1}(u;x_{kn}(u))^2 P_n(u;z)(z-x_{kn}(u))^{-1}$$

it follows that for $z \in \mathbf{C} \setminus [-1,1]$

$$\left| \frac{P_{n-1}(u;z)}{P_n(u;z)} \right| \le \rho(z)^{-1} \sum_{k=1}^{n} \lambda_{kn}(u) P_{n-1}(u;x_{kn}(u))^2 = \rho(z)^{-1},$$

which, coupled with (10.3.27) and (10.3.29), gives that for $z \in \mathbf{C} \setminus [-1,1]$

$$\begin{aligned} |a_n^{-1} Q_n(z) - \omega_n(w,m;z)| &\le 2\rho(z)^{-1} E_n(w,m) \Psi\left(\frac{\gamma_n(w,m)}{a_n} \|Q_n\|_{w,m} \right) \\ &\times |P_n(u;z)| [\|P_{n-1}(u) u w^{-1/m}\|_{L^q} + \rho(z)^{-1} \|P_n(u) u w^{-1/m}\|_{L^q}]. \end{aligned} \tag{10.3.30}$$

On the other hand, clearly

$$\begin{aligned} & \left| \frac{Q_n(z)}{P_n(w,m;z)} - 1 \right| \\ = & \left| \frac{Q_n(z) - a_n \omega_n(w,m;z) + a_n \omega_n(w,m;z)}{P_n(w,m;z)} - 1 \right| \\ \le & \; a_n \left| \frac{a_n^{-1} Q_n(z) - \omega_n(w,m;z)}{P_n(w,m;z)} \right| + \left| \frac{a_n}{\gamma_n(w,m)} - 1 \right|. \end{aligned}$$

Substituting (10.3.30) into the above inequality yields (10.3.26). □

The first main result in this section is as follows.

Theorem 10.3.2. *Let a weight w supported in $[-1,1]$ satisfy $w^{-1} \in L^q[-1,1]$ for every $q < \infty$. Let $1 < m < \infty$. Then*

$$\lim_{n \to \infty} \frac{\gamma_n(w,m)}{2^n} = \sigma_m^{-1} G(w)^{-1/m} \tag{10.3.31}$$

and for $z \in \mathbf{C} \setminus [-1,1]$

$$\lim_{n \to \infty} \frac{P_n(w,m;z)}{\phi(z)^n} = 2^{-1/m} \sigma_m^{-1} D(\bar{u}_m; \phi(z)^{-1})^{-2}. \tag{10.3.32}$$

Proof. The limit (10.3.31) follows from (10.3.13) and (10.3.20). Let us prove (10.3.32). We separate the cases when $2 \le m < \infty$ and $1 < m < 2$.

Case 1. $2 \le m < \infty$. In this case let R_m be a function defined by (10.3.5) satisfying the assumptions of Lemma 10.3.1. By (10.3.8), (10.3.10), and (10.3.11)

$$|P_n(R_m,m;z)R(x)| \le 2^{1-1/m} \sigma_m^{-1}, \quad x \in [-1,1], \tag{10.3.33}$$

and by (10.3.8), (10.3.10), and (10.3.12)

$$\left| \frac{P_n(R_m,m;z)}{2^{1-1/m} \sigma_m^{-1} \phi(z)^n D(R(\cos\theta); \phi(z)^{-1})^{-2}} - 1 \right| \le |\phi(z)|^{2q-2n-2}, \quad z \in \mathbf{C} \setminus [-1,1], \tag{10.3.34}$$

where $2q = \partial S$. Assume that $\Delta \subset [-1,1]$ is measurable and $\|w\|_\Delta < \infty$. Write $E = [-1,1] \setminus \Delta$. The inequality (10.3.22) with $Q_n = P_n(R_m, m)$ reads

$$\begin{aligned}
&\left| \frac{P_n(R_m, m; z)}{P_n(w, m; z)} - 1 \right| \\
\leq\; &\left| \frac{\gamma_n(R_m, m)}{\gamma_n(w, m)} - 1 \right| + \rho(z)^{-1} \left\{ \|P_n(R_m, m)\|_{w,m}^2 - \frac{\gamma_n(R_m, m)^2}{\gamma_n(w, m)^2} \right\}^{1/2}.
\end{aligned}$$

By (10.3.33)

$$\begin{aligned}
\|P_n(R_m, m)\|_{w,m} &\leq \|P_n(R_m, m)\|_{L^m_{R_m}(\Delta)} \|R_m^{-1} w\|_\Delta^{1/m} + 2^{1-1/m} \sigma_m^{-1} \|R^{-1}\|_{L^m_w(E)} \\
&\leq \|R_m^{-1} w\|_\Delta^{1/m} + 2^{1-1/m} \sigma_m^{-1} \|R^{-1}\|_{L^m_w(E)}.
\end{aligned}$$

Hence

$$\begin{aligned}
&\left| \tfrac{P_n(R_m, m; z)}{P_n(w, m; z)} - 1 \right| \leq \left| \tfrac{\gamma_n(R_m, m)}{\gamma_n(w, m)} - 1 \right| \\
&+\rho(z)^{-1} \left\{ \left[\|R_m^{-1} w\|_\Delta^{\frac{1}{m}} + 2^{1-\frac{1}{m}} \sigma_m^{-1} \|R^{-1}\|_{L^m_w(E)} \right]^2 - \tfrac{\gamma_n(R_m, m)^2}{\gamma_n(w, m)^2} \right\}^{\frac{1}{2}}. \qquad (10.3.35)
\end{aligned}$$

Assume that $g \in C[-1,1]$ and $g(x) > 0$, $x \in [-1,1]$. Then there exists a sequence $\{r_n\}$ of polynomials with

$$\partial r_n = q - 1 = q_n - 1 \leq (n-2)/2, \; r_n(x) > 0, \; x \in [-1,1], \quad n \in \mathbf{N},$$

such that $\lim_{n\to\infty} \|r_n - g\| = 0$. As in the proof of Theorem 10.3.1, let R and R_m be defined in (10.3.15) and (10.3.16), respectively. Put

$$h_n(f; z) = (2\sigma_m)^{-1} \phi(z)^n D(f(\cos\theta); \phi(z)^{-1})^{-2} \qquad (10.3.36)$$

and

$$\hat{u}_m(x) = w(x)^{1/m} (1 - x^2)^{1/(2m)}.$$

By (10.3.3) we have $\hat{u}_m(\cos\theta) = \bar{u}_m(\theta)$. Write

$$\frac{h_n(\hat{u}_m; z)}{P_n(w, m; z)} = \frac{P_n(R_m, m; z)}{P_n(w, m; z)} \cdot \frac{h_n(R; z)}{P_n(R_m, m; z)} \cdot \frac{h_n(\hat{u}_m; z)}{h_n(R; z)} = A_n(z) B_n(z) C_n(z). \qquad (10.3.37)$$

Remembering $q \leq n/2$, by (10.3.34) the inequality

$$\lim_{n\to\infty} B_n(z) = 1 \qquad (10.3.38)$$

holds uniformly in a compact subset of $\mathbf{C} \setminus [-1,1]$. We observe that

$$\lim_{n\to\infty} \|r_n^{-1} - g^{-1}\| = 0.$$

Hence by (1.2.17) and (10.3.36) the limit

$$\lim_{n\to\infty} C_n(z) = D((\hat{u}_m g)(\cos\theta); \phi(z)^{-1})^{-2} \qquad (10.3.39)$$

holds uniformly in a compact subset of $\mathbf{C}\setminus[-1,1]$. We see that by (10.3.8), (10.3.31), and (1.2.17)

$$\begin{aligned}\lim_{n\to\infty}\frac{\gamma_n(R_m,m)}{\gamma_n(w,m)} &= \lim_{n\to\infty}\left[\frac{G(w)}{G(R_m)}\right]^{1/m}\\ &= \lim_{n\to\infty} G(w^{1/m}R_m^{-1/m}) = G(\hat{u}_m g).\end{aligned}$$

Thus by (10.3.31) and (10.3.35) for any compact set $\Omega\subset\mathbf{C}\setminus[-1,1]$

$$\limsup_{n\to\infty}\|A_n-1\|_\Omega\le c(g,\Delta,\Omega)\|\rho^{-1}\|_\Omega \tag{10.3.40}$$

where

$$\begin{aligned}c(g,\Delta,\Omega) &= |G(\hat{u}_m g)-1|\ \|\rho(z)^{-1}\|_\Omega^{-1}\\ &+\left\{\left[\|\hat{u}_m g\|_\Delta+2^{1-1/m}\sigma_m^{-1}\|g\|_{L_w^m(E)}\right]^2-G(\hat{u}_m g)^2\right\}^{1/2}.\end{aligned}$$

It is easy to check that

$$\begin{aligned}&|A_n(z)B_n(z)C_n(z)-1|\le|A_n(z)-1|\ |B_n(z)-1|\ |C_n(z)-1|\\ &+|A_n(z)|\ |B_n(z)-1|+|B_n(z)|\ |C_n(z)-1|+|C_n(z)|\ |A_n(z)-1|.\end{aligned}$$

Thus by (10.3.37)–(10.3.40)

$$\begin{aligned}&\limsup_{n\to\infty}\left\|\frac{h_n(\hat{u}_m)}{P_n(w,m)}\right\|_\Omega\\ &\le \|D((\hat{u}_m g)(\cos\theta);\phi^{-1})^{-2}-1\|_\Omega\\ &+c(g,\Delta,\Omega)\ \|D((\hat{u}_m g)(\cos\theta);\phi^{-1})^{-2}\|_\Omega\ \|\rho^{-1}\|_\Omega.\end{aligned} \tag{10.3.41}$$

In a similar way as that of Theorem 10.3.1 we conclude that inequality (10.3.41) holds for any measurable function g with $0<c_1\le g\le c_2<\infty$. Then we choose Δ so that $0<c_1\le\hat{u}_m(x)\le c_2<\infty$, $x\in\Delta$, and

$$g(x)=\begin{cases}\hat{u}_m(x)^{-1}, & x\in\Delta,\\ 1, & x\in E.\end{cases}$$

It is easy to see that $|E|$ may be small enough. We conclude that $c(g,\Delta,\Omega)\to 0$ and $D((\hat{u}_m g)(\cos\theta);\phi(z)^{-1})^{-2}\to 1$ as $|E|\to 0$. Then relation (10.3.32) follows from (10.3.41).

Case 2. $1<m<2$. In this case we choose

$$u(x)=(1-x^2)^{1/2}\quad x\in[-1,1] \tag{10.3.42}$$

and hence obtain [24, p. 35]

$$\|P_n(u)u\|\le(2/\pi)^{1/2},\quad n\ge 2. \tag{10.3.43}$$

Let R_m be given by (10.3.5) satisfying the conditions of Lemma 10.3.1. Inserting $Q_n = P_n(R_m, m)$ into (10.3.26) and using (10.3.43), we get

$$\begin{aligned}\left|\frac{P_n(R_m,m;z)}{P_n(w,m;z)}-1\right| &\le \left|\frac{\gamma_n(R_m,m)}{\gamma_n(w,m)}-1\right| \\ &+2\rho(z)^{-1}\left|\frac{P_n(u;z)}{P_n(w,m;z)}\right|\cdot\left|\frac{\gamma_n(R_m,m)}{\gamma_n(w,m)}\right|\Psi\left[\|P_n(R_m,m)\|_{w,m}\frac{\gamma_n(w,m)}{\gamma_n(R_m,m)}\right] \\ &\times\left(\frac{2}{\pi}\right)^{1/2}[1+\rho(z)^{-1}]\|w^{-1/m}\|_{L^q}.\end{aligned}$$

In a similar way as above we choose a sequence of functions R_m such that

$$\lim_{n\to\infty}\frac{\gamma_n(R_m,m)}{\gamma_n(w,m)} = 1$$

and

$$\lim_{n\to\infty}\|P_n(R_m,m)\|_{w,m} = 1.$$

Recalling that $\Psi(1) = 0$ and (10.3.32) holds for $w = u$ and $m = 2$, by (10.3.8), (10.3.10), and (10.3.12) the above estimations give (10.3.32). □

Remark 10.3.1. We point out that the conditions on w^{-1} in the theorem imply $w \in \mathbf{S}$, that is $G(w) > 0$. In fact, assume that $w^{-1} \in L^q$ for some $q > 0$. Then by the arithmetic-geometric mean inequality [137, p. 2]

$$\begin{aligned}G(w)^{-q/4} &= \exp\left\{\frac{1}{\pi}\int_{-1}^{1}\ln w(x)^{-q/4}\frac{dx}{\sqrt{1-x^2}}\right\} \le \frac{1}{\pi}\int_{-1}^{1}w(x)^{-q/4}\frac{dx}{\sqrt{1-x^2}} \\ &\le \frac{1}{\pi}\|w^{-1}\|_{L^q}^{q/4}\left\|(1-x^2)^{-1/2}\right\|_{L^{4/3}} < \infty,\end{aligned}$$

which implies $G(w) > 0$.

For a Szegő measure we have a weaker conclusion, which is the second main result in this section.

Theorem 10.3.3. *Let $d\mu$ be a measure supported in $[-1,1]$. If $0 < m < \infty$ then the following statements are equivalent.*

(a) We have

$$d\mu \in \mathbf{S}. \tag{10.3.44}$$

(b) We have

$$\frac{\gamma_n(d\mu,m)}{2^n} \le C. \tag{10.3.45}$$

(c) We have

$$c \le \frac{\gamma_n(d\mu,m)}{2^n} \le C. \tag{10.3.46}$$

Moreover, if $2 \le m < \infty$ then Statements (a)–(c) and the following statements are equivalent.

(d) The inequalities

$$c \le \left|\frac{2^n\omega_n(d\mu,m;z)}{\phi(z)^n}\right| \le C, \quad z \in \Omega, \tag{10.3.47}$$

hold for every compact set $\Omega \subset \mathbf{C}\setminus[-1,1]$.

(e) The inequality

$$\left|\frac{P_n(d\mu,m;z)}{\phi(z)^n}\right| \le C, \quad z\in\Omega, \tag{10.3.48}$$

holds for every compact set $\Omega \subset \mathbf{C}\setminus[-1,1]$.

(f) The inequalities

$$c \le \left|\frac{P_n(d\mu,m;z)}{\phi(z)^n}\right| \le C, \quad z\in\Omega, \tag{10.3.49}$$

hold for every compact set $\Omega \subset \mathbf{C}\setminus[-1,1]$.

(g) We have

$$\left|\sum_{k=1}^{n} f(x_{kn}(d\mu,m)) - \frac{n}{\pi}\int_{-1}^{1} f(x)\frac{dx}{\sqrt{1-x^2}}\right| \le \|f\|_\Delta \tag{10.3.50}$$

whenever f *is analytic in an open set* $\Delta \supset [-1,1]$.

Proof. (a)$\Longrightarrow$(b) By (9.4.8)

$$\left[\frac{2^n}{\gamma_n(d\mu,m)}\right]^m = \int_{-1}^{1} |2^n\omega_n(d\mu,m;x)|^m d\mu(x) \ge \frac{\pi}{2}G(d\mu)$$

and hence

$$\frac{\gamma_n(d\mu,m)}{2^n} \le \left[\frac{\pi}{2}G(d\mu)\right]^{-1/m}. \tag{10.3.51}$$

(b)$\Longrightarrow$(c) Inequalities (10.3.46) follow from (10.3.45) and (10.1.14).

(c)$\Longrightarrow$(a) Let $m\ge 1$ and let $\rho_n(z) = z^n + a_{n-1}z^{n-1} + \cdots + a_0$ be a solution of (1.2.19), where $d\nu$ is defined by (1.2.21). Since $\nu(-\theta) = \nu(\theta)$, the polynomial

$$\overline{\rho_n(z^{-1})} = \bar{\rho}_n(z) = z^n + \bar{a}_{n-1}z^{n-1} + \cdots + \bar{a}_0$$

is also a solution of (1.2.19). Thus $\psi_n(z) = \frac{1}{2}[\rho_n(z) + \bar{\rho}_n(z)]$ is a solution of (1.2.19) and has real coefficients.

Let $z = e^{i\theta}$ and

$$\psi_n(z) = \sum_{k=0}^{n} b_k z^k = \sum_{k=0}^{n} b_k[\cos k\theta + i\sin k\theta], \quad b_n = 1.$$

For the kth Chebyshev polynomial of the first kind T_k we have

$$\begin{aligned} d_n(d\nu) &= \frac{1}{2\pi}\int_{-\pi}^{\pi} |\psi_n(e^{i\theta})|^m d\nu(\theta) \ge \frac{1}{2\pi}\int_{-\pi}^{\pi}\left|\sum_{k=0}^{n} b_k\cos k\theta\right|^m d\nu(\theta) \\ &= \frac{1}{\pi}\int_{-1}^{1}\left|\sum_{k=0}^{n} b_k T_k(x)\right|^m d\mu(x) \\ &= \frac{1}{\pi 2^m}\int_{-1}^{1} |2^n x^n + \cdots|^m\, d\mu(x) \\ &\ge \frac{1}{\pi 2^m} 2^n \gamma_n(d\mu,m)^{-1} \ge c. \end{aligned}$$

Thus, for $m \geq 1$, $d\mu \in \mathbf{S}$ follows at once from (1.2.20) and the above inequality.

If $m < 1$ then by (10.1.17) inequality (10.3.45) with $m = 1$ holds. By the previous conclusion we conclude $d\mu \in \mathbf{S}$.

Let us prove the second part of the theorem. First we point out that for $2 \leq m < \infty$ by (8.3.9) and (9.1.45)

$$c\gamma_n(d\mu, m) \leq \gamma_n(X(d\mu, m)) \leq C\gamma_n(d\mu, m). \qquad (10.3.52)$$

Thus by Theorem 9.4.2 we can conclude (d)$\Longleftrightarrow$(f)$\Longleftrightarrow$(g).

(b)$\Longrightarrow$(f) Inequality (10.3.45) means

$$\int_{-1}^{1} |2^n \omega_n(d\mu, m; x)|^m d\mu(x) \leq C,$$

which by Theorem 9.4.1 gives the second inequality of (10.3.49). To prove the first it remains to apply Theorem 9.4.3 and use (10.3.52).

(f)$\Longrightarrow$(e) Trivial.

(e)$\Longrightarrow$(b) Apply Theorem 9.4.4 and use (10.3.52). □

10.4 Asymptotics of Christoffel Type Functions

This section deals with the asymptotics of Christoffel type functions $\lambda_{jn}(d\mu, m; x)$. For a general measure we have the following

Theorem 10.4.1. *Let $d\mu$ be a measure supported on $[-1,1]$ and $j \in \mathbf{M}_2$. Then*

$$\limsup_{n\to\infty} n^{j+1} \lambda_{jn}(d\mu, m; x) \leq c\mu'(x)(1-x^2)^{(j+1)/2} \qquad (10.4.1)$$

holds for almost every $x \in [-1,1]$.

Proof. By (5.2.48) and (5.2.41)

$$\begin{aligned} &\lambda_{jn}(d\mu, m; x) \\ \leq\ & c \sum_{i=0}^{m-j-2} \Delta_n(x)^{-i} \int_{-1}^{1} |(t-x)^{j+i} \pi_{n-1}(x,t)^m| d\mu(t) \\ \leq\ & c \sum_{i=0}^{m-j-2} \Delta_n(x)^{-i} \int_{-1}^{1} \left| (t-x)^{j+i} \left[\frac{K_n(v; x, t)}{K_n(v; x, x)} \right]^{i+j+2} \right| d\mu(t). \end{aligned}$$

Further using (5.2.40) and (5.2.42) we obtain

$$\begin{aligned}
&\lambda_{jn}(d\mu,m;x)\\
\le\ & c\sum_{i=0}^{m-j-2} n^{-i-j-1}\Delta_n(x)^{-i}\\
&\times\lambda_n(v;x)\int_{-1}^{1}|(t-x)K_n(v;x,t)|^{i+j}K_n(v;x,t)^2d\mu(t)\\
\le\ & cn^{-j-1}\sum_{i=0}^{m-j-2}[n\Delta_n(x)]^{-i}\\
&\times\lambda_n(v;x)\int_{-1}^{1}\left[(1-x^2)^{1/2}+(1-t^2)^{1/2}\right]^{i+j}K_n(v;x,t)^2d\mu(t)\\
=\ & cn^{-j-1}\sum_{i=0}^{m-j-2}[n\Delta_n(x)]^{-i}\sum_{s=0}^{i+j}\binom{i+j}{s}(1-x^2)^{s/2}\\
&\times\lambda_n(v;x)\int_{-1}^{1}K_n(v;x,t)^2(1-t^2)^{(i+j-s)/2}d\mu(t)
\end{aligned}$$

or equivalently

$$\begin{aligned}
n^{j+1}\lambda_{jn}(d\mu,m;x)\le c\sum_{i=0}^{m-j-2}[n\Delta_n(x)]^{-i}\sum_{s=0}^{i+j}\binom{i+j}{s}(1-x^2)^{s/2}\\
\times\lambda_n(v;x)\int_{-1}^{1}K_n(v;x,t)^2(1-t^2)^{(i+j-s)/2}d\mu(t).
\end{aligned}$$

Here we need a formula given by Nevai in [65, Lemma 6.2.32, p. 93]

$$\lim_{n\to\infty}\lambda_n(v;x)\int_{-1}^{1}K_n(v;x,t)^2d\mu(t)=\mu'(x)(1-x^2)^{1/2},$$

which holds for almost every $x\in[-1,1]$. Hence for almost every $x\in[-1,1]$

$$\begin{aligned}
&\limsup_{n\to\infty} n^{j+1}\lambda_{jn}(d\mu,m;x)\\
\le\ & c\sum_{i=0}^{m-j-2}(1-x^2)^{-i/2}\sum_{s=0}^{i+j}\binom{i+j}{s}(1-x^2)^{s/2}(1-x^2)^{(i+j-s)/2}\mu'(x)(1-x^2)^{1/2}\\
\le\ & c\mu'(x)(1-x^2)^{(j+1)/2}.\quad\square
\end{aligned}$$

Lemma 10.4.1. [65, Lemma 5.1, p. 49] *Let $d\mu$ be a measure with a compact support and let f be continuous on $\Delta(d\mu)$ with the modulus of continuity ω. Then the inequality*

$$\left|\sum_{k=1}^{n}f(x_{kn}(d\mu))-\sum_{k=0}^{n-1}\int_{\mathbf{R}}f(x)P_k(d\mu;x)^2d\mu(x)\right|\le n\omega(n^{-\frac{1}{3}})\left[1+\frac{1}{2}|\Delta(d\mu)|^3\right]\tag{10.4.2}$$

holds for $n>|\Delta(d\mu)|^{-3}$.

For a regular measure we have the following two results.

Lemma 10.4.2. *Let $d\mu$ be a regular measure supported on $[-1,1]$. Then for almost every $x \in [-1,1]$*

$$\limsup_{n\to\infty} n\lambda_n(d\mu;x) = \pi\mu'(x)(1-x^2)^{1/2}. \tag{10.4.3}$$

Proof. Using the formula (1.2.5) inequality (10.4.2) becomes

$$\left|\frac{1}{n}\sum_{k=1}^{n} f(x_{kn}) - \sum_{k=0}^{n-1}\int_{\mathbf{R}} \frac{f(t)}{n\lambda_n(d\mu;t)}d\mu(t)\right| \le 5\omega(n^{-1/3}),$$

which, together with (10.1.10), implies that the relation

$$\lim_{n\to\infty}\int_{-1}^{1}\frac{f(t)}{n\lambda_n(d\mu;t)}d\mu(t) = \frac{1}{\pi}\int_{-1}^{1}\frac{f(t)dt}{\sqrt{1-t^2}} \tag{10.4.4}$$

holds for every $f \in C[-1,1]$. Using one-sided approximation we conclude that relation (10.4.4) remains true if f is the characteristic function of an interval. That is, the relation

$$\lim_{n\to\infty}\int_{c}^{d}\frac{1}{n\lambda_n(d\mu;t)}d\mu(t) = \frac{1}{\pi}\int_{c}^{d}\frac{1}{\sqrt{1-t^2}}dt$$

holds for every interval $[c,d] \subset [-1,1]$. According to Fatou lemma the inequality

$$\int_{c}^{d}\liminf_{n\to\infty}\frac{\mu'(t)}{n\lambda_n(d\mu;t)}dt \le \frac{1}{\pi}\int_{c}^{d}\frac{1}{\sqrt{1-t^2}}dt$$

holds for every interval $[c,d] \subset [-1,1]$. In particular, for each $x \in (-1,1)$ and for each positive number $\varepsilon \le \min\{1-x, 1+x\}$

$$\frac{1}{2\varepsilon}\int_{x-\varepsilon}^{x+\varepsilon}\liminf_{n\to\infty}\frac{\mu'(t)}{n\lambda_n(d\mu;t)}dt \le \frac{1}{2\pi\varepsilon}\int_{x-\varepsilon}^{x+\varepsilon}\frac{1}{\sqrt{1-t^2}}dt.$$

The left-hand side in the above inequality as $\varepsilon \to 0$ is equal to

$$\liminf_{n\to\infty}\frac{\mu'(x)}{n\lambda_n(d\mu;x)}$$

for almost every $x \in [-1,1]$. Thus the inequality

$$\liminf_{n\to\infty}\frac{\mu'(x)}{n\lambda_n(d\mu;x)} \le \frac{1}{\pi\sqrt{1-x^2}}$$

or equivalently, the inequality

$$\limsup_{n\to\infty} n\lambda_n(d\mu;x) \ge \pi\mu'(x)(1-x^2)^{1/2}$$

holds for almost every $x \in [-1,1]$. On the other hand, the inequality

$$\limsup_{n\to\infty} n\lambda_n(d\mu;x) \le \pi\mu'(x)(1-x^2)^{1/2}, \quad \text{a.e.}$$

is given by Nevai [65, Theorem 6.2.33, p. 93]. This completes the proof. □

Theorem 10.4.2. *Let $d\mu$ be a regular measure supported on $[-1,1]$ and let $m \in \mathbf{N}_2$. Then for almost every $x \in [-1,1]$*

$$\frac{2\pi}{m}\mu'(x)(1-x^2)^{1/2} \leq \limsup_{n\to\infty} n\lambda_{0n}(d\mu,m;x) \leq C\mu'(x)(1-x^2)^{1/2}. \tag{10.4.5}$$

Proof. This follows from (5.2.50), (10.4.3), and (10.4.1). □

For a generalized Jacobi weight we have the following

Theorem 10.4.3. *Let $m \in \mathbf{N}_2$. Then relation (5.3.2) is equivalent to*

$$\frac{c_1}{n}\widehat{W}_n(x) \leq \lambda_{0n}(u,m;x) \leq \frac{c_2}{n}\widehat{W}_n(x). \tag{10.4.6}$$

Proof. By Theorem 5.3.3 it suffices to show the implication (10.4.6) $\Longrightarrow$ (5.3.2). By (10.4.1) and (10.4.6)

$$u(x)(1-x^2)^{1/2} \geq c\limsup_{n\to\infty} n\lambda_{0n}(u,m;x) = c(1-x^2)^{1/2}\prod_{i=1}^{r}|x-t_i|^{p_i},$$

that is,

$$u(x) \geq c\prod_{i=1}^{r}|x-t_i|^{p_i}.$$

Thus $\lim_{n\to\infty}\gamma_n(u)^{1/n} = 2$. Then applying Theorem 10.4.2 and using (10.4.6), we obtain

$$u(x) \leq c\prod_{i=1}^{r}|x-t_i|^{p_i}. \quad \square$$

The asymptotics of the Christoffel numbers $\lambda_{kjn}(u,m)$ for a weight $u \sim W$ is as follows.

Theorem 10.4.4. *If (5.3.2) with (7.2.57) is true, then, with the constants associated with the symbol $\sim$ depending on u and m,*

$$\begin{cases} [sgn\ P_n'(u,m;x_{kn})^m]\,\lambda_{kjn}(u,m) \sim \frac{1}{n}\widehat{W}_n(x_{kn})\Delta_n(x_{kn})^j, & j \in \mathbf{M}_2, \\ |\lambda_{kjn}(u,m)| \leq \frac{c}{n}\widehat{W}_n(x_{kn})\Delta_n(x_{kn})^j, & j \in \mathbf{M}_1 \setminus \{0\}. \end{cases} \tag{10.4.7}$$

Proof. The first formula in (10.4.7) follows directly from (5.3.26) and (5.2.31); the second follows from the first, (7.2.65), and (5.2.31). □

The following theorem provides another form of the asymptotics of the Christoffel numbers $\lambda_{kjn}(v,m)$, where v is the Chebyshev weight.

Theorem 10.4.5. *Let $m \in \mathbf{N}_2$. Then with the notations of Section 6.6 and $s = [(j+1)/2]$, $0 \leq j \leq m-2$*

$$\lim_{n\to\infty}\max_{1\leq k\leq n}\left|n^{2s+1}\lambda_{kjn}(v,m) - \frac{\pi c_{s+1,m/2}}{2^{2s}[(m/2-1)!]^2}h_j(x_{kn})\right| = 0, \tag{10.4.8}$$

where $x_{kn} = x_{kn}(v)$ and

$$h_j(x) = \begin{cases} (1-x^2)^{j/2}, & j \in \mathbf{N}_{02} \\ -\frac{1}{2}j(j+1)x(1-x^2)^{(j-1)/2}, & j \in \mathbf{N}_1. \end{cases} \tag{10.4.9}$$

Moreover, the limit

$$\lim_{n\to\infty} n^{2s}\sum_{k=1}^{n}\lambda_{kjn}(v,m)f(x_{kn})=\frac{c_{s+1,m/2}}{2^{2s}[(m/2-1)!]^2}\int_{-1}^{1}f(x)h_j(x)\frac{dx}{\sqrt{1-x^2}} \tag{10.4.10}$$

holds for all $f\in C[-1,1]$.

Proof. Comparing (10.4.9) with (6.6.9) we see that $h_j=e_{sj}$. By (6.6.10) for each k, $1\le k\le n$,

$$\begin{aligned}&\left|n^{2s+1}\lambda_{kjn}(v,m)-\frac{\pi c_{s+1,m/2}}{2^{2s}[(m/2-1)!]^2}h_j(x_{kn})\right|\\ \le\ &\frac{1}{n^2}\left|\sum_{i=s+1}^{(m-2)/2}\frac{\pi c_{m/2,i+1}}{2^{2i}n^{2(i-s-1)}}e_{ij}(x_{kn})\right|\le\frac{c}{n^2}.\end{aligned} \tag{10.4.11}$$

Then relation (10.4.8) follows.

To prove (10.4.10) we use (10.4.11) to get

$$\begin{aligned}&\left|n^{2s}\sum_{k=1}^{n}\lambda_{kjn}(v,m)f(x_{kn})-\frac{c_{s+1,m/2}}{2^{2s}[(m/2-1)!]^2}\cdot\frac{\pi}{n}\sum_{k=1}^{n}f(x_{kn})h_j(x_{kn})\right|\\ \le\ &\frac{\|f\|}{n}\sum_{k=1}^{n}\left|n^{2s+1}\lambda_{kjn}(v,m)-\frac{\pi c_{s+1,m/2}}{2^{2s}[(m/2-1)!]^2}h_j(x_{kn})\right|\le\frac{c\|f\|}{n^2}.\end{aligned}$$

Thus

$$\begin{aligned}&\lim_{n\to\infty}n^{2s}\sum_{k=1}^{n}\lambda_{kjn}(v,m)f(x_{kn})\\ =\ &\frac{c_{s+1,m/2}}{2^{2s}[(m/2-1)!]^2}\lim_{n\to\infty}\frac{\pi}{n}\sum_{k=1}^{n}f(x_{kn})h_j(x_{kn})\\ =\ &\frac{c_{s+1,m/2}}{2^{2s}[(m/2-1)!]^2}\int_{-1}^{1}f(x)h_j(x)\frac{dx}{\sqrt{1-x^2}}.\quad\square\end{aligned}$$

Theorem 10.4.6. *Let a weight* $w\in C(\mathbf{R})$ *and* $w(x)>0$, $x\in\mathbf{R}$. *Let* $m\in\mathbf{N}_2$. *Then for every number* $a>0$

$$\lim_{n\to\infty}\{n\lambda_{0n}(w,m;x)\}^{-1}=0 \tag{10.4.12}$$

holds uniformly on $[-a,a]$.

Proof. First we prove the theorem for $m=2$. Let for $b>a$

$$w_b(x)=\begin{cases}w(x), & x\in[-b,b]\\ 0, & \text{otherwise.}\end{cases}$$

Then by [65, Example 6.2.10, p. 80] we see that

$$\lim_{n\to\infty}n\lambda_n(w_b;x)=\pi w_b(x)(b^2-x^2)^{\frac{1}{2}}$$

holds uniformly on $[-a,a]$. Since $w \geq w_b$, we conclude that $\lambda_n(w;x) \geq \lambda_n(w_b;x)$ and hence

$$\liminf_{n\to\infty} n\lambda_n(w;x) \geq \pi w_b(x)(b^2-x^2)^{\frac{1}{2}} = \pi w(x)(b^2-x^2)^{\frac{1}{2}}, \quad x \in [-a,a].$$

Thus

$$\limsup_{n\to\infty}[\pi\lambda_n(w;x)]^{-1} \leq \left[\pi w(x)(b^2-x^2)^{\frac{1}{2}}\right]^{-1}, \quad x \in [-a,a]$$

holds for any $b > a$. That is,

$$\limsup_{n\to\infty}[n\lambda_n(w;x)]^{-1} = 0$$

holds uniformly on $[-a,a]$.

Next for $m \geq 4$ it remains to apply (5.2.50). □

10.5 Notes

10.5.1. For ratio asymptotics we propose the following two conjectures.

Conjecture 10.5.1. *Statement (a) of Theorem 10.2.3 is the same for every m, $1 \leq m < \infty$.*

Conjecture 10.5.2. *Theorem 10.2.3 remains true if (9.3.16) in Statement (a) is replaced by the relations*

$$\lim_{n\to\infty} a_{1,1,n} = 0, \quad \lim_{n\to\infty} a_{2,1,n} = \frac{1}{4}.$$

We point out that Conjecture 10.5.2 is true for $m = 2$.

10.5.2. Theorems 10.3.1 and 10.3.2 are due to Lubinsky and Saff [52]. I believe that the relations (10.3.31) and (10.3.32) remain true for a weight $w \in \mathbf{S}$.

10.5.3. Nevai [65, Theorems 6.2.54 and 6.2.55, pp. 104–105] proved (10.4.3) for a measure $d\mu$ satisfying

$$\lim_{n\to\infty} \frac{\gamma_n(d\mu)}{\gamma_{n-1}(d\mu)} = 2 \text{ or } \mu'(x) > 0, \text{ a.e..}$$

Lemma 10.4.2 is an improvement of these results.

10.5.4. Theorem 10.4.5 was given by Micchelli and Sharma [59] in 1983. This theorem first gave an answer to the second part of Problem 26 of Turán (see Section 6.1).

Chapter 11

Interpolation and Power Orthogonal Series

11.1 Introduction

The bounds and inequalities given in Chapter 8, especially, (8.3.7), (8.3.20), and (8.5.1), make it possible to extend many important results previously obtained on convergence of power orthogonal series and on mean convergence of Lagrange and Hermite interpolation. The reason is that the inequalities of the forms (8.3.7), (8.3.20), and (8.5.1), etc. are just the crucial points to prove these results and these inequalities were previously obtained for certain "nice" measures only, since only for "nice" measures one can use asymptotics to get them.

Section 11.2 deals with absolute and mean convergence of power orthogonal series. In Sections 11.3 and 11.4 we discuss mean convergence of Lagrange interpolation on a matrix of nodes X and on the zeros of power orthogonal polynomials, respectively. The former is a base of the latter. We also discuss mean convergence of Hermite interpolation. This is given in Section 11.5. Using power orthogonal polynomials we can construct two positive operators $\mathcal{F}_n(d\mu,m,f;x)$ and $\mathcal{G}_n(d\mu,m,f;x)$, which are discussed in Section 11.6.

11.2 Absolute and Mean Convergence of Power Orthogonal Series

For $f \in L^1_{d\mu}$ and $2 \le m < \infty$ the nth partial sum of its *m orthogonal series* is defined by

$$\mathcal{S}_n(d\mu,m,f;x) = \sum_{k=0}^{n-1} P_k(d\mu,m;x) \int_{-1}^{1} f(t)P_k(d\mu,m;t)d\mu(t). \tag{11.2.1}$$

As usual, we write $\mathcal{S}_n(d\mu,2,f;x)=\mathcal{S}_n(d\mu,f;x)$, which is the nth partial sum of the orthogonal Fourier series.

We need a useful result, which is given by Nevai [65, Theorem 7.31, p. 138].

Lemma 11.2.1. *Let $d\mu$ be a measure and w a weight supported in $[-1,1]$. Let $2 \le p < \infty$. If*

$$\sup_{n\in\mathbf{N}_0} \|P_n(d\mu)\|_{w,p} < \infty,$$

then

$$\int_{-1}^{1} \left[\mu'(x)(1-x^2)^{1/2}\right]^{-p/2} w(x)dx < \infty. \tag{11.2.2}$$

First, for absolute convergence of power orthogonal series we have the following

Theorem 11.2.1. *Let $d\mu$ be a measure supported in $[-1,1]$ and let $2 \le m < \infty$. Then the series*

$$\sum_{n=0}^{\infty} |c_n P_n(d\mu, m; x)| \tag{11.2.3}$$

either diverges almost everywhere in $[-1,1]$ or converges almost everywhere in the set $\{x \in [-1,1] : \mu'(x) > 0\}$, and the latter case is equivalent to

$$\sum_{n=0}^{\infty} |c_n| < \infty. \tag{11.2.4}$$

Proof. Assume that the series (11.2.3) converges on a set $\Delta \subset [-1,1]$, $|\Delta| > 0$. By (8.5.2) with $p = 1$ applying the arguments used to prove Denjoy-Lusin theorem on absolute convergence of trigonometric series [162, p. 232] we obtain (11.2.4).

On the other hand, by Monotone convergence theorem it follows from (11.2.4) that

$$\begin{aligned}\int_{-1}^{1} \sum_{n=0}^{\infty} |c_n P_n(d\mu,m;x)|\mu'(x)dx &\le \int_{-1}^{1} \sum_{n=0}^{\infty} |c_n P_n(d\mu,m;x)| d\mu(x) \\ \le \sum_{n=0}^{\infty} |c_n| \int_{-1}^{1} |P_n(d\mu,m;x)| d\mu(x) &\le \sum_{n=0}^{\infty} |c_n| \left\{\int_{-1}^{1} d\mu(x)\right\}^{1/m} < \infty,\end{aligned}$$

which implies that the series

$$\sum_{n=0}^{\infty} |c_n P_n(d\mu,m;x)|\mu'(x) < \infty$$

holds almost everywhere in $[-1,1]$ and hence

$$\sum_{n=0}^{\infty} |c_n P_n(d\mu,m;x)| < \infty$$

holds almost everywhere in the set $\{x \in [-1,1] : \mu'(x) > 0\}$. □

Next, for mean convergence of power orthogonal series we have the following

Theorem 11.2.2. *Let $2 \le m < \infty$, $0 < p < \infty$, $1 < q < \infty$, $q' = q/(q-1)$, and $u^{1/(1-q)}, u, w \in L^1_{d\mu}$. Let $\int_{-1}^{1} w(x)\mu'(x)dx > 0$ and $\int_{-1}^{1} u(x)\mu'(x)dx > 0$. Suppose that the inequality*

$$\|S_n(d\mu,m,f)\|_{wd\mu,p} \le C\|f\|_{ud\mu,q} \tag{11.2.5}$$

holds for all $n \in \mathbf{N}$ and all $f \in L^q_{ud\mu}$, where C is independent of n and f. Then

$$\sup_{n\in\mathbf{N}_0} \|P_n(d\mu,m)\|_{wd\mu,p} < \infty \tag{11.2.6}$$

and

$$\sup_{n\in\mathbf{N}_0} \|P_n(d\mu,m)u^{-1}\|_{ud\mu,q'} < \infty. \tag{11.2.7}$$

Moreover, if $m = 2$ and $p \geq 2$ then

$$\int_{-1}^{1} \left[\mu'(x)(1-x^2)^{1/2}\right]^{-p/2} w(x)\mu'(x)dx < \infty \tag{11.2.8}$$

and if $m = 2$ and $1 < q \leq 2$ then

$$\int_{-1}^{1} \left[\mu'(x)(1-x^2)^{1/2}\right]^{q/(2(1-q))} u^{1/(1-q)}(x)\mu'(x)dx < \infty. \tag{11.2.9}$$

Proof. First, we point out that using Hölder inequality, $f \in L^q_{ud\mu}$, together with $u^{1/(1-q)} \in L^1_{d\mu}$, implies $f \in L^1_{d\mu}$. Next, inequality (11.2.5) gives

$$\|S_{n+1}(d\mu,m,f) - S_n(d\mu,m,f)\|_{wd\mu,p} \leq C\|f\|_{ud\mu,q}$$

i.e.,

$$\|P_n(d\mu,m)\|_{wd\mu,p} \left| \int_{-1}^{1} f(x)P_n(d\mu,m;x)d\mu(x) \right| \leq C\|f\|_{ud\mu,q}. \tag{11.2.10}$$

If we choose

$$f(x) = \left[|P_n(d\mu,m;x)|u(x)^{-1}\right]^{1/(q-1)} \operatorname{sgn} P_n(d\mu,m;x)$$

then we see that $f \in L^q_{ud\mu}$. So by (11.2.10)

$$\|P_n(d\mu,m)\|_{wd\mu,p}\|P_n(d\mu,m)u^{-1}\|_{ud\mu,q'} \leq C.$$

Applying Theorem 8.5.2 we conclude

$$\inf_{n\in\mathbf{N}_0} \|P_n(d\mu,m)\|_{wd\mu,p} > 0$$

and

$$\inf_{n\in\mathbf{N}_0} \|P_n(d\mu,m)u^{-1}\|_{ud\mu,q'} > 0.$$

Hence inequalities (11.2.6) and (11.2.7) follow. By means of Lemma 11.2.1 inequalities (11.2.8) and (11.2.9) must be valid. □

11.3 Mean Convergence of Lagrange Interpolation on a Matrix of Nodes X

As we know, the Lebesgue function $\sum_{k=1}^{n}|\ell_{kn}(x)|$ plays a decisive role in the uniform convergence of Lagrange interpolation. While, as it turns out from the following basic lemma, the Lebesgue function type sum $\Lambda_n(X;x)=\sum_{k=1}^{n}|(x-x_{kn})\ell_{kn}(x)|$ (defined in (9.1.4)) plays a fundamental role in mean convergence of Lagrange interpolation.

Lemma 11.3.1. *Let $d\mu$ be a measure supported in $[-1,1]$. Let $X\subset[-1,1]$ and $0<p\le\infty$. Then for $n\in\mathbf{N}$*

$$\|\mathcal{L}_n(X)\|_{d\mu,p}\ge 2^{-2-1/p}\|\Lambda_n(X)\|_{d\mu,p}, \tag{11.3.1}$$

where $\|\mathcal{L}_n(X)\|_{d\mu,p}$ is defined in (9.1.3).

Proof. Let n be fixed. Choose $f\in C[-1,1]$ so that $\|f\|=1$ and

$$f(x_k)=\operatorname{sgn}\omega_n'(x_k),\quad k=1,\dots,n.$$

Then applying the identities

$$\frac{\omega_n(x)}{\omega_n'(x_k)}=(x-x_k)\ell_k(x),\quad k=1,\dots,n, \tag{11.3.2}$$

we obtain

$$\Lambda_n(X;x)=\sum_{k=1}^{n}\left|\frac{\omega_n(x)}{\omega_n'(x_k)}\right|$$

and

$$\Lambda_n(X;x)=\left|\sum_{k=1}^{n}f(x_k)(x-x_k)\ell_k(x)\right|.$$

Thus by (5.3.6)

$$\begin{aligned}
&\|\Lambda_n(X)\|_{d\mu,p}\\
=&\left\|\sum_{k=1}^{n}f(x_k)(x-x_k)\ell_k(x)\right\|_{d\mu,p}\\
\le&\ 2^{1+1/p}\left\{\left\|x\sum_{k=1}^{n}f(x_k)\ell_k(x)\right\|_{d\mu,p}+\left\|\sum_{k=1}^{n}x_kf(x_k)\ell_k(x)\right\|_{d\mu,p}\right\}\\
\le&\ 2^{2+1/p}\|\mathcal{L}_n(X)\|_{d\mu,p}.
\end{aligned}$$

Here we use the obvious fact that inequality $\|f\|\le 1$ implies $\|xf(x)\|\le 1$. □

As an immediate consequence of this lemma we get a necessary condition for weighted mean convergence of $\mathcal{L}_n(X,f)$.

Corollary 11.3.1. *Let $d\mu$ be a measure supported in $[-1,1]$. Let $X\subset[-1,1]$ and $0<p<\infty$. If*

$$\sup_{n\in\mathbf{N}}\|\mathcal{L}_n(X)\|_{d\mu,p}<\infty, \tag{11.3.3}$$

then

$$\sup_{n\in\mathbf{N}} \|\Lambda_n(X)\|_{d\mu,p} < \infty. \tag{11.3.4}$$

Lemma 11.3.2. *Let $d\mu$ a measure supported in $[-1,1]$. Let $X \subset [-1,1]$ and $0 < p \le \infty$. Then*

$$\int_{-1}^{1} \Lambda_n(X;x)^p d\mu(x) \ge c(p) > 0. \tag{11.3.5}$$

Proof. By the same arguments as that of (8.1.42) we can obtain (11.3.5). □

The converse inequality of (11.3.1) is not true in general. We only have the following weaker result, in which

$$\|\mathcal{L}_n(X)\|_{C^1\to L^p_{d\mu}} = \sup_{\|f\|\le 1, \|f'\|\le 1} \|\mathcal{L}_n(X)\|_{d\mu,p}.$$

Obviously, we have

$$\|\mathcal{L}_n(X)\|_{C^1\to L^p_{d\mu}} \le \|\mathcal{L}_n(X)\|_{d\mu,p}.$$

Lemma 11.3.3. *Let $d\mu$ be a measure supported in $[-1,1]$. Let $X \subset [-1,1]$ and $0 < p \le \infty$. Then for $n \in \mathbf{N}$*

$$\|\mathcal{L}_n(X)\|_{C^1\to L^p_{d\mu}} \le c(d\mu,p)\|\Lambda_n(X)\|_{d\mu,p}. \tag{11.3.6}$$

Proof. Let n be fixed. By the mean value theorem for derivative and by (5.3.6)

$$\begin{aligned}
&\|\mathcal{L}_n(X)\|_{C^1\to L^p_{d\mu}} \\
&= \sup_{\|f\|\le 1, \|f'\|\le 1} \left\| f(x) + \sum_{k=1}^{n} [f(x_k) - f(x)]\ell_k(x) \right\|_{d\mu,p} \\
&= \sup_{\|f\|\le 1, \|f'\|\le 1} \left\| f(x) + \sum_{k=1}^{n} f'(\xi_k)(x_k - x)\ell_k(x) \right\|_{d\mu,p} \\
&\le \|1 + \Lambda_n(X)\|_{d\mu,p} \\
&\le 2^{1+1/p}\left\{ \left(\int_{-1}^{1} d\mu(x) \right)^{1/p} + \|\Lambda_n(X)\|_{d\mu,p} \right\}.
\end{aligned}$$

Then using (11.3.5) we obtain (11.3.6). □

As an immediate consequence of this lemma we get a sufficient condition for weighted mean convergence of $\mathcal{L}_n(X,f)$.

Corollary 11.3.2. *Let $d\mu$ be a measure supported in $[-1,1]$. Let $X \subset [-1,1]$ and $0 < p \le \infty$. If the inequality (11.3.4) is true, then the limit*

$$\lim_{n\to\infty} \|f - \mathcal{L}_n(X,f)\|_{d\mu,p} = 0 \tag{11.3.7}$$

holds for every function $f \in C^1[-1,1]$.

Proof. Use (11.3.6) and apply Banach-Steinhaus theorem. □

The following theorem gives a necessary condition of weighted mean convergence of Lagrange interpolation on X for a regular measure $d\mu$.

Theorem 11.3.1. *Let $d\mu$ be a regular measure supported in $[-1,1]$, $X \subset [-1,1]$, and $0 < p < \infty$. If limit (11.3.7) holds for every $f \in C[-1,1]$, then Statements (a)–(h) of Theorem 9.1.1 and Statements (i) and (j) of Theorem 9.1.2 are valid.*

Proof. If (11.3.7) holds for every $f \in C[-1,1]$, then applying Banach-Steinhaus theorem inequality (11.3.3) is valid. So relation (9.1.17) must hold, because

$$\|\mathcal{L}_n(X)\|_{d\mu,p} \geq \left\{\int_{-1}^{1} d\mu(x)\right\}^{1/p}.$$

Then applying Theorems 9.1.1 and 9.1.2 proves our theorem. □

The following theorem gives a necessary condition of weighted mean convergence of Lagrange interpolation on X for a measure $d\mu \in \mathbf{S}$.

Theorem 11.3.2. *Let $d\mu$ be a measure supported in $[-1,1]$ and $d\mu \in \mathbf{S}$. Let $X \subset [-1,1]$ and $0 < p < \infty$. If limit (11.3.7) holds for every $f \in C[-1,1]$, then inequality (9.4.18) and Statements (a)–(c) of Theorem 9.4.2 are valid.*

Proof. If (11.3.7) holds for every $f \in C[-1,1]$, then applying Banach-Steinhaus theorem inequality (11.3.3) is true and hence by Corollary 11.3.1 inequality (11.3.4), or equivalently, the inequality

$$\sup_{n\in\mathbf{N}} \gamma_n(X)\|\omega_n(X)\|_{d\mu,p} < \infty \tag{11.3.8}$$

is valid. Since $d\mu \in \mathbf{S}$, by (9.4.8) we get (9.4.18). On the other hand, by (9.1.23) it follows from (11.3.8) that

$$\sup_{n\in\mathbf{N}} 2^n\|\omega_n(X)\|_{d\mu,p} < \infty.$$

Then applying Theorems 9.4.1 and 9.4.3 yields Statement (a) of Theorem 9.4.2. It remains to apply Theorem 9.4.2. □

11.4 Mean Convergence of Lagrange Interpolation on the Zeros of Power Orthogonal Polynomials

A well known result proved by Erdős and Turán in [20, p. 146] is

Lemma 11.4.1. *Let $d\mu$ be a measure supported in $[-1,1]$. Then*

$$\lim_{n\to\infty} \|f - \mathcal{L}_n(d\mu, f)\|_{d\mu,2} = 0 \tag{11.4.1}$$

holds for every function $f \in C[-1,1]$.

As an extension of this theorem we state

Theorem 11.4.1. *Let $\{d\mu^{(n)}\}$ be a sequence of measures supported in $[-1,1]$ with uniformly bounded total masses*

$$\sup_{n\in\mathbf{N}} \int_{-1}^{1} d\mu^{(n)}(x) < \infty. \tag{11.4.2}$$

Then

$$\lim_{n\to\infty} \|f - \mathcal{L}_n(d\mu^{(n)}, f)\|_{d\mu^{(n)},2} = 0 \tag{11.4.3}$$

holds for every function $f \in C[-1,1]$.

In particular, If $d\mu$ *is a measure supported in* $[-1,1]$ *and* $2 \le m < \infty$, *then limit*

$$\lim_{n\to\infty} \int_{-1}^{1} |f(x) - \mathcal{L}_n(d\mu, m, f; x)|^2 \, |P_n(d\mu, m; x)|^{m-2} d\mu(x) = 0 \tag{11.4.4}$$

holds for every function $f \in C[-1,1]$.

Proof. Since

$$\sup_{\|f\|=1} \int_{-1}^{1} |\mathcal{L}_n(d\mu^{(n)}, f; x)|^2 d\mu^{(n)}(x) \le \int_{-1}^{1} \sum_{k=1}^{n} \ell_{kn}(d\mu^{(n)}; x)^2 d\mu^{(n)}(x) = \int_{-1}^{1} d\mu^{(n)}(x),$$

$\mathcal{L}_n(d\mu^{(n)}, f)$, as an operator from $C[-1,1]$ to $L^2[-1,1]$, is uniformly bounded in n. Thus limit (11.4.3) holds for every function $f \in C[-1,1]$. To verify (11.4.4) it suffices to apply (11.4.3) to the sequence of measures

$$d\mu^{(n)}(x) = |P_n(d\mu, m; x)|^{m-2} d\mu(x),$$

the total masses of which are uniformly bounded:

$$\int_{-1}^{1} d\mu^{(n)}(x) = \int_{-1}^{1} |P_n(d\mu, m; x)|^{m-2} d\mu(x) \le \left\{ \int_{-1}^{1} d\mu(x) \right\}^{2/m}.$$

This proves that limit (11.4.4) holds for every function $f \in C[-1,1]$. □

It is natural to propose

Problem 11.4.1. Is it true that

$$\lim_{n\to\infty} \|f - \mathcal{L}_n(d\mu, m, f)\|_{d\mu,m} = 0 \tag{11.4.5}$$

holds for every function $f \in C[-1,1]$?

Now we can find two possible answers to this problem. The first is the following corollary, which is an immediate consequence of Theorems 8.4.3 and 11.4.1

Corollary 11.4.1. *Let* $d\mu$ *be a measure supported in* $[-1,1]$ *and let* $2 \le m < \infty$. *Then for any given* $\varepsilon > 0$, *there exists a sequence of sets* $I_n \subset [-1,1]$ *such that* $|I_n| \le \varepsilon$ *and limit*

$$\lim_{n\to\infty} \int_{[-1,1]\setminus I_n} |f(x) - \mathcal{L}_n(d\mu, m, f; x)|^2 d\mu(x) = 0 \tag{11.4.6}$$

holds for every function $f \in C[-1,1]$.

To discuss the next result we formulate a useful lemma, which establishes relationship between $|P_n(d\mu, m; x)|$ and $\Lambda_n(d\mu, m; x)$.

Lemma 11.4.2. *Let* $d\mu$ *be a measure supported in* $[-1,1]$ *and* $2 \le m < \infty$. *Then*

$$|P_n(d\mu, m; x)| \sim \Lambda_n(d\mu, m; x). \tag{11.4.7}$$

Proof. Since

$$\Lambda_n(d\mu,m;x) = \sum_{k=1}^{n} \left| \frac{P_n(d\mu,m;x)}{P_n'(d\mu,m;x_k)} \right| = |P_n(d\mu,m;x)| \sum_{k=1}^{n} \frac{1}{|P_n'(d\mu,m;x_k)|},$$

by (8.3.9) we obtain (11.4.7). □

The second answer to the above problem is as follows..

Theorem 11.4.2. *Let $d\mu$ be a measure supported in $[-1,1]$ and $2 \le m < \infty$. Then limit (11.4.5) holds for every function $f \in C^1[-1,1]$.*

Proof. By (11.4.7) we have

$$\sup_{n \in \mathbf{N}} \|\Lambda_n(d\mu,m)\|_{d\mu,m} < \infty. \tag{11.4.8}$$

The remainder of the proof is to apply Corollary 11.3.2 □

Applying Corollary 11.3.1 and Theorem 8.4.1 we can get a necessary condition for mean convergence of Lagrange interpolation based on the zeros of $P_n(d\mu,m;x)$.

Theorem 11.4.3. *Let $d\mu$ and $d\nu$ be measures supported in $[-1,1]$. Let $2 \le m < \infty$ and $0 < p < \infty$. If*

$$\lim_{n \to \infty} \|f - \mathcal{L}_n(d\mu,m,f)\|_{d\nu,p} = 0 \tag{11.4.9}$$

holds for every function $f \in C[-1,1]$, then

$$\sup_{n \in \mathbf{N}_0} \|P_n(d\mu,m)\|_{d\nu,p} < \infty. \tag{11.4.10}$$

Moreover, if $m = 2$ and $p \ge 2$ then

$$\int_{-1}^{1} \left[\mu'(x)(1-x^2)^{1/2}\right]^{-p/2} \nu'(x)dx < \infty. \tag{11.4.11}$$

Proof. Using (11.4.7) and applying Corollary 11.3.1 we obtain (11.4.10).

To prove the second part of the theorem we apply Lemma 11.2.1. □

11.5 Mean Convergence of Hermite Interpolation

Let $A_{kj} \in \mathbf{P}_{qn-1}$ be the fundamental polynomials defined by (2.2.9) with $m_k \equiv q \in \mathbf{N}$ on the nodes (9.1.1). Then for a fixed integer r, $0 \le r \le q-1$, and $f \in C^r[-1,1]$ the unique *truncated Hermite interpolation polynomial* is defined by

$$\mathcal{H}_{nqr}(f;x) = \mathcal{H}_{nqr}(X,f;x) = \sum_{j=0}^{r} \sum_{k=1}^{n} f^{(j)}(x_{kn}) A_{kj}(x). \tag{11.5.1}$$

We recognize that $\mathcal{H}_{n10} = \mathcal{L}_n$ is the Lagrange interpolation and $\mathcal{H}_{n20} = \mathcal{H}_n$ is the classical Hermite-Fejér interpolation. According to [136], $\mathcal{H}_{nqr}$ is called *Lagrange type interpolation* for $q \in \mathbf{N}_1$ and *Hermite-Fejér type interpolation* for $q \in \mathbf{N}_2$, respectively.

We also need the following notations:

$$d_k = d_{kn} = \max\{|x_{kn} - x_{k-1,n}|, |x_{kn} - x_{k+1,n}|, \quad k = 1,2,\dots,n,$$

$$\Gamma_{nqr}(X) = \sum_{k=1}^{n} \frac{1}{d_k^{q-1-r}|\omega_n'(x_k)|^q}, \tag{11.5.2}$$

and

$$\Lambda_{nqr}(X;x) = \Gamma_{nqr}(x)|\omega_n(x)|^q. \tag{11.5.3}$$

Clearly, we have

$$\Gamma_n = \Gamma_{n10}, \quad \Lambda_n(x) = \Lambda_{n10}(x). \tag{11.5.4}$$

Again we write $\mathcal{H}_{nqr}(X,f;x) = \mathcal{H}_{nqr}(d\mu,m,f;x)$, etc. if $x_{kn} = x_{kn}(d\mu,m)$. As it turns out, the *Lebesgue function type sum* $\sum_{k=1}^{n}|(x-x_k)A_{kj}(x)|$ (as well as $\Lambda_{nqj}(x)$) will play a fundamental role for mean convergence of truncated Hermite interpolation.

Lemma 11.5.1. *Let $X \subset [-1,1]$. Then*

$$\Gamma_{nqr}(X) \geq 4^{-q}n^{-r}\Gamma_n(X)^q \tag{11.5.5}$$

and

$$\Lambda_{nqr}(X;x) \geq 4^{-q}n^{-r}\Lambda_n(X;x)^q. \tag{11.5.6}$$

If $q-r \in \mathbf{N}_1$, then

$$\sum_{k=1}^{n}|(x-x_k)A_{kr}(X;x)| \geq \frac{1}{r!}\Lambda_{nqr}(X;x). \tag{11.5.7}$$

Proof. The conclusion for $q=1$ is trivial, because in this case $r=0$. For $q>1$ applying Hölder inequality twice we have

$$\begin{aligned}
\Gamma_n(X)^q &= \left\{\sum_{k=1}^{n}\frac{1}{|\omega_n'(x_k)|}\right\}^q \\
&\leq \left\{\sum_{k=1}^{n} d_k^{(q-1-r)/(q-1)}\right\}^{q-1}\sum_{k=1}^{n}\frac{1}{d_k^{q-1-r}|\omega_n'(x_k)|^q} \\
&\leq \left\{\sum_{k=1}^{n} d_k\right\}^{q-1-r}\left\{\sum_{k=1}^{n}1\right\}^r\sum_{k=1}^{n}\frac{1}{d_k^{q-1-r}|\omega_n'(x_k)|^q} \\
&\leq 4^q n^r\Gamma_{nqr}(X),
\end{aligned}$$

which is equivalent to (11.5.5). Inequality (11.5.6) follows from (11.5.3) and (11.5.5). Inequality (11.5.7) follows from (2.2.9), (2.4.20), (11.5.2), and (11.5.3). □

Lemma 11.5.2. *There exists a function $G \in C^r[-1,1]$ such that*

$$\begin{cases}
G^{(r)}(x_k) = sgn\ \omega_n'(x_k), & k=1,2,\ldots,n, \\
G^{(j)}(x_k) = 0, & k=1,2,\ldots,n,\ j=0,1,\ldots,r-1, \\
\|G^{(j)}\| \leq h_r = \frac{[(2r+1)!]^2}{r![(r+1)!]^2}, & j=0,1,\ldots,r.
\end{cases} \tag{11.5.8}$$

Proof. First we show that the polynomial

$$g(x) = g(a,b;x) = \frac{(x-a)^{r+1}(x-b)^r}{r!(b-a)^{r+1}} \quad (a \neq b,\ a,b \in [-1,1])$$

satisfies

$$\begin{cases} g^{(j)}(a) = 0, & j = 0,1\ldots,r, \\ g^{(j)}(b) = 0, & j = 0,1\ldots,r-1, \\ g^{(r)}(b) = 1 \end{cases} \tag{11.5.9}$$

and for $a' = \min\{a,b\}$, $b' = \max\{a,b\}$

$$\|g^{(j)}\|_{[a',b']} \leq h_r, \quad j = 0,1,\ldots,r. \tag{11.5.10}$$

In fact, relation (11.5.9) is obvious. Let us show (11.5.10) for $r \geq 1$, since inequality (11.5.10) is trivial for $r = 0$. Suppose without loss of generality that $a < b$. It is easy to check that $|g|$ attains its maximum on $[a,b]$ at $x = [ar + b(r+1)]/(2r+1)$ and

$$\|g\|_{[a,b]} = \frac{r^r(r+1)^{r+1}(b-a)^r}{r!(2r+1)^{2r+1}} \leq \frac{1}{r!} \leq h_r. \tag{11.5.11}$$

Hence by Markov inequality for each j, $1 \leq j \leq r$,

$$\begin{aligned} \|g^{(j)}\|_{[a,b]} &\leq \left(\frac{2}{b-a}\right)^j (2r+1)^2(2r)^2 \cdots (2r-j+2)^2 \|g\|_{[a,b]} \\ &\leq \frac{[(2r+1)!]^2}{r![(r+1)!]^2} = h_r. \end{aligned}$$

Next put

$$G(x) = \begin{cases} g\left(\frac{x_{k-1}+x_k}{2}, x_k; x\right) \operatorname{sgn} \omega_n'(x_k), & x \in \left[\frac{x_{k-1}+x_k}{2}, x_k\right],\ 1 \leq k \leq n+1, \\ g\left(\frac{x_{k+1}+x_k}{2}, x_k; x\right) \operatorname{sgn} \omega_n'(x_k), & x \in \left[x_k, \frac{x_{k+1}+x_k}{2}\right],\ 0 \leq k \leq n. \end{cases} \tag{11.5.12}$$

We assume $1 \leq k \leq n$ in the first line of (11.5.12) when $x_n = 1$ and in the second when $x_1 = -1$. It is easy to see that G satisfies (11.5.8). □

The following basic theorem shows that the function $\sum_{k=1}^n |(x-x_k)A_{kr}(x)|$ (as well as $\Lambda_{nqr}(x)$) plays a fundamental role for mean convergence of Lagrange type interpolation.

Theorem 11.5.1. *Let $d\mu$ be a measure supported in $[-1,1]$. Let $X \subset [-1,1]$, $0 < p \leq \infty$, $q \in \mathbf{N}_1$, and $r \in \mathbf{N}_{02}$. Then*

$$\|\mathcal{H}_{nqr}(X)\|_{d\mu,p} \geq c(p,r) \left\| \sum_{k=1}^n |(x-x_k)A_{kr}(X;x)| \right\|_{d\mu,p}, \tag{11.5.13}$$

$$\|\mathcal{H}_{nqr}(X)\|_{d\mu,p} \geq c(p,r) \|\Lambda_{nqr}(X)\|_{d\mu,p}, \tag{11.5.14}$$

and

$$\|\mathcal{H}_{nqr}(X)\|_{d\mu,p} \geq \frac{c(p,q,r)}{n^r} \|\Lambda_n(X)^q\|_{d\mu,p}. \tag{11.5.15}$$

Proof. Let G be defined as in Lemma 11.5.3. Then using the identities (11.3.2) we obtain

$$|(x-x_k)\ell_k(x)| = G^{(r)}(x_k)\frac{|\omega_n(x)|}{\omega_n'(x_k)}, \quad k=1,\dots,n.$$

Hence by (2.2.9), (2.2.20), and (11.5.8)

$$\begin{aligned}
&\sum_{k=1}^{n}|(x-x_k)A_{kr}(x)| \\
&= \frac{1}{r!}\sum_{k=1}^{n}G^{(r)}(x_k)\frac{|\omega_n(x)|}{\omega_n'(x_k)}(x-x_k)^r\ell_k(x)^{q-1}B_{kr}(x) \\
&= \left|\frac{1}{r!}\sum_{k=1}^{n}G^{(r)}(x_k)\frac{\omega_n(x)}{\omega_n'(x_k)}(x-x_k)^r\ell_k(x)^{q-1}B_{kr}(x)\right| \\
&= \left|\sum_{k=1}^{n}G^{(r)}(x_k)(x-x_k)A_{kr}(x)\right| \\
&= \left|\sum_{j=0}^{r}\sum_{k=1}^{n}G^{(j)}(x_k)(x-x_k)A_{kj}(x)\right|.
\end{aligned}$$

Thus by (5.3.6)

$$\begin{aligned}
\left\|\sum_{k=1}^{n}|(x-x_k)A_{kr}(x)|\right\|_{d\mu,p} &= \left\|\sum_{j=0}^{r}\sum_{k=1}^{n}G^{(j)}(x_k)(x-x_k)A_{kj}(x)\right\|_{d\mu,p} \\
\leq\ 2^{1+1/p}&\left\{\left\|x\sum_{j=0}^{r}\sum_{k=1}^{n}G^{(j)}(x_k)A_{kj}(x)\right\|_{d\mu,p}+\left\|\sum_{j=0}^{r}\sum_{k=1}^{n}x_kG^{(j)}(x_k)A_{kj}(x)\right\|_{d\mu,p}\right\}.
\end{aligned}$$

We observe that by (11.5.8)

$$x_kG^{(j)}(x_k) = [xG(x)]^{(j)}_{x=x_k} - jG^{(j-1)}(x_k) = [xG(x)]^{(j)}_{x=x_k}$$

and that inequality $\max_{0\leq j\leq r}\|f^{(j)}\| \leq h_r$ implies $\max_{0\leq j\leq r}\|\{xf(x)\}^{(j)}\| \leq 2^r h_r$. Thus we have

$$\begin{aligned}
&\left\|\sum_{k=1}^{n}|(x-x_k)A_{kr}(x)|\right\|_{d\mu,p} \\
&\leq\ 2^{2+1/p}\sup_{\|f\|_r\leq 2^r h_r}\left\|\sum_{j=0}^{r}\sum_{k=1}^{n}f^{(j)}(x_k)A_{kj}\right\|_{d\mu,p} \\
&=\ 2^{2+r+1/p}h_r\left\|\mathcal{H}_{nqr}(X)\right\|_{d\mu,p}.
\end{aligned}$$

This proves (11.5.13). Inequality (11.5.14) follows from (11.5.13) and (11.5.7); inequality (11.5.15) follows from (11.5.14) and (11.5.6). □

As consequences of this theorem we get necessary conditions for weighted mean convergence of $\mathcal{H}_{nqr}(X,f)$.

Corollary 11.5.1. *Let $d\mu$ be a measure supported in $[-1,1]$. Let $X \subset [-1,1]$, $0 < p < \infty$, $q \in \mathbf{N}_1$, and $r \in \mathbf{N}_{02}$. If*

$$\sup_{n\in\mathbf{N}} \|\mathcal{H}_{nqr}(X)\|_{d\mu,p} < \infty, \tag{11.5.16}$$

then

$$\sup_{n\in\mathbf{N}} \left\| \sum_{k=1}^{n} |(x-x_k)A_{kr}(X;x)| \right\|_{d\mu,p} < \infty, \tag{11.5.17}$$

$$\sup_{n\in\mathbf{N}} \|\Lambda_{nqr}(X)\|_{d\mu,p} < \infty, \tag{11.5.18}$$

and

$$\|\Lambda_n(X)\|_{d\mu,p} \le cn^{r/q}. \tag{11.5.19}$$

Proof. Apply Theorem 11.5.1. □

Corollary 11.5.2. *Let $d\mu$ and $d\nu$ be arbitrary measures supported in $[-1,1]$. Let $2 \le m < \infty$, $0 < p < \infty$, $q \in \mathbf{N}_1$, and $r \in \mathbf{N}_{02}$. If*

$$\sup_{n\in\mathbf{N}} \|\mathcal{H}_{nqr}(d\mu,m)\|_{d\nu,p} < \infty, \tag{11.5.20}$$

then

$$\|P_n(d\mu,m)\|_{d\nu,p} \le cn^{r/q}. \tag{11.5.21}$$

Proof. Use (11.5.19) and (11.4.7). □

Now let us discuss mean convergence of Hermit-Fejér type interpolation. To this end put

$$f_j(x) = x^j, \quad j = 0,1,\ldots$$

and

$$\mathcal{R}_{nqr}(X,f;x) = f(x) - \mathcal{H}_{nqr}(X,f;x).$$

In this case the following basic theorem shows that $\sum_{k=1}^{n} |(x-x_k)A_{k,r+1}(x)|$ (as well as $\Lambda_{n,m,r+1}(x)$) plays an important role for mean convergence of Hermit-Fejér type interpolation.

Theorem 11.5.2. *Let $X \subset [-1,1]$, $q \in \mathbf{N}_2$, and $r \in \mathbf{N}_{02}$. Then*

$$|\mathcal{R}_{nqr}(X,f_{r+1};x)| + |\mathcal{R}_{nqr}(X,f_{r+2};x)| \ge \frac{1}{2}\sum_{k=1}^{n} |(x-x_k)A_{k,r+1}(x)|, \tag{11.5.22}$$

$$|\mathcal{R}_{nqr}(X,f_{r+1};x)| + |\mathcal{R}_{nqr}(X,f_{r+2};x)| \ge \frac{1}{2}\Lambda_{n,q,r+1}(X;x), \tag{11.5.23}$$

and

$$|\mathcal{R}_{nqr}(X,f_{r+1};x)| + |\mathcal{R}_{nqr}(X,f_{r+2};x)| \ge 2^{-2q-1}n^{-r-1}\Lambda_n(X;x)^q. \tag{11.5.24}$$

Proof. Here we agree to replace $1/p!$ by 0 if $p<0$. Then by definition we have

$$x^p = \sum_{j=0}^{q-1}\sum_{k=1}^{n} \frac{p!}{(p-j)!} x_k^{p-j} A_{kj}(x), \quad 0 \le p \le qn-1.$$

In particular, for $p=r+1$ and $p=r+2$ we get

$$\mathcal{R}_{nqr}(f_{r+1};x) = (r+1)!\sum_{k=1}^{n} A_{k,r+1}(x)$$

and

$$\mathcal{R}_{nqr}(f_{r+2};x) = (r+2)!\sum_{k=1}^{n} [x_k A_{k,r+1}(x) + A_{k,r+2}(x)],$$

respectively. Hence

$$\begin{aligned}
&\sum_{k=1}^{n} [(x-x_k)A_{k,r+1}(x) - A_{k,r+2}(x)] \\
\le\ & \left|\sum_{k=1}^{n} A_{k,r+1}(x)\right| + \left|\sum_{k=1}^{n} [x_k A_{k,r+1}(x) + A_{k,r+2}(x)]\right| \\
=\ & \frac{1}{(r+1)!}|\mathcal{R}_{nqr}(f_{r+1};x)| + \frac{1}{(r+2)!}|\mathcal{R}_{nqr}(f_{r+2};x)| \\
\le\ & \frac{1}{(r+1)!}[|\mathcal{R}_{nqr}(f_{r+1};x)| + |\mathcal{R}_{nqr}(f_{r+2};x)|]. \qquad (11.5.25)
\end{aligned}$$

On the other hand, by (2.2.9) we see

$$\begin{aligned}
&\sum_{k=1}^{n} [(x-x_k)A_{k,r+1}(x) - A_{k,r+2}(x)] \\
=\ & \sum_{k=1}^{n} (x-x_k)^{r+2}\ell_k(x)^q \left[\frac{B_{k,r+1}(x)}{(r+1)!} - \frac{B_{k,r+2}(x)}{(r+2)!}\right]. \qquad (11.5.26)
\end{aligned}$$

According to (2.2.6) and (2.2.20)

$$B_{k,r+1}(x) - B_{k,r+2}(x) = b_{k,q-2-r}(x-x_k)^{q-2-r} \ge 0. \qquad (11.5.27)$$

Thus by (11.5.26) and (2.2.9)

$$\sum_{k=1}^{n} [(x-x_k)A_{k,r+1}(x) - A_{k,r+2}(x)] \ge \frac{r+1}{r+2}\sum_{k=1}^{n} |(x-x_k)A_{k,r+1}(x)|.$$

This inequality, coupled with (11.5.25), gives

$$|\mathcal{R}_{nqr}(X,f_{r+1};x)| + |\mathcal{R}_{nqr}(X,f_{r+2};x)| \ge \frac{(r+1)(r+1)!}{r+2}\sum_{k=1}^{n} |(x-x_k)A_{k,r+1}(x)|,$$

which implies (11.5.22). Inequality (11.5.23) follows from the above inequality and (11.5.7); inequality (11.5.24) follows from (11.5.23) and (11.5.6). □

This theorem has many applications. Combining (11.5.24) and (11.4.7) gives

Corollary 11.5.3. *Let $d\mu$ and $d\nu$ be measures supported in $[-1,1]$. Let $2 \le m < \infty$, $0 < p \le \infty$, $q \in \mathbf{N}_2$, and $r \in \mathbf{N}_{02}$. Then*

$$\|\mathcal{R}_{nqr}(d\mu,m,f_{r+1})\|_{d\nu,p} + \|\mathcal{R}_{nqr}(d\mu,m,f_{r+2})\|_{d\nu,p} \ge \frac{c(d\mu,p,q)}{n^{r+1}}\|P_n(d\mu,m)^q\|_{d\nu,p}. \tag{11.5.28}$$

Moreover, if

$$\lim_{n\to\infty}\{\|\mathcal{R}_{nqr}(d\mu,m,f_{r+1})\|_{d\nu,p} + \|\mathcal{R}_{nqr}(d\mu,m,f_{r+2})\|_{d\nu,p}\} = 0, \tag{11.5.29}$$

then

$$\|P_n(d\mu,m)\|_{d\nu,p} = o(n^{(r+1)/q}). \tag{11.5.30}$$

Corollary 11.5.4. *Let $d\mu$ be a measure supported in $[-1,1]$. Let $X \subset [-1,1]$, $0 < p \le \infty$, $q \in \mathbf{N}_2$, and $r \in \mathbf{N}_{02}$. If*

$$\|\mathcal{R}_{nqr}(X,f_{r+1})\|_{d\mu,p} + \|\mathcal{R}_{nqr}(X,f_{r+2})\|_{d\mu,p} \le cn^{-r-1}, \tag{11.5.31}$$

then

$$\sup_{n\in\mathbf{N}}\|\Lambda_n(X)\|_{d\mu,p} < \infty. \tag{11.5.32}$$

Moreover, if (11.5.31) is true and $d\mu$ is regular then Statements (a)–(h) of Theorem 9.1.1 and Statements (i) and (j) of Theorem 9.1.2 are valid.

If (11.5.31) is true and $d\mu \in \mathbf{S}$ then relation (9.4.18) and Statements (a)–(c) of Theorem 9.4.2 are valid.

Corollary 11.5.5. *Let $d\mu$ and $d\nu$ be measures supported in $[-1,1]$. Let $2 \le m < \infty$, $0 < p \le \infty$, $q \in \mathbf{N}_2$, and $r \in \mathbf{N}_{02}$. If*

$$\|\mathcal{R}_{nqr}(d\mu,m,f_{r+1})\|_{d\nu,p} + \|\mathcal{R}_{nqr}(d\mu,m,f_{r+2})\|_{d\nu,p} \le cn^{-r-1}, \tag{11.5.33}$$

then

$$\sup_{n\in\mathbf{N}_0}\|P_n(d\mu,m)^q\|_{d\nu,p} < \infty. \tag{11.5.34}$$

Furthermore, if $m = 2$ and $pq \ge 2$ then

$$\int_{-1}^{1}\left[\mu'(x)(1-x^2)^{1/2}\right]^{-pq/2}\nu'(x)dx < \infty. \tag{11.5.35}$$

Proof. Inequality (11.5.34) follows from (11.5.28) and (11.5.33); inequality (11.5.35) follows by Lemma 11.2.1 from (11.5.34). □

11.6 Two Positive Operators

In this section we deal with mean convergence of the two operators

$$\mathcal{F}_n(d\mu,m,f;x) = \lambda_n(d\mu_n;x)\sum_{k=1}^{n} f(x_{kn})\frac{\ell_{kn}(d\mu_n;x)^2}{\lambda_{kn}(d\mu_n)} \tag{11.6.1}$$

and

$$\mathcal{G}_n(d\mu,m,f;x)=\lambda_n(d\mu_n;x)\int_{-1}^{1}f(t)K_n(d\mu_n;x,t)^2d\mu_n(t), \tag{11.6.2}$$

where $x_{kn}=x_{kn}(d\mu_n)=x_{kn}(d\mu,m)$, $d\mu_n$ is defined in (8.2.1), and

$$K_n(d\mu_n;x,t)=\frac{\gamma_{n-1}(d\mu_n)}{\gamma_n(d\mu_n)}\cdot\frac{P_{n-1}(d\mu_n;t)P_n(d\mu_n;x)-P_{n-1}(d\mu_n;x)P_n(d\mu_n;t)}{x-t}. \tag{11.6.3}$$

It is easy to prove the following

Lemma 11.6.1. *Let $d\mu$ be a measure supported in $[-1,1]$ and $2\le m<\infty$.*
(a) If $f(x)\equiv 1$ then $\mathcal{F}_n(d\mu,m,f;x)\equiv 1$.
(b) If $f(x)\ge 0$ for $x\in[-1,1]$ then $\mathcal{F}_n(d\mu,m,f;x)\ge 0$ for $x\in[-1,1]$.
(c) $\mathcal{F}_n(d\mu,m,f;x_{kn})=f(x_{kn})$ for $k=1,2,\ldots,n$.
(d) $\mathcal{F}_n'(d\mu,m,f;x_{kn})=0$ for $k=1,2,\ldots,n$.
(e) $\mathcal{F}_n(d\mu,m,f)$ is a rational function of degree $(2n-2,2n-2)$, only the numerator depends on f.
(f) If $f(x)\equiv 1$ then $\mathcal{G}_n(d\mu,m,f;x)\equiv 1$.
(g) If $f(x)\ge 0$ for $x\in[-1,1]$ then $\mathcal{G}_n(d\mu,m,f;x)\ge 0$ for $x\in[-1,1]$.
(h) $\mathcal{G}_n(d\mu,m,f)$ is a rational function of degree $(2n-2,2n-2)$, only the numerator depends on f.

We accept the notations $\mu(1+0)=\mu(1)$ and $\mu(-1-0)=\mu(-1)$. As auxiliary lemmas we give several results which are of independent interest.

Lemma 11.6.2. *Let $d\mu$ be a measure supported in $[-1,1]$ and $0<m<\infty$. Then $\mu(t)$ is continuous at $t=x$ if and only if*

$$\lim_{n\to\infty}\lambda_n(d\mu,m;x)=0, \tag{11.6.4}$$

where $\lambda_n(d\mu,m;x)$ is defined in (5.1.7).

Proof. ($\Longrightarrow$) We give the proof for $x\in(-1,1)$ only; the proof for $x\in\{-1,1\}$ is similar.

Assume that $\varepsilon>0$ is arbitrary. By continuity there exists a positive number $\delta\le\min\{\varepsilon,1+x,1-x,1/2\}$ such that $|\mu(t)-\mu(x)|\le\varepsilon/2$ whenever $|t-x|\le\delta$. Choose $f\in C[-1,1]$ so that

$$f(t)=\begin{cases}1, & t\in[x-\delta/2,x+\delta/2],\\ 0, & t\in[-1,x-\delta]\cup[x+\delta,1],\\ \text{a linear function}, & t\in[x-\delta,x-\delta/2],\\ \text{a linear function}, & t\in[x+\delta/2,x+\delta].\end{cases}$$

By Weierstrass theorem there is a polynomial $P\in\mathbf{P}_{N-1}$ for N large enough so that $\|P-$

$f\| \leq \delta$. Then by (5.1.7) we have

$$\begin{aligned}
&\lambda_N(d\mu, m; x) \\
\leq\ & \frac{1}{|P(x)|^m} \int_{-1}^{1} |P(t)|^m d\mu(t) \\
\leq\ & \frac{1}{|f(x)-\delta|^m} \int_{-1}^{1} [f(t)+\delta]^m d\mu(t) \\
=\ & \frac{1}{(1-\delta)^m} \left\{ \int_{-1}^{x-\delta} \delta^m d\mu(t) + \int_{x+\delta}^{1} \delta^m d\mu(t) + \int_{x-\delta}^{x+\delta} [f(t)+\delta]^m d\mu(t) \right\} \\
\leq\ & \frac{1}{(1-\delta)^m} \left\{ \delta^m \int_{-1}^{1} d\mu(t) + (1+\delta)^m \int_{x-\delta}^{x+\delta} d\mu(t) \right\} \\
=\ & \frac{1}{(1-\delta)^m} \{\delta^m [\mu(1)-\mu(-1)] + (1+\delta)^m [\mu(x+\delta)-\mu(x-\delta)]\} \\
\leq\ & 2^m \{\varepsilon^m [\mu(1)-\mu(-1)] + (3/2)^m \varepsilon\}.
\end{aligned}$$

Since $\lambda_n(d\mu, m; x) \leq \lambda_N(d\mu, m; x)$ for $n \geq N$, we obtain

$$\lim_{n\to\infty} \lambda_n(d\mu, m; x) \leq 2^m \{\varepsilon^m [\mu(1)-\mu(-1)] + (3/2)^m \varepsilon\}.$$

Letting $\varepsilon \to 0$ we get (11.6.4).

($\Longleftarrow$) It suffices to show

$$\lambda_n(d\mu, m; x) \geq \mu(x+0) - \mu(x-0). \tag{11.6.5}$$

To this end let $P \in \mathbf{P}_{n-1}$ with $P(x) = 1$ satisfy

$$\lambda_n(d\mu, m; x) = \int_{-1}^{1} |P(t)|^m d\mu(t).$$

Then for $\varepsilon > 0$

$$\begin{aligned}
\lambda_n(d\mu, m; x) &\geq \int_{[-1,1]\cap[x-\varepsilon, x+\varepsilon]} |P(t)|^m d\mu(t) \\
&\geq \left\{ \min_{t\in[-1,1]\cap[x-\varepsilon, x+\varepsilon]} |P(t)|^m \right\} [\mu(x+0) - \mu(x-0)],
\end{aligned}$$

which yields (11.6.5) by letting $\varepsilon \to 0$, because $P(x) = 1$. □

Lemma 11.6.3. *Let $d\mu$ be a measure supported in $[-1,1]$ and $0 < m < \infty$. Then the following statements are equivalent.*

(a) $\mu \in C[-1,1]$.

(b) $\lim_{n\to\infty} \lambda_n(d\mu, m; x) = 0, \quad \forall x \in [-1,1]$.

(c) $\lim_{n\to\infty} \|\lambda_n(d\mu, m)\| = 0$.

(d) $\lim_{n\to\infty} \int_{-1}^{1} \lambda_n(d\mu, m; x)\mu(x) = 0$.

Proof. (a)$\Longleftrightarrow$(b) Apply Lemma 11.6.2.

(b)$\Longleftrightarrow$(c) Apply Dini theorem, since by (5.1.7) the function $\lambda_n(d\mu,m;x)$ is monotonically decreasing with respect to n for each fixed x.

(c)$\Longrightarrow$(d) Trivial.

(d)$\Longrightarrow$(a) By (11.6.5) we get that for $x \in [-1,1]$

$$\int_{-1}^{1} \lambda_n(d\mu,m;x)d\mu(x) \geq [\mu(x+0)-\mu(x-0)]^2, \tag{11.6.6}$$

which by Statement (d) implies Statement (a). □

Lemma 11.6.4. *Let $d\mu$ be a measure supported in $[-1,1]$, $2 \leq m < \infty$, and $0 < p < \infty$. If $\mu \in C[-1,1]$, then*

$$\lim_{n\to\infty} \int_{-1}^{1} [\lambda_n(d\mu_n;x)P_n(d\mu_n;x)^2]^p d\mu_n(x) = 0 \tag{11.6.7}$$

and

$$\lim_{n\to\infty} \int_{-1}^{1} [\lambda_n(d\mu_n;x)P_{n-1}(d\mu_n;x)^2]^p d\mu_n(x) = 0. \tag{11.6.8}$$

Proof. It is enough to prove (11.6.7). If $p \geq 1$ then by (8.4.3) and (8.2.16) we have with $c = c(d\mu,m)$ defined in (8.4.3)

$$\begin{aligned}
&\int_{-1}^{1} [\lambda_n(d\mu_n;x)P_n(d\mu_n;x)^2]^p d\mu_n(x) \\
= \ & c^p \int_{-1}^{1} \left[\frac{\lambda_n(d\mu_n;x)P_n(d\mu_n;x)^2}{c}\right]^p d\mu_n(x) \\
\leq \ & c^p \int_{-1}^{1} \frac{\lambda_n(d\mu_n;x)P_n(d\mu_n;x)^2}{c} d\mu_n(x) \\
\leq \ & c^{p-1} \int_{-1}^{1} \lambda_n(d\mu,m;x)^{2/m} P_n(d\mu_n;x)^2 d\mu_n(x) \\
\leq \ & c^{p-1} \|\lambda_n(d\mu,m)\|^{2/m},
\end{aligned}$$

which by Lemma 11.6.3 implies (11.6.7).

If $p < 1$ then by Hölder inequality

$$\begin{aligned}
&\int_{-1}^{1} [\lambda_n(d\mu_n;x)P_n(d\mu_n;x)^2]^p d\mu_n(x) \\
\leq \ & \left[\int_{-1}^{1} \lambda_n(d\mu_n;x)P_n(d\mu_n;x)^2 d\mu_n(x)\right]^p \left[\int_{-1}^{1} d\mu_n(x)\right]^{1-p} \\
\leq \ & c\left[\int_{-1}^{1} \lambda_n(d\mu_n;x)P_n(d\mu_n;x)^2 d\mu_n(x)\right]^p,
\end{aligned}$$

which by the previous conclusion again implies (11.6.7). □

Now we can give our main result in this section as follows.

Theorem 11.6.1. *Let a measure $d\mu$ supported in $[-1,1]$ satisfy $\mu \in C[-1,1]$. Let $2 \le m < \infty$ and $0 < p < \infty$. If $f \in C[-1,1]$, then*

$$\lim_{n\to\infty} \|\mathcal{F}_n(d\mu,m,f) - f\|_{d\mu_n,p} = 0 \tag{11.6.9}$$

and

$$\lim_{n\to\infty} \|\mathcal{G}_n(d\mu,m,f) - f\|_{d\mu_n,p} = 0. \tag{11.6.10}$$

Proof. First we point out that since $\mathcal{F}_n(d\mu,m,f)$ and $\mathcal{G}_n(d\mu,m,f)$ are linear positive operators and $\mathcal{F}_n(d\mu,m,1;x) \equiv \mathcal{G}_n(d\mu,m,1;x) \equiv 1$, in order to prove (11.6.9) and (11.6.10), it suffices to show

$$\lim_{n\to\infty} \|\mathcal{F}_n(d\mu,m,\psi_x;x)\|_{d\mu_n,p} = 0 \tag{11.6.11}$$

and

$$\lim_{n\to\infty} \|\mathcal{G}_n(d\mu,m,\psi_x;x)\|_{d\mu_n,p} = 0, \tag{11.6.12}$$

respectively, where $\psi_x(t) = (x-t)^2$.

To prove (11.6.11) by definition

$$\|\mathcal{F}_n(d\mu,m,\psi_x;x)\|_{d\mu_n,p}^p = \int_{-1}^1 \left[\lambda_n(d\mu_n;x)\sum_{k=1}^n \frac{(x-x_{kn})^2\ell_{kn}(d\mu_n;x)^2}{\lambda_{kn}(d\mu_n)}\right]^p d\mu_n(x). \tag{11.6.13}$$

By (8.2.11), (8.3.7), and (1.2.8), we get

$$\begin{aligned}
&\sum_{k=1}^n \frac{(x-x_{kn})^2\ell_{kn}(d\mu_n;x)^2}{\lambda_{kn}(d\mu_n)}\\
=\;& \frac{\gamma_{n-1}(d\mu_n)^2}{\gamma_n(d\mu_n)^2} P_n(d\mu_n;x)^2 \sum_{k=1}^n \lambda_{kn}(d\mu_n)P_{n-1}(d\mu_n;x_{kn})^2\\
\le\;& P_n(d\mu_n;x)^2.
\end{aligned}$$

Using this inequality equation (11.6.13) gives

$$\|\mathcal{F}_n(d\mu,m,\psi_x;x)\|_{d\mu_n,p}^p \le \int_{-1}^1 \left[\lambda_n(d\mu_n;x)P_n(d\mu_n;x)^2\right]^p d\mu_n(x),$$

which by (11.6.7) yields (11.6.11).

To prove (11.6.12) by definition

$$\begin{aligned}
&\|\mathcal{G}_n(d\mu,m,\psi_x;x)\|_{d\mu_n,p}^p\\
=\;& \int_{-1}^1 \left\{\lambda_n(d\mu_n;x)\int_{-1}^1 (x-t)^2K_n(d\mu_n;x,t)^2 d\mu_n(t)\right\}^p d\mu_n(x).
\end{aligned} \tag{11.6.14}$$

By (11.6.3) and (8.3.7) we obtain

$$\begin{aligned}
&\int_{-1}^1 (x-t)^2K_n(d\mu_n;x,t)^2 d\mu_n(t)\\
=\;& \int_{-1}^1 \frac{\gamma_{n-1}(d\mu_n)^2}{\gamma_n(d\mu_n)^2}\left[P_{n-1}(d\mu_n;t)P_n(d\mu_n;x) - P_{n-1}(d\mu_n;x)P_n(d\mu_n;t)\right]^2 d\mu_n(t)\\
\le\;& P_n(d\mu_n;x)^2 + P_{n-1}(d\mu_n;x)^2.
\end{aligned}$$

Using this inequality and (5.3.6) equation (11.6.14) yields

$$\begin{aligned}&\|\mathcal{G}_n(d\mu,m,\psi_x;x)\|^p_{d\mu_n,p}\\ \le\ & 2^p\int_{-1}^{1}\left\{\left[\lambda_n(d\mu_n;x)P_n(d\mu_n;x)^2\right]^p+\left[\lambda_n(d\mu_n;x)P_{n-1}(d\mu_n;x)^2\right]^p\right\}d\mu_n(x),\end{aligned}$$

which by (11.6.7) and (11.6.8) gives (11.6.12). □

11.7 Notes

11.7.1. Theorem 11.2.1 with $m=2$ was proved by Máté, Nevai and Totik [56] for a measure $d\mu$ with $\mu'(x)>0$, a.e. in $[-1,1]$, and then by Shi [85] for an arbitrary measure $d\mu$. Theorem 11.2.1 in the present form appears in [99].

11.7.2. Theorem 11.2.2 with $m=2$ was given by Nevai [65, Theorem 8.13, p. 154] for a measure $d\mu\in\mathbf{S}$, by Máté, Nevai and Totik [56] for a measure $d\mu$ with $\mu'(x)>0$, a.e. in $[-1,1]$. Theorem 11.2.2 with $m=2$ and an arbitrary measure $d\mu$ is due to Shi [85]. For other related results of weighted mean convergence of orthogonal Fourier series we refer to [56] and its references.

Instead of (11.2.1) we can also construct the nth partial sum of m orthogonal series in the form

$$S_n^*(d\mu,m,f;x)=\sum_{k=0}^{n-1}P_k(d\mu,m;x)\int_{-1}^{1}f(t)P_k(d\mu,m;t)|P_k(d\mu,m;t)|^{m-2}d\mu(t).$$

It is interesting to have a convergence result here.

11.7.3. Dealing with mean convergence of $\mathcal{L}_n(X,f)$ for $X\subset[-1,1]$ Turán [142, pp. 33-34] proposed

Problem 8. Given $p>1$, what is a necessary and sufficient condition that

$$\lim_{n\to\infty}\int_{-1}^{1}|f(x)-L_n(X,f,x)|^p dx=0$$

for every $f\in C[-1,1]$?

Mastroianni and Vértesi [55] and Shi [89] discussed this problem and obtained some necessary conditions for weighted mean convergence of Lagrange interpolation on a matrix of nodes X. Later Shi [94] proved Lemma 11.3.1, gave a necessary condition for weighted mean convergence of Lagrange interpolation (Theorem 11.3.1), and found the fundamental role of the Lebesgue function type sum $\Lambda(X;x)$ in weighted mean convergence of Lagrange interpolation.

11.7.4. Comparing Theorem 10.1.1 with Theorems 9.1.1 and 9.1.2, Theorem 11.3.1 shows that under its assumptions the asymptotic behaviour of $\omega_n(X;x)$ is very closely related to the regular (nth root) asymptotic behaviour of $\omega_n(d\mu,m;x)$.

Similarly, comparing Theorem 10.3.3 with Theorem 9.4.2, Theorem 11.3.2 shows that under its assumptions the asymptotic behaviour of $\omega_n(X;x)$ is very closely related to the power asymptotic behaviour of $\omega_n(d\mu,m;x)$.

Unfortunately, we have no a similar result for a ratio asymptotic behaviour.

11.7.5. The main results in Section 11.4 appears in [99]. It is interesting to have a result of the form (11.4.11) for the case when $m > 2$ here.

11.7.6. The main results in Section 11.5 are due to Shi [96].

11.7.7. The two operators (11.6.1) and (11.6.2) for $m = 2$ were introduced by Nevai [65, p. 58 and p. 74] and extended to the case when $m > 2$ by Shi [99].

Chapter 12

Gaussian Quadrature Formulas for Extended Chebyshev Spaces

12.1 Chebyshev Spaces and Weak Chebyshev Spaces

This chapter deals with the existence, characterizations, uniqueness, and extremal properties of a generalized Gaussian quadrature formula for an *extended Chebyshev space*. We begin with some definitions and lemmas.

Definition 12.1.1. *A system $\{u_0, \ldots, u_N\}$ is said to be an extended Chebyshev system and a space $U = \text{span}\,\{u_0, \ldots, u_N\}$ an extended Chebyshev space (of order $N+1$) on $[a,b]$ if $\{u_0, \ldots, u_N\} \subset C^N[a,b]$ and if each nonzero function $u \in U$ has at most N zeros on $[a,b]$, counting multiplicities up to order N.*

Definition 12.1.2. *An n-dimensional space U is said to be a weak Chebyshev space on $[a,b]$ if $U \subset C[a,b]$ and each function $u \in U$ has at most $n-1$ sign changes on $[a,b]$.*

Definition 12.1.3. *A space $U \subset C[a,b]$ is said to be nondegenerate on $[a,b]$ if $u \in U$ with the property that $u(x) = 0$ on a nontrivial interval $[c,d] \subset [a,b]$ implies $u = 0$ on $[a,b]$.*

A weak Chebyshev space has the following basic property.

Lemma 12.1.1. [15, Lemma 2.2] *If U is a weak Chebyshev space of dimension n on $[a,b]$ then for each r, $0 \le r \le n-1$, and each set of r points*

$$a = x_0 < x_1 < \cdots < x_r < x_{r+1} = b \tag{12.1.1}$$

there exists a nonzero function $u \in U$ such that

$$u(x) \prod_{k=1}^{r} (x - x_k) \ge 0, \quad x \in [a,b], \tag{12.1.2}$$

here we agree to replace $\prod_{k=1}^{r}(x-x_k)$ by 1 if $r = 0$.

We need a topological tool (Borsuk [11], see also Amann [2]) in which a mapping $\mathcal{T}$ is said to be odd if $\mathcal{T}(\mathbf{x}) = -\mathcal{T}(-\mathbf{x})$.

Theorem 12.1.1. (Brosuk antipodality theorem) *Let Δ be a bounded, open, symmetric neighborhood of 0 in $\mathbf{R}^{n+1}$ and $\mathcal{T}:\partial\Delta\to\mathbf{R}^n$ be an odd, continuous mapping. Then there exists some $\mathbf{x}\in\partial\Delta$ such that $\mathcal{T}(\mathbf{x})=0$.*

Lemma 12.1.2. (Hobby and Rice [40]) *Let a measure $d\mu$ supported in $[a,b]$ satisfy $\mu\in C[a,b]$. Let U be an n-dimensional space in $L^1_{d\mu}[a,b]$. Then there exist $r\le n$ points (12.1.1) such that*

$$\sum_{k=0}^{r}(-1)^k\int_{x_k}^{x_{k+1}}u(x)d\mu(x)=0,\quad u\in U. \tag{12.1.3}$$

Proof. Let $\mathbf{T}\subset\mathbf{R}^{n+1}$ be defined by

$$\mathbf{T}=\left\{\mathbf{t}=(t_0,t_1,\ldots,t_n):\sum_{k=0}^{n}|t_k|=b-a\right\}. \tag{12.1.4}$$

For $\mathbf{t}\in\mathbf{T}$ we associate to $\mathbf{t}$ the vector $\mathbf{x}=\mathbf{x}(\mathbf{t})$ specified by $x_0=a,\ x_{k+1}=x_k+|t_k|,\ k=0,1,\ldots,n$. Define $\mathcal{T}:\mathbf{T}\to\mathbf{R}^n$ by

$$\mathcal{T}_j(\mathbf{t})=\sum_{k=0}^{n}(\operatorname{sgn}t_k)\int_{x_k(\mathbf{t})}^{x_{k+1}(\mathbf{t})}u_j(x)d\mu(x),\quad j=1,2,\ldots,n. \tag{12.1.5}$$

Since $\mu\in C[a,b]$, $\mathcal{T}$ is continuous. Clearly $\mathcal{T}$ is odd. By Brosuk antipodality theorem we can find a vector $\mathbf{t}^*\in\mathbf{T}$ such that $\mathcal{T}(\mathbf{t}^*)=0$ and the points $x_k(\mathbf{t}^*)$ with $t^*_kt^*_{k+1}\le 0,\ k=1,2,\ldots,n$ give a solution for (12.1.3). □

Lemma 12.1.3. *Let $d\mu$ be a measure supported in $[a,b]$ and U an n-dimensional weak Chebyshev space on $[a,b]$. Assume that for the system of nodes (12.1.1) relation (12.1.3) holds. Then for each k, $0\le k\le r$,*

$$\mu(x_{k+1})-\mu(x_k)\ge\delta_1=\frac{1}{2}\min_{u\in U,\|u\|=1}\int_a^b|u(x)|d\mu(x)>0. \tag{12.1.6}$$

Furthermore, if $\mu\in C[a,b]$ then there exists a number $\delta=\delta(d\mu,U)>0$ such that for each k, $0\le k\le r$,

$$x_{k+1}-x_k\ge\delta. \tag{12.1.7}$$

Proof. If $r=0$ then the lemma is trivial. Now let $r\ge1$ and let k, $0\le k\le r$, be fixed. By Lemma 12.1.1 we can find a function $u\in U$ with $\|u\|=1$ such that

$$u(x)\prod_{i=1}^{k-1}(x-x_i)\prod_{i=k+2}^{r}(x-x_i)\ge0,\quad x\in[a,b],$$

here we agree to replace $\prod_{i=p}^{q}(x-x_i)$ by 1 if $q<p$. Inserting u into (12.1.3), we get

$$\int_a^b|u(x)|\operatorname{sgn}\left[(x-x_k)(x-x_{k+1})\right]d\mu(x)=0,$$

which implies

$$\int_{x_k}^{x_{k+1}}|u(x)|d\mu(x)=\int_{[a,b]\setminus[x_k,x_{k+1}]}|u(x)|d\mu(x),=\frac{1}{2}\int_a^b|u(x)|d\mu(x)\ge\delta_1.$$

Hence

$$\int_{x_k}^{x_{k+1}} d\mu(x) \geq \int_{x_k}^{x_{k+1}} |u(x)|d\mu(x) \geq \delta_1,$$

which is equivalent to (12.1.6). Inequality (12.1.7) follows from (12.1.6) by the uniform continuity of μ on $[a,b]$. □

The following lemma will play a crucial role in this section.

Lemma 12.1.4. ([46, Theorems B and 1]) *Let $\mu \in C[a,b]$ be strictly increasing on $[a,b]$. Let $U = span\,\{u_1,\dots,u_n\}$ be a nondegenerate weak Chebyshev space of dimension n on $[a,b]$. Then there exists a unique set of points (12.1.1) with $r = n$ such that (12.1.3) with $r = n$ holds.*

Proof. By Lemma 12.1.2 there exist $r \leq n$ points (12.1.1) such that (12.1.3) holds. We need to prove $r = n$. Suppose not and let $r < n$. In this case we rewrite (12.1.3) in the form

$$\sum_{k=0}^{r} s_k \int_{I'_k} u_j(x)d\mu(x) = 0, \quad j = 1,2,\dots,n, \tag{12.1.8}$$

where $s_k = (-1)^k$, $I'_k = [x_k, x_{k+1}]$, $k = 0,1,\dots,r$. We define the intervals I_k, $k = 0,1,\dots,n-1$ as follows. Put $I_k = I'_k$ for $k = 0,1,\dots,r-1$. If $r = n-1$ then $I_{n-1} = I'_{n-1}$; if $r < n-1$ then we substitute I'_r by the union of its $n-r$ disjoint nontrivial subintervals $I'_r = \cup_{k=r}^{n-1} I_k$. Then put $s_k = s_r$, $k = r+1,\dots,n-1$. In this case the relation (12.1.8) becomes

$$\sum_{k=0}^{n-1} s_k \int_{I_k} u_j(x)d\mu(x) = 0, \quad j = 1,2,\dots,n. \tag{12.1.9}$$

This means that the system of equations with unknowns $y_0,\dots,y_{n-1}$

$$\sum_{k=0}^{n-1} y_k \int_{I_k} u_j(x)d\mu(x) = 0, \quad j = 1,2,\dots,n, \tag{12.1.10}$$

has a nontrivial solution $s_0,\dots,s_{n-1}$. Thus the system of equations with n unknowns, the matrix of coefficients for which is the transpose of the matrix in (12.1.10), has a nontrivial solution, say, $a_1,\dots,a_n$, that is,

$$\sum_{j=1}^{n} a_j \int_{I_k} u_j(x)d\mu(x) = 0, \quad k = 0,1,\dots,n-1.$$

If we denote $u = \sum_{j=1}^{n} a_j u_j$ then u is a nonzero function and the above relation reads

$$\int_{I_k} u(x)d\mu(x) = 0, \quad k = 0,1,\dots,n-1. \tag{12.1.11}$$

This shows that the function u must change sign in each interval I_k, $k = 0,1,\dots,n-1$, and hence must change sign at least n times in all, a contradiction. This contradiction proves $r = n$.

To prove the uniqueness of nodes x_k suppose to the contrary that there exists another set of nodes $a = y_0 < y_1 < \cdots < y_n < y_{n+1} = b$ such that

$$\sum_{k=0}^{n}(-1)^k \int_{y_k}^{y_{k+1}} u_j(x)d\mu(x) = 0, \quad j = 1,2,\ldots,n.$$

Then

$$\sum_{k=0}^{n}(-1)^k \left[\int_{x_k}^{x_{k+1}} u_j(x)d\mu(x) - \int_{y_k}^{y_{k+1}} u_j(x)d\mu(x)\right] = 0, \quad j = 1,2,\ldots,n.$$

This implies that there exist $r+1 \leq n$ disjoint nontrivial subintervals I_k' and $r+1$ numbers $s_k = \pm 1$ such that the relations (12.1.8) hold. By the same argument as above we again obtain (12.1.11) for some nontrivial function $u \in U$, which leads to a contradiction and proves the uniqueness. □.

To state the following results we need some notations. In what follows let $m_k \in \mathbf{N}$, $k = 1,2,\ldots,n$ and $N = \sum_{k=1}^{n} m_k - 1$. Let $U = \text{span}\,\{u_0, u_1, \ldots, u_N\}$ be an extended Chebyshev space on $[a,b]$. For $\mathbf{y} \in \mathbf{Y}$, where the set $\mathbf{Y}$ is defined in Chapter 3, denote by $u_{kj} = u_{kj}(\mathbf{y}) \in U$, $j = 0,1,\ldots,m_k - 1$, $k = 1,\ldots,n$, the fundamental functions for the Hermite interpolation, i.e., u_{kj} satisfy the interpolation conditions:

$$u_{kj}^{(p)}(y_q) = \delta_{jp}\delta_{kq}, \quad p = 0,1,\ldots,m_q - 1,\ q = 1,\ldots,n. \tag{12.1.12}$$

Put for $\mathbf{y} \in \overline{\mathbf{Y}}$

$$V(\mathbf{y}) = \{v \in U : v^{(j)}(y_k) = 0,\ j = 0,1,\ldots,m_k - 2,\ k = 1,\ldots,n\} \tag{12.1.13}$$

and

$$W(\mathbf{y}) = \left\{ v(x)\text{sgn} \prod_{k=1}^{n}(x - y_k)^{m_k - 1} : v \in V(\mathbf{y}) \right\}. \tag{12.1.14}$$

Here and later the case $(y_{k-1} <) y_k = \cdots = y_r (< y_{r+1})$ is equivalent to the node y_k with multiplicity $m_k + \cdots + m_r$.

Lemma 12.1.5. *Let U be an $(N+1)$-dimensional extended Chebyshev space on $[a,b]$. Then for each $\mathbf{y} \in \overline{\mathbf{Y}}$ the set $V(\mathbf{y})$ is a space of dimension n and the set $W(\mathbf{y})$ is a nondegenerate weak Chebyshev space of dimension n on $[a,b]$.*

Moreover, for each positive constant $\delta < (b-a)/(n+1)$ the space $W(\mathbf{y})$ has a basis that depends continuously on $\mathbf{y}$ in the set

$$\mathbf{Y}_\delta := \{\mathbf{y} = (y_1, \ldots, y_n) : y_{k+1} - y_k \geq \delta,\ k = 0,1,\ldots,n,\ y_0 = a,\ y_{n+1} = b\}.$$

Proof. First, we point out that by definition $V(\mathbf{y}) \subset C[a,b]$ and $W(\mathbf{y}) \subset C[a,b]$. Next, choose n points $a < z_1 < \cdots < z_n < a + \delta$ so that $z_k \neq y_j$, $k,j = 1,\ldots,n$. Let $v_k \in V(\mathbf{y})$, $k = 1,\ldots,n$, satisfy further interpolation conditions:

$$v_k(z_j) = \delta_{kj}, \quad j = 1,2,\ldots,n.$$

It is easy to see that

$$V(\mathbf{y}) = \text{span}\,\{v_1, \ldots, v_n\}. \tag{12.1.15}$$

This implies that both $V(\mathbf{y})$ and $W(\mathbf{y})$ are spaces of dimension n on $[a,b]$. Meanwhile $W(\mathbf{y})$ has a basis

$$\left\{ v_j(x)\text{sgn} \prod_{k=1}^{n} (x-y_k)^{m_k-1}, \quad j = 1,2,\ldots,n \right\}, \tag{12.1.16}$$

which depends continuously on $\mathbf{y}$ in $\mathbf{Y}_\delta$.

To prove that $W(\mathbf{y})$ is a weak Chebyshev space of dimension n on $[a,b]$ suppose to the contrary that there is a vector $w \in W(\mathbf{y})$ which changes sign at n points $a < t_1 < \cdots < t_n < b$. By (12.1.14) there is a function $v \in V(\mathbf{y})$ such that

$$w(x) = v(x)\text{sgn} \prod_{k=1}^{n} (x-y_k)^{m_k-1}. \tag{12.1.17}$$

If $t_i = y_k$ with $y_{k-1} < y_k = \cdots = y_r < y_{r+1}$, $1 \le k \le r \le n$, then by (12.1.13) t_i is a zero of at least $(m_k-1)+\cdots+(m_r-1)$ multiplicities of v; hence further by (12.1.17) t_i is a zero of at least $(m_k-1)+\cdots+(m_r-1)+1$ multiplicities of v, because w changes sign at t_i. If $t_i \notin \{y_k : 1 \le k \le n\}$ then t_i is a new zero of v. To sum up v would have at least

$$\sum_{k=1}^{n} (m_k-1) + n = N+1$$

zeros counting multiplicities. This means $v = 0$ and hence $w = 0$, a contradiction. This contradiction proves that $W(\mathbf{y})$ is a weak Chebyshev space of dimension n on $[a,b]$. □

Remark 12.1.2. If U is an extended Chebyshev space on $[c,d] \supset [a,b]$ with $c < a$ or $b < d$ then choosing n points $z_1 < \cdots < z_n$ in $[c,d] \setminus [a,b]$ we can prove that $W(\mathbf{y})$ has a basis (12.1.16) on $[a,b]$ that depends continuously on $\mathbf{y}$ in $\overline{\mathbf{Y}}$.

12.2 Existence and Characterizations

A Gaussian quadrature formula for an $(N+1)$-dimensional extended Chebyshev space on $[a,b]$ with respect to a measure $d\mu$ is of the form

$$\int_a^b u(x)S(\mathbf{x};x)d\mu(x) = \sum_{k=1}^{n} \sum_{j=0}^{m_k-2} c_{kj}(\mathbf{x})u^{(j)}(x_k), \tag{12.2.1}$$

which is exact for all $u \in U$, where $\mathbf{x} = (x_1, \ldots, x_n)$ with

$$a = x_0 < x_1 < x_2 < \cdots < x_n < x_{n+1} = b, \tag{12.2.2}$$

and

$$S(\mathbf{x};x) = \text{sgn} \prod_{k=1}^{n} (x-x_k)^{m_k}. \tag{12.2.3}$$

Lemma 12.2.1. *Let $u_{kj}(\mathbf{y};x)$, $1 \le k, j \le n$, be defined in (12.1.12). Then*

$$\frac{\partial u_{k,m_k-1}(\mathbf{y};x)}{\partial y_j} = \begin{cases} -u^{(m_j)}_{k,m_k-1}(\mathbf{y};y_j)u_{j,m_j-1}(\mathbf{y};x), & j \neq k, \\ -u^{(m_k)}_{k,m_k-1}(\mathbf{y};y_k)u_{k,m_k-1}(\mathbf{y};x) - u_{k,m_k-2}(\mathbf{y};x), & j = k. \end{cases} \tag{12.2.4}$$

Here $u_{k,-1}(\mathbf{y};\cdot) := 0$.

Proof. According to (12.1.12) it follows that the function

$$v_{kj}(x) = \frac{\partial u_{k,m_k-1}(\mathbf{y};x)}{\partial y_j} \tag{12.2.5}$$

satisfies the interpolation conditions:

$$v^{(p)}_{kj}(y_q) = 0, \quad p = \begin{cases} 0, \dots m_q - 1, & q \neq j, \\ 0, \dots, m_q - 2, & q = j,\ j \neq k, \\ 0, \dots, m_q - 3, & q = j,\ j = k. \end{cases} \tag{12.2.6}$$

It is easy to see that (12.2.6) with $j \neq k$ by (12.1.12) yields the first line of (12.2.4). Similarly, (12.2.6) with $j = k$ by (12.1.12) gives

$$v_{kk} = c u_{k,m_k-1} + d u_{k,m_k-2}$$

with constants c and d to be determined. Again using (12.1.12) we can further determine these two constants: $c = -u^{(m_k)}_{k,m_k-1}(\mathbf{y};y_k)$, $d = -u^{(m_k-1)}_{k,m_k-1}(\mathbf{y};y_k) = -1$. This gives the second line of (12.2.4). □

Lemma 12.2.2. *Let $d\mu$ be a measure supported in $[a,b]$ and when $\min_{1\le k\le n} m_k = 1$ assume further that $\mu \in C^1[a,b]$. Let $1 \le k, j \le n$ and write*

$$\varphi_k(\mathbf{y}) = \int_a^b u_{k,m_k-1}(\mathbf{y};x)S(\mathbf{y};x)d\mu(x). \tag{12.2.7}$$

Then

$$\frac{\partial \varphi_k(\mathbf{y})}{\partial y_j} = \begin{cases} -u^{(m_j)}_{k,m_k-1}(\mathbf{y};y_j)c_{j,m_j-1}(\mathbf{y}), & j \neq k, \\ -u^{(m_k)}_{k,m_k-1}(\mathbf{y};y_k)c_{k,m_k-1}(\mathbf{y}) - c_{k,m_k-2}(\mathbf{y}), & j = k. \end{cases} \tag{12.2.8}$$

Here we use the notation

$$c_{k,-1}(\mathbf{y}) = 2S(\mathbf{y};y_k+0)\mu'(y_k).$$

Moreover, if a set of nodes (12.2.2) is a solution of the system of normal equations:

$$\varphi_k(\mathbf{x}) = 0, \quad k = 1,2,\dots,n, \tag{12.2.9}$$

then

$$\frac{\partial \varphi_k(\mathbf{x})}{\partial y_j} = \begin{cases} 0, & j \neq k, \\ -c_{k,m_k-2}(\mathbf{x}), & j = k. \end{cases} \tag{12.2.10}$$

Proof. Denote by $\mathbf{e}_j$ the jth unit vector of dimension n. Then

$$\begin{aligned}
&\frac{\partial \varphi_k(\mathbf{y})}{\partial y_j} \\
&= \lim_{h\to 0}\frac{1}{h}\int_a^b [u_{k,m_k-1}(\mathbf{y}+h\mathbf{e}_j;x)S(\mathbf{y}+h\mathbf{e}_j;x)-u_{k,m_k-1}(\mathbf{y};x)S(\mathbf{y};x)]d\mu(x) \\
&= \lim_{h\to 0}\frac{1}{h}\int_a^b [u_{k,m_k-1}(\mathbf{y}+h\mathbf{e}_j;x)-u_{k,m_k-1}(\mathbf{y};x)]S(\mathbf{y}+h\mathbf{e}_j;x)d\mu(x) \\
&\quad +\lim_{h\to 0}\frac{1}{h}\int_a^b u_{k,m_k-1}(\mathbf{y};x)[S(\mathbf{y}+h\mathbf{e}_j;x)-S(\mathbf{y};x)]d\mu(x) \\
&= S_1+S_2.
\end{aligned}$$

It is easy to see that

$$S_1=\int_a^b \frac{\partial u_{k,m_k-1}(\mathbf{y};x)}{\partial y_j}S(\mathbf{y};x)d\mu(x).$$

To compute S_2 we separate two cases.

Case 1. $j\neq k$ or $m_k\geq 2$. In this case recalling $m_k\geq 2$ for $j=k$, we have $u_{k,m_k-1}(\mathbf{y};y_j)=0$ and hence

$$\begin{aligned}
&|S_2| \\
&= \left|\lim_{h\to 0}\frac{1}{h}\int_{y_j}^{y_j+h}[u_{k,m_k-1}(\mathbf{y};x)-u_{k,m_k-1}(\mathbf{y};y_j)][S(\mathbf{y}+h\mathbf{e}_j;x)-S(\mathbf{y};x)]d\mu(x)\right| \\
&\leq \lim_{h\to 0}2\max_{|x-y_j|\leq h}|u'_{k,m_k-1}(\mathbf{y};x)|\;|\mu(y_j+h)-\mu(y_j)|=0.
\end{aligned}$$

Case 2. $j=k$ and $m_k=1$. In this case we have that for any h with $|h|>0$ small enough

$$S_2=\lim_{h\to 0}(\operatorname{sgn} h)\frac{1}{h}\int_{y_k}^{y_k+h}u_{k,0}(\mathbf{y};x)\,[S(\mathbf{y}+h\mathbf{e}_k;x)-S(\mathbf{y};x)]\,d\mu(x).$$

For $x\in(y_k,y_k+h)$ with $h>0$ we have

$$S(\mathbf{y}+h\mathbf{e}_k;x)=(-1)^{m_k}S(\mathbf{y};x)=-S(\mathbf{y};x)$$

and

$$S(\mathbf{y};x)=S(\mathbf{y};y_k+0);$$

for $x\in(y_k+h,y_k)$ with $h<0$ we have

$$S(\mathbf{y}+h\mathbf{e}_k;x)=-S(\mathbf{y};x)$$

and

$$S(\mathbf{y};x)=S(\mathbf{y};y_k-0).$$

Since $m_k=1$, $S(\mathbf{y};y_k-0)=-S(\mathbf{y};y_k+0)$ and hence

$$S(\mathbf{y}+h\mathbf{e}_k;x)-S(\mathbf{y};x)=-2(\operatorname{sgn} h)S(\mathbf{y};y_k+0)$$

for $x \in (\min\{y_k, y_k+h\}, \max\{y_k, y_k+h\})$. Thus

$$\begin{aligned} S_2 &= -2S(\mathbf{y}; y_k+0) \lim_{h\to 0} \frac{1}{h} \int_{y_k}^{y_k+h} u_{k,0}(\mathbf{y}; x) d\mu(x) \\ &= -2S(\mathbf{y}; y_k+0) \lim_{h\to 0} \frac{1}{h} [\mu(y_k+h) - \mu(y_k)] \\ &= -2S(\mathbf{y}; y_k+0)\mu'(y_k). \end{aligned}$$

Using (12.2.4) we obtain (12.2.8).

By (12.2.1), (12.1.12), and (12.2.7)

$$c_{k,m_k-1}(\mathbf{y}) = \varphi_k(\mathbf{y}), \quad k = 1, 2, \ldots, n. \tag{12.2.11}$$

Thus equations (12.2.9) are equivalent to

$$c_{k,m_k-1}(\mathbf{x}) = 0, \quad k = 1, 2, \ldots, n. \tag{12.2.12}$$

Relation (12.2.10) follows directly from (12.2.8) and (12.2.12). □

Theorem 12.2.1. *Let $d\mu$ a measure supported in $[a,b]$. Then the following statements are equivalent.*

(a) There exists a set of nodes (12.2.2) such that the relation (12.2.1) is exact for every $u \in U$.

(b) A set of nodes (12.2.2) is a solution of the system of normal equations (12.2.9).

(c) A set of nodes (12.2.2) satisfies the orthogonality relation

$$\int_a^b v(x) S(\mathbf{x}; x) d\mu(x) = 0, \quad v \in V(\mathbf{x}). \tag{12.2.13}$$

Proof. (a)⟺(b) By (12.2.11) and (12.2.12) we obtain equivalence of Statements (a) and (b).

(b)⟺(c) According to (12.1.13)

$$V(\mathbf{y}) = \text{span}\,\{u_{1,m_1-1}(\mathbf{y}; \cdot), \ldots, u_{n,m_n-1}(\mathbf{y}; \cdot)\}$$

and hence by (12.2.7) we obtain equivalence of Statements (b) and (c). □

Theorem 12.2.2. *Let a measure $d\mu$ supported in $[a,b]$ satisfy that $\mu \in C[a,b]$. Then there exists a set of nodes (12.2.2) such that the relation (12.2.1) is exact for every $u \in U$.*

Moreover, Statements (a)–(c) of Theorem 12.2.1 are equivalent.

Proof. First assume that $\mu(x)$ is strictly increasing. To prove the theorem according to Theorem 12.2.1 it suffices to show that there exists a set of nodes (12.2.2) satisfying (12.2.13), or equivalently, by (12.1.14) satisfying the orthogonality relation

$$\int_a^b w(x) \left[\text{sgn} \sum_{k=1}^{n} (x - x_k)\right] d\mu(x) = 0, \quad w \in W(\mathbf{x}). \tag{12.2.14}$$

By Lemma 12.1.3 if $\mathbf{x} \in \overline{\mathbf{Y}}$ satisfies (12.2.14) then

$$\begin{aligned} \mu(x_{k+1}) - \mu(x_k) &\geq \frac{1}{2} \min_{w \in W(\mathbf{x}), \|w\|=1} \int_a^b |w(x)| d\mu(x) \\ &\geq \delta_1 = \frac{1}{2} \min_{u \in U, \|u\|=1} \int_a^b |u(x)| d\mu(x). \end{aligned}$$

By uniform continuity we can choose $\delta > 0$ so small that inequality $0 < y - x < \delta$ implies $\mu(y) - \mu(x) < \delta_1$. Then for each k , $1 \leq k \leq n$, the inequality $\mu(x_{k+1}) - \mu(x_k) \geq \delta_1$ implies $x_{k+1} - x_k \geq \delta$, which gives $\mathbf{x} \in \mathbf{Y}_\delta$. By Lemma 12.1.4 for each $\mathbf{y} \in \mathbf{Y}_\delta$ there exists a unique vector $\mathbf{z} \in \mathbf{Y}_\delta$ such that

$$\int_a^b w(x) \left[\operatorname{sgn} \sum_{k=1}^{n} (x - z_k) \right] d\mu(x) = 0, \quad w \in W(\mathbf{y}). \tag{12.2.15}$$

Then we define a map $\mathcal{T} : \mathbf{Y}_\delta \to \mathbf{Y}_\delta$ by $\mathbf{z} = \mathcal{T}\mathbf{y}$, which is a continuous map of $\mathbf{Y}_\delta$ into itself. Now applying the Brouwer fixed-point theorem it has a fixed point $\mathbf{x} \in \mathbf{Y}_\delta$, i.e., $\mathcal{T}\mathbf{x} = \mathbf{x}$, which satisfies (12.2.14).

Next, for a measure $d\mu$ satisfying the assumption of the theorem only let us consider the measure $\mu_q(x) = \mu(x) + qx$, $q > 0$, which is strictly increasing. According to the previous conclusion, for each $q > 0$ there is a solution $\mathbf{x}_q \in \mathbf{Y}_\delta$ satisfying

$$\varphi_k(\mathbf{x}_q) = 0, \quad k = 1, 2, \ldots, n. \tag{12.2.16}$$

Let $\mathbf{x} \in \mathbf{Y}_\delta$ be a limit point of $\mathbf{x}_q$ as $q \to 0$. Then equations (12.2.16) lead to (12.2.9). Then the first part of the theorem is proved. The second may be proved by applying Theorem 12.2.1. □

Remark 12.2.1. The proof of the first part of Theorem 12.2.2 proceeds passing from $q > 0$ to $q = 0$. The reason is that under the present assumptions of Theorem 12.2.2 the solution of (12.2.15) is not necessarily unique (see Remark 3.2.3).

Remark 12.2.2. The example in Remark 3.2.1 shows that the condition $\mu \in C[a,b]$ can not be dropped for the existence in the case when $m = 1$ in general.

12.3 Uniqueness

This section deals with the uniqueness of a solution $\mathbf{x}$ of a Gaussian quadrature formula. To this end we determine the signs of the leading coefficients in (12.2.1).

Lemma 12.3.1. *Let U be an $(N+1)$-dimensional extended Chebyshev space on $[a,b]$ and let a measure $d\mu$ supported on $[a,b]$ satisfy that $\mu \in C[a,b]$. When $\min_{1 \leq k \leq n} m_k = 1$ assume further that $\mu \in C^1[a,b]$ and $\mu'(x) > 0$ in (a,b). If $c_{k,m_k-1}(\mathbf{x}) = 0$ then*

$$sgn\ c_{k,m_k-2}(\mathbf{x}) = S(\mathbf{x}; x_k + 0). \tag{12.3.1}$$

Proof. If $m_k = 1$ then by Lemma 12.2.2

$$c_{k,-1}(\mathbf{x}) = 2S(\mathbf{x}; x_k + 0)\mu'(x_k),$$

which proves (12.3.1). Let $m_k \geq 2$ and let $u \in U$ be defined by the $N+1$ interpolation conditions

$$\begin{cases} u^{(j)}(x_k) = \delta_{j,m_k-2}, & j = 0,\dots,m_k-2, \\ u^{(j)}(x_i) = 0, & j = 0,\dots,m_i-1,\ i = 1,\dots,n,\ i \neq k, \\ u(b) = 0. \end{cases} \tag{12.3.2}$$

Then the function u has no more zeros than the N zeros specified by (12.3.2) and the function $uS(\mathbf{x};\cdot)$ does not change sign in $[a,b]$. Clearly, $u(x_k+0) > 0$ and hence

$$\text{sgn } c_{k,m_k-2}(\mathbf{x}) = \text{sgn} \int_a^b u(x)S(\mathbf{x};x)d\mu(x) = S(\mathbf{x};x_k+0). \quad \square$$

The main result in this section is as follows.

Theorem 12.3.1. *Let U be an $(N+1)$-dimensional extended Chebyshev space on $[a,b]$ and let a measure $d\mu$ supported on $[a,b]$ satisfy that $\mu \in C[a,b]$. When $\min_{1\leq k\leq n} m_k = 1$ assume further that $\mu \in C^1[a,b]$ and $\mu'(x) > 0$ in (a,b). Then there exists a unique set of nodes (12.2.2) such that relation (12.2.1) is exact for every $u \in U$.*

Proof. By Theorem 12.2.2 there exists a set of nodes (12.2.2) satisfies the system of normal equations (12.2.9). Again by Theorem 12.2.2 to prove the uniqueness it suffices to show that the system of normal equations (12.2.9) has a unique solution (12.2.2). This may be proved by an argument similar to that of Theorem 3.2.2. We omit the details. $\square$

Remark 12.3.1. The example in Remark 3.2.3 shows that the condition that $\mu \in C^1[a,b]$ and $\mu(x) > 0$ in (a,b) can not be dropped for the uniqueness in the case when $m = 1$ in general.

12.4 Extremal Properties of Gaussian Quadrature Formulas

In this section we discuss extremal properties of Gaussian quadrature formulas for an extended Chebyshev space. To this end choose a function u_{N+1} so that the system $\{u_0,\dots,u_N,u_{N+1}\}$ is an extended Chebyshev system of order $N+2$. For $\mathbf{y} \in \overline{\mathbf{Y}}$ we also choose a number $\rho = 1$ or -1 and a function

$$P(\mathbf{y};x) = u_{N+1}(x) + \sum_{i=0}^{N} d_i u_i(x) \tag{12.4.1}$$

so that

$$P^{(j)}(\mathbf{y};y_k) = 0, \quad j = 0,1,\dots,m_k-1, \quad k = 1,\dots,n, \tag{12.4.2}$$

and

$$\rho S(\mathbf{y};x)P(\mathbf{y};x) \geq 0, \quad x \in [a,b]. \tag{12.4.3}$$

Theorem 12.4.1. *Let both $\{u_0,\dots,u_N\}$ and $\{u_0,\dots,u_N,u_{N+1}\}$ be extended Chebyshev systems on $[a,b]$. Assume that $d\mu$ is a measure on $[a,b]$ satisfying $\mu \in C[a,b]$. When*

$\min_{1\le k\le n} m_k = 1$ *assume further that* $\mu \in C^1[a,b]$ *and* $\mu'(x) > 0$ *in* (a,b). *Then there exists a unique vector* $\mathbf{x} \in \mathbf{Y}$ *satisfying*

$$\int_a^b |P(\mathbf{x};x)|d\mu(x) = \min_{\mathbf{y}\in\overline{\mathbf{Y}}} \int_a^b |P(\mathbf{y};x)|d\mu(x) \tag{12.4.4}$$

and the solution $\mathbf{x}$ *is independent of the particular choice of* u_{N+1} *from the set of functions*

$$\mathbf{K}(U) = \{v:\ u_0,\ldots,u_N,v \text{ is an extended Chebyshev system on } [a,b]\}.$$

Moreover, if $\mathbf{x} \in \mathbf{Y}$ *is a solution of (12.4.4) then* $\mathbf{x}$ *satisfies the relation (12.2.1).*

Proof. The extremal problem (12.4.4) subject to the conditions (12.4.2) is equivalent to the minimization of the functional

$$\begin{aligned}\mathcal{E}(\mathbf{y},\mathbf{d}) = &\int_a^b \left[u_{N+1}(x) + \sum_{i=0}^N d_i u_i(x)\right] S(\mathbf{y};x)d\mu(x)\\ &- \sum_{k=1}^n \sum_{j=0}^{m_k-1} \lambda_{kj} \left[u_{N+1}^{(j)}(y_k) + \sum_{i=0}^N d_i u_i^{(j)}(y_k)\right],\end{aligned}$$

where the λ_{kj} are the corresponding Lagrange multipliers. If $(\mathbf{x},\mathbf{c})$ minimizes the functional $\mathcal{E}(\mathbf{y},\mathbf{d})$ then we have the equations

$$\frac{\partial \mathcal{E}}{\partial d_i}(\mathbf{x},\mathbf{c}) = 0, \quad i = 0,1,\ldots,N, \tag{12.4.5}$$

and

$$\frac{\partial \mathcal{E}}{\partial y_k}(\mathbf{x},\mathbf{c}) = 0, \quad k = 1,\ldots,n. \tag{12.4.6}$$

These equations by (12.4.1) yield

$$\int_a^b u_i(x)S(\mathbf{x};x)d\mu(x) = \sum_{k=1}^n \sum_{j=0}^{m_k-1} \lambda_{kj} u_i^{(j)}(x_k), \quad i = 0,1,\ldots,N, \tag{12.4.7}$$

and

$$\sum_{j=0}^{m_k-1} \lambda_{kj} P^{(j+1)}(\mathbf{x};x_k) = 0, \quad k = 1,\ldots,n. \tag{12.4.8}$$

By (12.4.2) and (12.4.8)

$$\lambda_{k,m_k-1} P^{(m_k)}(\mathbf{x};x_k) = 0, \quad k = 1,\ldots,n.$$

Recalling that $P(\mathbf{x};x)$ satisfies (12.4.2) and $Z'(P(\mathbf{x})) \le N+1$, we conclude

$$P^{(m_k)}(\mathbf{x};x_k) \ne 0, \quad k = 1,\ldots,n,$$

and hence

$$\lambda_{k,m_k-1} = 0, \quad k = 1,\ldots,n.$$

Thus (12.4.7) becomes (12.2.1).

By Theorem 12.3.1 the solution of (12.2.1) is unique. So we can also conclude that the solution $\mathbf{x}$ of (12.4.4) is independent of the particular choice of u_{N+1} from the set $\mathbf{K}(U)$. □

12.5 Notes

12.5.1. Lemma 12.1.2 was first given by Hobby and Rice [40] in 1965. The simple proof here is from Pinkus [75].

12.5.2. The topic of Chebyshev spaces and weak Chebyshev spaces is the subject of several treatises. The book of Karlin and Studden [43] treats Chebyshev spaces exhaustively; in important papers Nürnberger and Sommer [68], and Deutsch, Nürenberger, and Singer [15] have studied weak Chebyshev spaces in detail.

12.5.3. A Gaussian quadrature formula was extended to Chebyshev systems by Krein in his famous paper [47]. The existence, uniqueness, and relation to best L^1 best approximation of the Hermite Gaussian quadrature formula for complete extended Chebyshev systems was given by Karlin and Pinkus [41, 42]. Later Barrow [4] provided a simple elegant proof of the existence and uniqueness based on the topological degree dropping the condition of completeness of extended Chebyshev systems. A generalized Gaussian quadrature formula for an extended Chebyshev system at prescribed nodes was investigated by Bojanov, Braess and Dyn [10] under the assumption:

$$\mu \in C^1[a,b], \quad \mu' > 0. \tag{12.5.1}$$

They proved the existence using the Borsuk antipodality theorem and Shi using the Brouwer fixed-point theorem [106], respectively. Using a modification of ideas of Bojanov [6] and Bojanov, Braess and Dyn [10], Shi [106] proved the uniqueness for a measure $d\mu$ satisfying that $\mu \in C[a,b]$ in the case when $m > 1$, omitting the assumption (12.5.1), and that $\mu \in C[a,b]$ is strictly increasing in the case when $m = 1$, respectively.

12.5.4. The results in Section 12.4 are due to Bojanov, Braess and Dyn [10].

Chapter 13

Construction Methods

13.1 Construction of Power Orthogonal Polynomials

In this chapter we use the notations of Chapters 2–4. Let $n \in \mathbf{N}$ be fixed and let $m_1, \ldots, m_n \geq 2$.

The construction method of power orthogonal polynomials is based on Theorem 3.2.2, Theorem 3.4.2, (3.2.3), (3.2.4), and (3.2.7). A basic idea for finding the zeros of $\mathbf{m}$ orthogonal polynomials is to solve the system of normal equations

$$\varphi_k(\mathbf{y}) := -m_k \int_{\mathbf{R}} \frac{\Omega_n(\mathbf{y};x)}{x - y_k} d\mu(x) = 0, \quad k = 1, 2, \ldots, n, \tag{13.1.1}$$

by the *Newton iterative method*.

(1) **A Newton Iteration Step**. We use the Newton iterative formula

$$\mathbf{x}^{(i+1)} = \mathbf{x}^{(i)} - J_n(\mathbf{x}^{(i)})^{-1} F(\mathbf{x}^{(i)}), \quad i = 0, 1, \ldots, \tag{13.1.2}$$

where $\mathbf{x}^{(i)} = (x_1^{(i)}, x_2^{(i)}, \ldots, x_n^{(i)})^T$, $F(\mathbf{y}) = (\varphi_1(\mathbf{y}), \varphi_2(\mathbf{y}), \ldots, \varphi_n(\mathbf{y}))^T$ and $J_n(\mathbf{y})$ is the *Jacobian matrix* of (13.1.1):

$$J_n(\mathbf{y}) = \left(\frac{\partial \varphi_k(\mathbf{y})}{\partial y_j} \right)_{k,j=1}^{n}. \tag{13.1.3}$$

Here

$$\frac{\partial \varphi_k(\mathbf{y})}{\partial y_j} = \begin{cases} m_k m_j \int_{\mathbf{R}} \frac{\Omega_n(\mathbf{y};x)}{(x-y_k)(x-y_j)} d\mu(x) = \frac{m_k \varphi_j(\mathbf{y}) - m_j \varphi_k(\mathbf{y})}{y_k - y_j}, & j \neq k, \\ m_k(m_k - 1) \int_{\mathbf{R}} \frac{\Omega_n(\mathbf{y};x)}{(x-y_k)^2} d\mu(x), & j = k. \end{cases} \tag{13.1.4}$$

The above integrals can be computed by, say, the classical Gaussian quadrature formula with respect to the measure $d\mu$

$$\int_{\mathbf{R}} f(x) d\mu(x) = \sum_{k=1}^{M} \lambda_k f(y_k) \tag{13.1.5}$$

with $M \geq \sum_{k=1}^{n} m_k$. If all m_k are even, this formula is exact except for rounding errors, otherwise we choose $M = 2^p, 2^{p+1}, \ldots$, such that the difference of the successive approximated integration values is within a given error control bound.

(2) **Error Estimation**. Let $\mathbf{x}, \mathbf{y} \in \mathbf{Y}$ satisfy

$$\varphi_k(\mathbf{x}) = 0, \quad \varphi_k(\mathbf{y}) = r_k, \quad k = 1, 2, \ldots, n, \tag{13.1.6}$$

respectively. Then by Taylor expansion of multivariable functions, we have

$$\mathbf{y} - \mathbf{x} = J_n(\mathbf{x})^{-1}\mathbf{r} + O(\|\mathbf{r}\|^2) \tag{13.1.7}$$

and

$$\mathbf{y} - \mathbf{x} = J_n(\mathbf{y})^{-1}\mathbf{r} + O(\|\mathbf{r}\|^2), \tag{13.1.8}$$

where $\mathbf{r} = (r_1, r_2, \ldots, r_n)^T$ and $\|\mathbf{r}\| = \sum_{k=1}^n |r_k|$.

(3) **Progressive Computation of the Zeros**. Based on Theorem 3.4.2 the system of equations (13.1.1) is solved progressively from a lower order to the higher order target to make each Newton iteration is surely successful. Let

$$\mathbf{m}^{(i)} = (m_1^{(i)}, m_2^{(i)}, \ldots, m_n^{(i)}), \ i = 0, 1, \ldots.$$

Here $\mathbf{m}^{(0)}$ may be chosen so that $\mathbf{x}(d\mu, \mathbf{m}^{(0)})$ is known, say, $\mathbf{m}^{(0)} = (2, 2, \ldots, 2)$. Meanwhile

$$m_k^{(i+1)} = \min\{m_k, m_k^{(i)} + h_{ki}\}, \quad h_{ki} > 0, \quad k = 1, 2, \ldots, n.$$

Use the zeros $\mathbf{x}^{(i)} = \mathbf{x}(d\mu, \mathbf{m}^{(i)})$ as initial values to calculate the zeros $\mathbf{x}^{(i+1)} = \mathbf{x}(d\mu, \mathbf{m}^{(i+1)})$ using the Newton iteration step described above. If this iterative process is not convergent, we decrease the $\mathbf{m}$-steplengths h_{ki} so that it becomes convergent. The failure of the Newton iteration is recognized by an increased residual $\sum_{k=1}^n \varphi_k(\mathbf{y})^2$ or disordered $\mathbf{x}^{(i+1)}$.

Take the case $\mathbf{m} = (4, \ldots, 4)$ as an example. Using the solution for $\mathbf{m}^{(0)} = (2, \ldots, 2)$ as an initial value, we solve the system of equations (13.1.1) successively for $\mathbf{m}^{(i)} = (2 + ih, \ldots, 2 + ih)$, for $i = 1, \ldots, 2^p$, where $h = 2^{1-p}$ is referred to as $\mathbf{m}$-steplength. We choose $p = 1, 2, \ldots$, until the Newton iteration is successful.

According to the continuity of the zeros of the nth $\mathbf{m}$ orthogonal polynomial with respect to $\mathbf{m}$ such an iteration strategy is guaranteed to be effective. Note that all the intermediate steps are merely for computing initial values for the next step, each such a Newton iteration is unnecessarily very accurate. This observation will significantly reduce computational cost.

Since the elements of $J_n(\mathbf{y})$ are the integrations of piecewise polynomials of high degree over finite or infinite domains, the values of these elements could be very large or very small. This feature could lead to the failure of the numerical inversion of $J_n(\mathbf{y})$. To overcome this difficulty, $J_n(\mathbf{y})$ is normalized by multiplying a factor $\rho(\mathbf{y})^{-1}$, where $\rho(\mathbf{y})$ is taken to be the maximal element of $\varphi_k(\mathbf{y})$ and $F_k(\mathbf{y})$ in magnitude.

Remark 13.1.1. The proposed method has the following advantages.

(1) This method directly calculates the zeros $\mathbf{x}$ of the $\mathbf{m}$ orthogonal polynomial rather than first computing its coefficients and then determining its zeros.

(2) The calculation of $J_n(\mathbf{y})$ needs only computing $2n$ integrals

$$\varphi_k(\mathbf{y}) = -m_k \int_{\mathbf{R}} \frac{\Omega_n(\mathbf{y}; x)}{x - y_k} d\mu(x), \ F_k(\mathbf{y}) = m_k(m_k - 1) \int_{\mathbf{R}} \frac{\Omega_n(\mathbf{y}; x)}{(x - y_k)^2} d\mu(x)$$

for $k = 1, 2, \ldots, n$, rather than n^2 integrals.

(3) The Jacobian matrix $J_n(\mathbf{y})$ has good properties: it is symmetric; further, all the elements in the main diagonal are positive and the other elements are small if $\mathbf{y}$ is close to the solution of (13.1.1) according to (3.2.7).

(4) We may adjust the **m**-steplength h_{ki} to obtain a convergent iterative process.

(5) In each Newton iteration step we can obtain an error estimation from the solution.

13.2 Construction of Christoffel Numbers

The construction method of Christoffel numbers is based on the formulas (2.2.11), (2.2.12), and (4.1.7). The Christoffel Numbers $\lambda_{kj} = \lambda_{kj}(\mathbf{x})$ are calculated in the following steps.

(1) Calculation of the numbers

$$a_{kj} = \sum_{i \neq k} \frac{m_i}{(x_i - x_k)^j}, \quad j \geq 1. \tag{13.2.1}$$

(2) Calculation of the numbers

$$b_{kj} = \frac{1}{j} \sum_{i=1}^{j} a_{ki} b_{k,j-i}, \quad b_{k0} = 1, \quad j \geq 1. \tag{13.2.2}$$

(3) Calculation of the integrals

$$c_{kj} = \int_{\mathbf{R}} (x - x_k)^j L_k(x) S_n(\mathbf{x}; x) d\mu(x), \quad j = 0, 1, \ldots, m_k - 2, \tag{13.2.3}$$

using Gaussian quadrature formula (13.1.5), where $L_k(x)$ and $S_n(\mathbf{x}; x)$ are defined by (2.2.4) and (4.1.1), respectively.

(4) Calculation of the Christoffel numbers

$$\lambda_{kj} = \frac{1}{j!} \sum_{i=0}^{m_k - j - 2} b_{ki} c_{k,i+j}, \quad j = 0, 1, \ldots, m_k - 2. \tag{13.2.4}$$

13.3 Numerical Results

Our intention is to provide double precision numerical results. To make sure the provided numerical results are significant till last digits, we use long double precision arithmetic (which has accuracy $2^{-108} \approx 3 \times 10^{-33}$) operation to conduct all the computation. Hence, the computed numerical results are much more accurate than the required double precision($\approx 10^{-16}$). The numerical results are finally rounded off to 16 decimal digits.

Example 13.3.1. *Jacobi weight*

$$w(x) = v_m(x) = (1 - x)^{-1/2}(1 + x)^{(m-1)/2}, \quad x \in [-1, 1], \quad m \in \mathbf{N}_2,$$

given by (6.2.14). By (6.3.57) and (1.3.10) we have

$$2^n P_n(v_m, m; x) = V_n(x) = \frac{\cos[(2n+1)\theta/2]}{\cos(\theta/2)}, \quad \theta \in [0, \pi].$$

Meanwhile the Christoffel numbers $\lambda_{kj}(v_m, m)$ are given by (6.4.70) with (6.4.62). We use this weight as the first example to show the numerical results coincide with the exact ones within the accuracy of double precision. In the following table the maximal absolute error for the zeros of $\mathbf{m}$ orthogonal polynomials and the maximal relative error for λ_{kj} are given for $n = 10$ and $m = 4, 6, 8, 10$.

m	$\max_k \lvert x_k - x_k^* \rvert$	$\max_{k,j} \lvert(\lambda_{kj} - \lambda_{kj}^*)/\lambda_{kj}\rvert$
4	3.41499E-31	1.18939E-29
6	2.92641E-31	3.73973E-28
8	5.32472E-31	4.47021E-28
10	8.30516E-31	4.97258E-28

Here x_k^* and λ_{kj}^* are the computed values of the exact values x_k and λ_{kj}, respectively.

Example 13.3.2. *Legendre weight*

$$w(x) = \begin{cases} 1, & |x| \le 1, \\ 0, & |x| > 1. \end{cases}$$

We calculate the zeros and Christoffel numbers for $n = 2, \ldots, 16$ and $m = 2, 4, \ldots, 10$. We choose the $\mathbf{m}$-steplength to be the maximal, i.e., $h = 2$. Each step needs around 5–7 Newton iterations.

Example 13.3.3. *Hermite weight*

$$w(x) = e^{-x^2}, \quad x \in \mathbf{R}.$$

Again, we choose $h = 2$ for $n = 2, \ldots, 10$, $m = 2, 4, 6, 8$. Each step needs about 5 Newton iterations.

Example 13.3.4. *Laguerre weight*

$$w(x) = \begin{cases} e^{-x}, & x \ge 0, \\ 0, & x < 0. \end{cases}$$

We choose the $\mathbf{m}$-steplength $h = 0.5$ for $n = 2, \ldots, 7$, and $h = 0.25$ for $n = 8, 9, 10$. Each step needs 4–7 Newton iterations.

Example 13.3.5. *Generalized Laguerre weight*

$$w(x) = \begin{cases} x^{-1/2} e^{-x}, & x > 0, \\ 0, & x \le 0. \end{cases}$$

For $n = 8$, $\mathbf{m} = (8, 8, 8, 10, 10, 10, 10, 10)$, we take $h = 2$ from $\mathbf{m}_2 = (2, 2, 2, 2, 2, 2, 2, 2)$ to $\mathbf{m}_4$, $\mathbf{m}_6$, and $\mathbf{m}_8 = (8, 8, 8, 8, 8, 8, 8, 8)$. These three steps require 6, 6, and 5 Newton iterations, respectively. From $\mathbf{m}_8$ to $\mathbf{m}$, we use the four steps

$$\mathbf{m}_8^{(l)} = \mathbf{m}_8 + l(0, 0, 0, 0.5, 0.5, 0.5, 0.5, 0.5), \; l = 1, 2, 3, 4$$

to arrive at the result for $\mathbf{m} = \mathbf{m}_8^{(4)}$. These four steps require 6, 6, 6, and 7 Newton iterations, respectively. Hence the total number of iterations is 42 from $\mathbf{m}_2$ to $\mathbf{m}$. This number is much smaller than the number of iterations 180 required in [10]. The zeros we obtained are

x_1	2.68359 22430 12328E−01	x_2	2.43080 10306 07161E+00
x_3	6.85565 84519 19513E+00	x_4	1.45478 47160 11330E+01
x_5	2.67958 39682 64774E+01	x_6	4.39580 11697 97213E+01
x_7	6.77657 34844 62150E+01	x_8	1.02919 75077 35820E+02

13.4 Notes

13.4.1. Construction of power orthogonal polynomials and Christoffel numbers for the case when all m_k are even integers is discussed by several authors (see [61] for references). Here we mention two of them. Milovanović [60] gave a stable method for numerical construction of power orthogonal polynomials with $m_k \equiv m$ using a reinterpretation of the orthogonality relations (3.2.6); this method first calculates $2n$ coefficients of a three term recurrence relation to get the nth power orthogonal polynomial and then finds its zeros. In a very recent paper [63] Milovanović and Spalević directly solve the system of nonlinear equations with n unknowns $x_1, x_2, ..., x_n$ consisting of the orthogonality relations (3.2.6) with $q(x) = x^i$, $i = 0, 1, ..., n-1$. The construction method for general power orthogonal polynomial using solving the system of normal equations (13.1.1) here is given by Shi and Xu [127]. The numerical tables in Section 13.5 are given by Xu and Shi [160].

13.4.2. Although our construction methods are given under the restriction $m_k \geq 2$, they remain valid for a general case ($m_k \geq 1$). But it should do some modifications. We need to use the formulas (3.2.4) instead of (13.1.4) and in this case the integrals in them may be singular; some of the **m**-steplengths h_{ki} may be negative.

13.5 Numerical Tables

Numerical results of Gaussian quadrature formulas for Legendre, Hermite, and Laguerre weights are tabled in the following. For Legendre weight, we provide results for $n = 2, 3, \ldots, 10, 12, 16$, and for $m = 2, 4, 6$. For Hermite and Laguerre weights, we provide results for $n = 2, 3, \ldots, 10$, and for $m = 2, 4$. Since Gaussian quadrature formulas for Legendre weight are most often used in the scientific computing, we provide more results for them.

The integers following x_k in the first column of the tables are the indices j of the Christoffel number λ_{kj}. In each box of the second column, the first row is the value of x_k, the following values are the corresponding Christoffel number λ_{kj} for $j = 0, \cdots, m-2$. Since the quadrature formulas for the Legendre and Hermite weights are symmetric about the origin, only half of results are given. In the tables, $x_k = \mp z$ means $x_k = -z$, $x_{n+1-k} = z$. Correspondingly, $\lambda_{kj} = \pm c$ means $\lambda_{kj} = c$ and $\lambda_{n+1-k,j} = -c$.

	Legendre Weight
	n = 2, m = 2
x_1	∓5.77350 26918 96258E−01
0	1.00000 00000 00000E+00
	n = 2, m = 4
x_1	∓6.29211 12834 99094E−01
0	1.00000 00000 00000E+00
1	±9.62917 68337 98977E−02
2	2.93011 96853 73820E−02
	n = 2, m = 6
x_1	∓6.50027 75347 58294E−01
0	1.00000 00000 00000E+00
1	±1.28870 46065 24314E−01
2	3.91680 02549 54509E−02
3	±2.61996 18925 38938E−03
4	2.21667 30957 74221E−04
	n = 3, m = 2
x_1	∓7.74596 66924 14834E−01
0	5.55555 55555 55556E−01
x_2	0.00000 00000 00000E+00
0	8.88888 88888 88889E−01
	n = 3, m = 4
x_1	∓8.14439 18557 77632E−01
0	5.33316 40592 17778E−01
1	±3.11646 66597 13555E−02
2	4.10748 07292 59521E−03
x_2	0.00000 00000 00000E+00
0	9.33367 18815 64444E−01
1	0.00000 00000 00000E+00
2	2.21270 84978 17585E−02
	n = 3, m = 6
x_1	∓8.29232 73806 92652E−01
0	5.24188 93974 74567E−01
1	±4.02639 79240 75502E−02
2	5.58568 67762 41462E−03
3	±2.03107 14206 26517E−04
4	7.32688 54482 32734E−06
x_2	0.00000 00000 00000E+00
0	9.51622 12050 50865E−01
1	0.00000 00000 00000E+00
2	2.84919 45868 18255E−02
3	0.00000 00000 00000E+00
4	1.46478 18704 20226E−04
	n = 4, m = 2
x_1	∓8.61136 31159 40526E−01
0	3.47854 84513 74539E−01
x_2	∓3.39981 04358 48563E−01
0	6.52145 15486 25461E−01
	n = 4, m = 4
x_1	∓8.89676 81360 55590E−01
0	3.24686 89761 89835E−01
1	±1.20175 28374 47594E−02
2	8.99742 55652 69651E−04
x_2	∓3.58858 52703 88840E−01
0	6.75313 10238 10165E−01
1	±1.01906 72094 71714E−02
2	8.13355 64909 82577E−03
	n = 4, m = 6
x_1	∓8.99829 21267 59577E−01
0	3.15604 20599 52367E−01
1	±1.51791 92721 96107E−02
2	1.21397 65322 11799E−03

	Legendre Weight(Cont.)
3	±2.67403 74323 26841E−05
4	5.42643 51770 10564E−07
x_2	∓3.65924 35475 89185E−01
0	6.84395 79400 47633E−01
1	±1.35560 93536 57987E−02
2	1.04801 63839 33739E−02
3	±1.12802 51014 77524E−04
4	2.63642 35509 78017E−05
	n = 5, m = 2
x_1	∓9.06179 84593 86640E−01
0	2.36926 88505 61891E−01
x_2	∓5.38469 31010 56831E−01
0	4.78628 67049 93665E−01
x_3	0.00000 00000 00000E+00
0	5.68888 88888 88889E−01
	n = 5, m = 4
x_1	∓9.27117 86960 98907E−01
0	2.16953 77170 38250E−01
1	±5.46114 33814 36805E−03
2	2.64610 23711 45917E−04
x_2	∓5.60867 41916 16367E−01
0	4.87860 66287 09246E−01
1	±7.50689 05096 56810E−03
2	3.02254 11252 25670E−03
x_3	0.00000 00000 00000E+00
0	5.90371 13085 05007E−01
1	0.00000 00000 00000E+00
2	5.35650 45934 41792E−03
	n = 5, m = 6
x_1	∓9.34368 97529 91895E−01
0	2.09350 62105 25997E−01
1	±6.79621 28545 49237E−03
2	3.53574 34356 19558E−04
3	±5.17555 44865 96472E−06
4	6.75368 85967 22694E−08
x_2	∓5.69032 43875 82429E−01
0	4.91360 69424 32071E−01
1	±9.83418 16637 82983E−03
2	3.87087 33280 43485E−03
3	±4.14354 84729 49138E−05
4	4.87180 22178 87692E−06
x_3	0.00000 00000 00000E+00
0	5.98577 36940 83864E−01
1	0.00000 00000 00000E+00
2	6.90255 37165 42489E−03
3	0.00000 00000 00000E+00
4	1.30717 45403 72660E−05
	n = 6, m = 2
x_1	∓9.32469 51420 31520E−01
0	1.71324 49237 91703E−01
x_2	∓6.61209 38646 62645E−01
0	3.60761 57304 81386E−01
x_3	∓2.38619 18608 31969E−01
0	4.67913 93457 26910E−01
	n = 6, m = 4
x_1	∓9.48342 05761 50594E−01
0	1.54752 77018 01832E−01
1	±2.80502 58766 47973E−03
2	9.52689 10733 87759E−05
x_2	∓6.82436 57430 97147E−01
0	3.63002 91500 21001E−01

	Legendre Weight(Cont.)
1	±4.78340 27205 47237E−03
2	1.23512 68879 19864E−03
x_3	∓2.47915 39222 31809E−01
0	4.82244 31481 77167E−01
1	±2.30962 16172 47315E−03
2	2.89603 65493 23533E−03
	n = 6, m = 6
x_1	∓9.53739 48171 93577E−01
0	1.48571 49434 75303E−01
1	±3.45483 61570 26248E−03
2	1.26222 94164 29491E−04
3	±1.31217 14444 09668E−06
4	1.19485 01648 18873E−08
x_2	∓6.90031 39443 30314E−01
0	3.63823 48552 31565E−01
1	±6.19501 22047 40081E−03
2	1.57066 92566 12427E−03
3	±1.41704 47484 39937E−05
4	1.06541 49428 81127E−06
x_3	∓2.51276 35056 12515E−01
0	4.87605 02012 93131E−01
1	±3.02481 41529 77250E−03
2	3.71817 34063 88870E−03
3	±1.24281 39272 16579E−05
4	4.60734 29117 90451E−06
	n = 7, m = 2
x_1	∓9.49107 91234 27585E−01
0	1.29484 96616 88697E−01
x_2	∓7.41531 18559 93944E−01
0	2.79705 39148 92767E−01
x_3	∓4.05845 15137 73972E−01
0	3.81830 05050 51189E−01
x_4	0.00000 00000 00000E+00
0	4.17959 18367 34694E−01
	n = 7, m = 4
x_1	∓9.61503 03455 39719E−01
0	1.15776 03149 02838E−01
1	±1.57894 39683 17105E−03
2	3.96972 71671 85798E−05
x_2	∓7.60412 11616 39885E−01
0	2.78594 63973 99043E−01
1	±3.03521 55794 27089E−03
2	5.55580 76862 99488E−04
x_3	∓4.18914 50189 49846E−01
0	3.90499 66894 86395E−01
1	±2.34485 26121 31758E−03
2	1.53007 14465 83882E−03
x_4	0.00000 00000 00000E+00
0	4.30259 31964 23447E−01
1	0.00000 00000 00000E+00
2	2.04664 19580 06152E−03
	n = 7, m = 6
x_1	∓9.65662 66156 33674E−01
0	1.10738 03614 31444E−01
1	±1.93004 89767 36421E−03
2	5.22270 60561 97953E−05
3	±4.04819 17064 89110E−07
4	2.71970 32829 37477E−09
x_2	∓7.67075 18014 66233E−01
0	2.78174 08874 30212E−01
1	±3.89719 62776 93681E−03

	Legendre Weight(Cont.)
2	7.01969 89806 96297E−04
3	±5.18009 73706 01230E−06
4	2.75420 40419 44965E−07
x_3	∓4.23579 11332 18298E−01
0	3.93690 46161 77698E−01
1	±3.04704 54067 49855E−03
2	1.95485 00050 37456E−03
3	±8.11245 15727 79383E−06
4	1.56397 68379 28687E−06
x_4	0.00000 00000 00000E+00
0	4.34794 82699 21292E−01
1	0.00000 00000 00000E+00
2	2.62481 10181 11459E−03
3	0.00000 00000 00000E+00
4	2.56968 70785 39801E−06
	n = 8, m = 2
x_1	∓9.60289 85649 75362E−01
0	1.01228 53629 03763E−01
x_2	∓7.96666 47741 36267E−01
0	2.22381 03445 33745E−01
x_3	∓5.25532 40991 63290E−01
0	3.13706 64587 78873E−01
x_4	∓1.83434 64249 56498E−01
0	3.62683 78337 83620E−01
	n = 8, m = 4
x_1	∓9.70215 58612 97783E−01
0	8.98034 94114 60722E−02
1	±9.53494 60921 97774E−04
2	1.84663 81788 44586E−05
x_2	∓8.13129 67933 25268E−01
0	2.19718 29808 57071E−01
1	±1.97474 76704 17815E−03
2	2.71650 94821 40515E−04
x_3	∓5.39670 08230 85669E−01
0	3.18604 36773 98684E−01
1	±1.90134 83044 11887E−03
2	8.28300 53685 68778E−04
x_4	∓1.88910 45950 29638E−01
0	3.71873 84005 98173E−01
1	±7.76897 68663 14359E−04
2	1.31710 70684 02638E−03
	n = 8, m = 6
x_1	∓9.73513 41389 22066E−01
0	8.56520 81089 15241E−02
1	±1.15881 92286 57935E−03
2	2.41547 53908 61337E−05
3	±1.44841 33395 42559E−07
4	7.47678 79163 75232E−10
x_2	∓8.18878 93183 94185E−01
0	2.18741 89271 79282E−01
1	±2.51879 42092 10430E−03
2	3.41320 55881 26384E−04
3	±2.06205 99321 59935E−06
4	8.22290 92609 85562E−08
x_3	∓5.44664 90130 32453E−01
0	3.20383 56186 79556E−01
1	±2.45486 02289 10768E−03
2	1.05313 86712 01147E−03
3	±4.31144 60015 26879E−06
4	5.54317 75163 83516E−07
x_4	∓1.90852 90843 98490E−01

	Legendre Weight(Cont.)
0	3.75222 46432 49638E−01
1	±1.00749 37652 80005E−03
2	1.68441 62711 78414E−03
3	±2.42703 66208 16640E−06
4	1.22139 05918 73061E−06
	n = 9, m = 2
x_1	∓9.68160 23950 76261E−01
0	8.12743 88361 57441E−02
x_2	∓8.36031 10732 66358E−01
0	1.80648 16069 48574E−01
x_3	∓6.13371 43270 05904E−01
0	2.60610 69640 29355E−01
x_4	∓3.24253 42340 38089E−01
0	3.12347 07704 00028E−01
x_5	0.00000 00000 00000E+00
0	3.30239 35500 12598E−01
	n = 9, m = 4
x_1	∓9.76277 27457 24844E−01
0	7.16537 67594 24625E−02
1	±6.08568 98486 37640E−04
2	9.35938 31787 33040E−06
x_2	∓8.50325 39868 87252E−01
0	1.77329 40342 17334E−01
1	±1.32483 83212 15100E−03
2	1.42486 36124 84003E−04
x_3	∓6.27321 74179 19216E−01
0	2.63102 00977 60689E−01
1	±1.45078 35747 78876E−03
2	4.65396 76805 43477E−04
x_4	∓3.32661 91628 20943E−01
0	3.18838 22112 20131E−01
1	±9.32379 31086 48005E−04
2	8.28253 74332 65852E−04
x_5	0.00000 00000 00000E+00
0	3.38153 19617 18768E−01
1	0.00000 00000 00000E+00
2	9.88081 88489 24777E−04
	n = 9, m = 6
x_1	∓9.78953 17435 88977E−01
0	6.81889 79800 34725E−02
1	±7.36275 53260 65714E−04
2	1.21837 51797 25566E−05
3	±5.81697 75238 54387E−08
4	2.37967 03266 22398E−10
x_2	∓8.55276 31117 65375E−01
0	1.76128 15443 65298E−01
1	±1.68101 68881 18938E−03
2	1.78181 65652 18554E−04
3	±8.89806 18355 87377E−07
4	2.76952 83281 95165E−08
x_3	∓6.32210 36624 57233E−01
0	2.63996 89397 75812E−01
1	±1.86328 22557 96621E−03
2	5.89120 60942 56344E−04
3	±2.21591 49320 27833E−06
4	2.09554 58465 06155E−07
x_4	∓3.35622 22864 62853E−01
0	3.21180 42908 61552E−01
1	±1.20349 95395 88889E−03
2	1.05573 03716 47811E−03
3	±2.11846 58300 62217E−06

	Legendre Weight(Cont.)
4	5.58535 99022 91360E−07
x_5	0.00000 00000 00000E+00
0	3.41011 08539 87731E−01
1	0.00000 00000 00000E+00
2	1.26216 25325 96015E−03
3	0.00000 00000 00000E+00
4	7.53613 37957 70683E−07
	n = 10, m = 2
x_1	∓9.73906 52851 71717E−01
0	6.66713 44308 68814E−02
x_2	∓8.65063 36668 89845E−01
0	1.49451 34915 05806E−01
x_3	∓6.79409 56829 90244E−01
0	2.19086 36251 59820E−01
x_4	∓4.33395 39412 92472E−01
0	2.69266 71930 99964E−01
x_5	∓1.48874 33898 16312E−01
0	2.95524 22471 47529E−01
	n = 10, m = 4
x_1	∓9.80662 59593 65866E−01
0	5.84827 98111 42158E−02
1	±4.06140 53858 65962E−04
2	5.08058 07401 80610E−06
x_2	∓8.77500 22098 48185E−01
0	1.45927 62204 14931E−01
1	±9.15780 01355 31670E−04
2	7.92748 59676 39646E−05
x_3	∓6.92624 42514 00476E−01
0	2.20060 17419 82068E−01
1	±1.09051 82361 30125E−03
2	2.71873 49351 57937E−04
x_4	∓4.43200 99195 06441E−01
0	2.73698 97591 60047E−01
1	±8.67956 79178 78477E−04
2	5.23074 32592 66366E−04
x_5	∓1.52470 58767 94166E−01
0	3.01830 42973 28738E−01
1	±3.29291 69627 45838E−04
2	7.01508 84861 52027E−04
	n = 10, m = 6
x_1	∓9.82875 89957 28033E−01
0	5.55548 44924 35525E−02
1	±4.89579 02507 50776E−04
2	6.58711 97410 26815E−06
3	±2.56243 03120 19941E−08
4	8.51254 80670 85129E−11
x_2	∓8.81779 37678 72088E−01
0	1.44663 73900 72849E−01
1	±1.15709 94569 32595E−03
2	9.87343 74743 76455E−05
3	±4.12389 77879 08487E−07
4	1.03168 48274 09576E−08
x_3	∓6.97224 70402 37363E−01
0	2.20405 69172 14339E−01
1	±1.39451 84833 08066E−03
2	3.42808 42299 38869E−04
3	±1.15371 54664 32990E−06
4	8.47038 70294 09334E−08
x_4	∓4.46631 37154 63451E−01
0	2.75285 31640 15096E−01
1	±1.11580 52192 22343E−03

	Legendre Weight(Cont.)
2	6.64548 64845 36479E−04
3	±1.44007 09997 80959E−06
4	2.57458 42642 16434E−07
x_5	∓1.53731 23413 37648E−01
0	3.04090 40794 54165E−01
1	±4.24255 54091 61337E−04
2	8.94277 66630 59607E−04
3	±6.68132 28237 74083E−07
4	4.23453 62594 97274E−07
	n = 12, m = 2
x_1	∓9.81560 63424 67193E−01
0	4.71753 36386 51183E−02
x_2	∓9.04117 25637 04749E−01
0	1.06939 32599 53184E−01
x_3	∓7.69902 67419 43047E−01
0	1.60078 32854 33462E−01
x_4	∓5.87317 95428 66174E−01
0	2.03167 42672 30659E−01
x_5	∓3.67831 49899 81802E−01
0	2.33492 53653 83548E−01
x_6	∓1.25233 40851 14689E−01
0	2.49147 04581 34028E−01
	n = 12, m = 4
x_1	∓9.86445 07017 79790E−01
0	4.10636 49884 72770E−02
1	±2.00708 20647 22860E−04
2	1.75500 94854 48719E−06
x_2	∓9.13674 51106 16951E−01
0	1.03570 74778 34508E−01
1	±4.73491 78824 00275E−04
2	2.82815 49167 63860E−05
x_3	∓7.81193 00834 88735E−01
0	1.59505 14338 83510E−01
1	±6.23732 91439 78491E−04
2	1.03308 27430 74085E−04
x_4	∓5.97620 33998 46387E−01
0	2.05018 29042 85820E−01
1	±6.13356 24975 07094E−04
2	2.19376 53812 00127E−04
x_5	∓3.74959 01286 73543E−01
0	2.37123 35470 10316E−01
1	±4.45103 31400 55431E−04
2	3.39418 52929 12434E−04
x_6	∓1.27772 15307 27657E−01
0	2.53718 81381 38569E−01
1	±1.62290 83994 09749E−04
2	4.15786 72113 98467E−04
	n = 12, m = 6
x_1	∓9.88030 17157 65983E−01
0	3.89018 99784 49516E−02
1	±2.40613 76652 52119E−04
2	2.26099 16841 97244E−06
3	±6.15945 40180 79665E−09
4	1.42673 40800 64894E−11
x_2	∓9.16930 26137 96710E−01
0	1.02377 90862 71641E−01
1	±5.94483 13107 80530E−04
2	3.49970 34407 30914E−05
3	±1.05842 10817 07640E−07
4	1.82304 05876 71435E−09
x_3	∓7.85083 86997 12541E−01

	Legendre Weight(Cont.)
0	1.59300 51440 06876E−01
1	±7.92323 72392 85896E−04
2	1.29428 27246 24285E−04
3	±3.41550 09475 46853E−07
4	1.66299 45211 81643E−08
x_4	∓6.01189 12354 99201E−01
0	2.05672 17716 12473E−01
1	±7.83397 64440 10801E−04
2	2.77083 32763 71351E−04
3	±5.62926 04999 73276E−07
4	5.96602 68357 19667E−08
x_5	∓3.77434 52644 80556E−01
0	2.38408 87145 80945E−01
1	±5.70120 34723 18231E−04
2	4.30692 94125 41651E−04
3	±5.50464 64950 81218E−07
4	1.24859 01607 10260E−07
x_6	∓1.28654 92268 73433E−01
0	2.55338 62856 83114E−01
1	±2.08136 15704 91179E−04
2	5.28713 12653 08373E−04
3	±2.30514 71656 41451E−07
4	1.75950 51147 14388E−07
	n = 16, m = 2
x_1	∓9.89400 93499 16499E−01
0	2.71524 59411 75409E−02
x_2	∓9.44575 02307 32326E−01
0	6.22535 23938 64789E−02
x_3	∓8.65631 20238 78317E−01
0	9.51585 11682 49278E−02
x_4	∓7.55404 40835 50030E−01
0	1.24628 97125 55339E−01
x_5	∓6.17876 24440 26437E−01
0	1.49595 98881 65767E−01
x_6	∓4.58016 77765 72274E−01
0	1.69156 51939 50025E−01
x_7	∓2.81603 55077 92589E−01
0	1.82603 41504 49236E−01
x_8	∓9.50125 09837 63744E−02
0	1.89450 61045 50685E−01
	n = 16, m = 4
x_1	∓9.92287 58337 56924E−01
0	2.34037 85448 38437E−02
1	±6.53509 83090 53079E−05
2	3.24217 42237 07533E−07
x_2	∓9.50619 89245 67715E−01
0	5.96685 66338 66670E−02
1	±1.61176 17636 85674E−04
2	5.39623 47240 99992E−06
x_3	∓8.73562 07525 27428E−01
0	9.38244 96290 94070E−02
1	±2.32988 43171 95737E−04
2	2.09807 76012 75952E−05
x_4	∓7.63959 38969 61797E−01
0	1.24491 29054 90308E−01
1	±2.70371 54011 39192E−04
2	4.90107 10522 28600E−05
x_5	∓6.25892 80724 99503E−01
0	1.50520 78507 27011E−01
1	±2.67827 72832 44131E−04
2	8.66290 80297 63061E−05

	Legendre Weight(Cont.)
x_6	∓4.64506 00758 26291E−01
0	1.70942 26643 36583E−01
1	±2.25736 89798 91017E−04
2	1.26888 62625 41788E−04
x_7	∓2.85812 00421 35018E−01
0	1.84994 62001 37826E−01
1	±1.50315 03573 02256E−04
2	1.60824 28547 66953E−04
x_8	∓9.64687 94937 08162E−02
0	1.92154 18985 28354E−01
1	±5.26987 11022 55119E−05
2	1.80228 68525 68416E−04
	n = 16, m = 6
x_1	∓9.93213 30957 80071E−01
0	2.20958 96508 97760E−02
1	±7.78034 40660 86895E−05
2	4.14168 44786 90130E−07
3	±6.40905 28467 02880E−10
4	8.39615 81327 72686E−13
x_2	∓9.52653 57178 92294E−01
0	5.87671 87340 62048E−02
1	±2.00770 68616 38062E−04
2	6.61984 01910 06675E−06
3	±1.17479 49874 32524E−08
4	1.13096 83385 74493E−10
x_3	∓8.76260 57799 33663E−01
0	9.33588 73371 54774E−02
1	±2.93485 25336 18545E−04
2	2.60535 35634 23402E−05
3	±4.33411 43669 17385E−08
4	1.14444 00249 07228E−09
x_4	∓7.66885 34957 62099E−01
0	1.24442 55170 94300E−01
1	±3.42390 17936 55589E−04
2	6.13714 41116 77962E−05
3	±8.98385 22108 95071E−08
4	4.81572 86642 59386E−09
x_5	∓6.28642 66276 48348E−01
0	1.50842 84820 76196E−01
1	±3.40217 74707 77242E−04
2	1.09060 48554 66068E−04
3	±1.31162 58720 34424E−07
4	1.26020 57030 52818E−08
x_6	∓4.66735 91033 07185E−01
0	1.71565 17980 56725E−01
1	±2.87296 97035 49262E−04
2	1.60292 24570 48624E−04
3	±1.43282 42505 68533E−07
4	2.39863 21328 88150E−08
x_7	∓2.87259 66433 28458E−01
0	1.85829 29188 58835E−01
1	±1.91522 78232 12973E−04
2	2.03578 53725 03117E−04
3	±1.12060 45147 54351E−07
4	3.57592 81287 43014E−08
x_8	∓9.69699 83011 86319E−02
0	1.93098 17117 02486E−01
1	±6.71811 54026 49346E−05
2	2.28361 78688 77437E−04
3	±4.24431 26082 71405E−08
4	4.33220 23297 24119E−08

	Hermite Weight
	n = 2, m = 2
x_1	∓7.07106 78118 65475E−01
0	8.86226 92545 27580E−01
	n = 2, m = 4
x_1	∓1.02643 74224 55781E+00
0	8.86226 92545 27580E−01
1	±3.21587 85692 17385E−01
2	8.47938 15438 81906E−02
	n = 3, m = 2
x_1	∓1.22474 48713 91589E+00
0	2.95408 97515 09193E−01
x_2	0.00000 00000 00000E+00
0	1.18163 59006 03677E+00
	n = 3, m = 4
x_1	∓1.76999 13663 17813E+00
0	1.55120 36199 79434E−01
1	±5.19577 51904 51674E−02
2	7.83728 70088 93461E−03
x_2	0.00000 00000 00000E+00
0	1.46221 31269 09629E+00
1	0.00000 00000 00000E+00
2	1.25396 59214 22954E−01
	n = 4, m = 2
x_1	∓1.65068 01238 85785E+00
0	8.13128 35447 24518E−02
x_2	∓5.24647 62327 52903E−01
0	8.04914 09000 55128E−01
	n = 4, m = 4
x_1	∓2.37892 83495 39110E+00
0	1.56583 71567 08298E−02
1	±4.61275 31514 94507E−03
2	5.13642 63694 48319E−04
x_2	∓7.54289 18997 68490E−01
0	8.70568 55388 56750E−01
1	±1.46849 76507 13885E−01
2	5.08200 37277 39906E−02
	n = 5, m = 2
x_1	∓2.02018 28704 56086E+00
0	1.99532 42059 04591E−02
x_2	∓9.58572 46461 38185E−01
0	3.93619 32315 22412E−01
x_3	0.00000 00000 00000E+00
0	9.45308 72048 29419E−01
	n = 5, m = 4
x_1	∓2.90576 12455 72040E+00
0	1.15012 85243 98021E−03
1	±3.02085 17726 17314E−04
2	2.73948 77464 73663E−05
x_2	∓1.37534 10771 26207E+00
0	2.68976 47220 41771E−01
1	±5.76315 68926 53326E−02
2	1.08759 06530 74376E−02
x_3	0.00000 00000 00000E+00
0	1.23220 06494 48366E+00
1	0.00000 00000 00000E+00
2	6.30915 48223 85307E−02
	n = 6, m = 2
x_1	∓2.35060 49736 74492E+00
0	4.53000 99055 08846E−03
x_2	∓1.33584 90740 13697E+00

	Hermite Weight(Cont.)
0	1.57067 32032 28566E−01
x_3	∓4.36077 41192 76165E−01
0	7.24629 59522 43925E−01
	n = 6, m = 4
x_1	∓3.37605 42313 02644E+00
0	6.86671 24122 95233E−05
1	±1.63464 81992 75554E−05
2	1.27100 02549 75848E−06
x_2	∓1.91388 33000 94388E+00
0	5.29562 12469 82014E−02
1	±1.16515 85305 23187E−02
2	1.57417 66454 25220E−03
x_3	∓6.24276 68016 68255E−01
0	8.33202 04585 88149E−01
1	±8.21729 00074 45389E−02
2	3.38972 61878 62648E−02
	n = 7, m = 2
x_1	∓2.65196 13568 35233E+00
0	9.71781 24509 95192E−04
x_2	∓1.67355 16287 67471E+00
0	5.45155 82819 12703E−02
x_3	∓8.16287 88285 89647E−01
0	4.25607 25261 01278E−01
x_4	0.00000 00000 00000E+00
0	8.10264 61755 68073E−01
	n = 7, m = 4
x_1	∓3.80445 92703 94727E+00
0	3.54091 24629 66712E−06
1	±7.74414 70587 59668E−07
2	5.32745 90471 16060E−08
x_2	∓2.39508 94914 53728E+00
0	7.57345 98519 23196E−03
1	±1.61904 27132 24390E−03
2	1.74319 46495 25240E−04
x_3	∓1.16718 93912 57087E+00
0	3.38210 49973 52895E−01
1	±4.91361 96733 89993E−02
2	1.08071 03598 79694E−02
x_4	0.00000 00000 00000E+00
0	1.08087 88499 06165E+00
1	0.00000 00000 00000E+00
2	3.93636 75865 12016E−02
	n = 8, m = 2
x_1	∓2.93063 74202 57244E+00
0	1.99604 07221 13676E−04
x_2	∓1.98165 67566 95843E+00
0	1.70779 83007 41348E−02
x_3	∓1.15719 37124 46780E+00
0	2.07802 32581 48919E−01
x_4	∓3.81186 99020 73221E−01
0	6.61147 01255 82413E−01
	n = 8, m = 4
x_1	∓4.20027 14338 58624E+00
0	1.63634 48943 64315E−07
1	±3.32337 10565 85344E−08
2	2.06586 48904 73658E−09
x_2	∓2.83357 34679 09654E+00
0	8.54957 10026 89954E−04
1	±1.74624 92513 61741E−04
2	1.58592 89017 30733E−05
x_3	∓1.65314 74919 04498E+00

	Hermite Weight(Cont.)
0	9.40062 39863 52047E−02
1	±1.54906 22522 75267E−02
2	2.40011 28506 07768E−03
x_4	∓5.44358 19969 79902E−01
0	7.91365 56485 44791E−01
1	±5.21984 18111 15754E−02
2	2.45191 97288 29617E−02
	n = 9, m = 2
x_1	∓3.19099 32017 81528E+00
0	3.96069 77263 26438E−05
x_2	∓2.26658 05845 31843E+00
0	4.94362 42755 36947E−03
x_3	∓1.46855 32892 16668E+00
0	8.84745 27394 37657E−02
x_4	∓7.23551 01875 28376E−01
0	4.32651 55900 25558E−01
x_5	0.00000 00000 00000E+00
0	7.20235 21560 60510E−01
	n = 9, m = 4
x_1	∓4.56981 34387 27062E+00
0	6.94273 08140 16587E−09
1	±1.32064 35753 33371E−09
2	7.53055 42783 61847E−11
x_2	∓3.23868 91762 56225E+00
0	8.04668 53776 27220E−05
1	±1.56346 83497 15651E−05
2	1.24027 25458 11537E−06
x_3	∓2.09644 36491 11669E+00
0	1.95369 02469 68726E−02
1	±3.34420 80731 75786E−03
2	4.07891 27970 06466E−04
x_4	∓1.03246 68908 45192E+00
0	3.79930 99462 58787E−01
1	±3.99722 93313 99399E−02
2	9.87996 79609 24689E−03
x_5	0.00000 00000 00000E+00
0	9.73357 10912 13699E−01
1	0.00000 00000 00000E+00
2	2.74863 96836 83820E−02
	n = 10, m = 2
x_1	∓3.43615 91188 37738E+00
0	7.64043 28552 32621E−06
x_2	∓2.53273 16742 32790E+00
0	1.34364 57467 81233E−03
x_3	∓1.75668 36492 99882E+00
0	3.38743 94455 48106E−02
x_4	∓1.03661 08297 89514E+00
0	2.40138 61108 23147E−01
x_5	∓3.42901 32722 37046E−01
0	6.10862 63373 53258E−01
	n = 10, m = 4
x_1	∓4.91761 13588 81648E+00
0	2.75015 22329 24451E−10
1	±4.93345 59313 29764E−11
2	2.60980 41098 40104E−12
x_2	∓3.61684 04950 53145E+00
0	6.55581 64756 40347E−06
1	±1.21198 46325 26528E−06
2	8.59761 95345 10362E−08
x_3	∓2.50628 83553 80640E+00
0	3.22809 10035 96067E−03

	Hermite Weight(Cont.)
1	±5.53012 05419 81994E−04
2	5.63445 10835 66073E−05
x_4	∓1.47826 33993 58294E+00
0	1.31366 56307 40137E−01
1	±1.66936 21720 22682E−02
2	2.87753 09640 03399E−03
x_5	∓4.88897 95606 84079E−01
0	7.51625 71528 36574E−01
1	±3.60119 41471 04198E−02
2	1.87529 47668 26694E−02
	Laguerre Weight
	n = 2, m = 2
x_1	5.85786 43762 69050E−01
0	8.53553 39059 32738E−01
x_2	3.41421 35623 73095E+00
0	1.46446 60940 67262E−01
	n = 2, m = 4
x_1	8.93499 36017 64848E−01
0	9.64611 86872 03390E−01
1	−3.76871 83665 63099E−02
2	2.17213 65580 39585E−01
x_2	6.38779 82740 32094E+00
0	3.53881 31279 66099E−02
1	−5.02451 47764 07492E−02
2	3.03829 66736 97636E−02
	n = 3, m = 2
x_1	4.15774 55678 34791E−01
0	7.11093 00992 91730E−01
x_2	2.29428 03602 79042E+00
0	2.78517 73356 92408E−01
x_3	6.28994 50829 37479E+00
0	1.03892 56501 58614E−02
	n = 3, m = 4
x_1	6.21277 35419 98259E−01
0	8.82464 24711 92333E−01
1	5.30801 93722 54686E−02
2	9.35609 10877 27375E−02
x_2	4.19360 31302 98466E+00
0	1.17328 06456 25582E−01
1	−9.55572 08631 52833E−02
2	6.06878 07044 02238E−02
x_3	1.21938 09738 60654E+01
0	2.07688 31820 84594E−04
1	−3.37888 37576 70129E−04
2	1.89513 41944 33898E−04
	n = 4, m = 2
x_1	3.22547 68961 93923E−01
0	6.03154 10434 16336E−01
x_2	1.74576 11011 58347E+00
0	3.57418 69243 77997E−01
x_3	4.53662 02969 21128E+00
0	3.88879 08515 00538E−02
x_4	9.39507 09123 01133E+00
0	5.39294 70556 13275E−04
	n = 4, m = 4
x_1	4.76295 51555 90591E−01
0	7.94973 96840 11831E−01
1	6.36612 81860 67082E−02
2	4.82716 61104 39498E−02
x_2	3.15240 69445 57096E+00

	Laguerre Weight(Cont.)
0	2.02467 98138 59689E−01
1	−9.97551 19566 49862E−02
2	6.77119 47690 89374E−02
x_3	8.65854 88020 60417E+00
0	2.55745 47883 32455E−03
1	−2.96399 10545 58945E−03
2	1.67456 39939 17743E−03
x_4	1.84749 50322 43054E+01
0	5.95424 51553 08978E−07
1	−1.02547 90558 52397E−06
2	5.58934 87823 68760E−07
	n = 5, m = 2
x_1	2.63560 31971 81409E−01
0	5.21755 61058 28087E−01
x_2	1.41340 30591 06517E+00
0	3.98666 81108 31759E−01
x_3	3.59642 57710 40722E+00
0	7.59424 49681 70760E−02
x_4	7.08581 00058 58838E+00
0	3.61175 86799 22048E−03
x_5	1.26408 00844 27578E+01
0	2.33699 72385 77623E−05
	n = 5, m = 4
x_1	3.86192 11523 01377E−01
0	7.16286 42415 73014E−01
1	5.81559 49779 38570E−02
2	2.80265 41038 29104E−02
x_2	2.53268 08971 66395E+00
0	2.74194 78145 27249E−01
1	−8.61224 28169 98978E−02
2	6.30631 68825 73960E−02
x_3	6.80559 64648 13736E+00
0	9.49512 63831 90954E−03
1	−8.01968 03267 47451E−03
2	4.61567 21371 42154E−03
x_4	1.37705 43148 95434E+01
0	2.36668 60784 08890E−05
1	−3.18263 58185 86023E−05
2	1.72946 33055 85475E−05
x_5	2.50390 67879 50020E+01
0	1.14599 86145 53192E−09
1	−2.03720 20689 30169E−09
2	1.09362 54865 45233E−09
	n = 6, m = 2
x_1	2.22846 60417 92607E−01
0	4.58964 67394 99636E−01
x_2	1.18893 21016 72623E+00
0	4.17000 83077 21210E−01
x_3	2.99273 63260 59314E+00
0	1.13373 38207 40450E−01
x_4	5.77514 35691 04511E+00
0	1.03991 97453 14907E−02
x_5	9.83746 74183 82590E+00
0	2.61017 20281 49321E−04
x_6	1.59828 73980 60170E+01
0	8.98547 90642 96212E−07
	n = 6, m = 4
x_1	3.24757 29547 85770E−01
0	6.48701 74032 55691E−01
1	5.00884 55297 63713E−02
2	1.76753 82486 40956E−02

	Laguerre Weight(Cont.)
x_2	2.11905 14050 38185E+00
0	3.29542 73624 06996E−01
1	−6.88387 75330 39359E−02
2	5.48343 77614 27702E−02
x_3	5.63131 22369 28527E+00
0	2.15689 05741 41581E−02
1	−1.35822 77128 37239E−02
2	7.99140 99645 83722E−03
x_4	1.11496 26596 13132E+01
0	1.86488 53704 98548E−04
1	−1.98848 53472 72210E−04
2	1.08799 59021 28928E−04
x_5	1.93013 31385 59176E+01
0	1.29153 56130 74535E−07
1	−1.88593 64440 38168E−07
2	1.00555 36783 54472E−07
x_6	3.17937 84130 14327E+01
0	1.70436 03254 47506E−12
1	−3.09191 26312 22564E−12
2	1.64410 93757 89997E−12
	n = 7, m = 2
x_1	1.93043 67656 03624E−01
0	4.09318 95170 12739E−01
x_2	1.02666 48953 39192E+00
0	4.21831 27786 17198E−01
x_3	2.56787 67449 50746E+00
0	1.47126 34865 75053E−01
x_4	4.90035 30845 26485E+00
0	2.06335 14468 71694E−02
x_5	8.18215 34445 62861E+00
0	1.07401 01432 80746E−03
x_6	1.27341 80291 79781E+01
0	1.58654 64348 56420E−05
x_7	1.93957 27862 26254E+01
0	3.17031 54789 95581E−08
	n = 7, m = 4
x_1	2.80184 39130 00106E−01
0	5.91257 67338 40351E−01
1	4.25854 80400 74389E−02
2	1.18487 12456 11058E−02
x_2	1.82258 50338 88525E+00
0	3.70180 36084 71202E−01
1	−5.30761 40578 77899E−02
2	4.64069 01816 11444E−02
x_3	4.81222 11159 13568E+00
0	3.78342 04946 73292E−02
1	−1.81669 42767 61649E−02
2	1.09474 89400 10344E−02
x_4	9.41872 85564 57246E+00
0	7.25693 74743 41352E−04
1	−6.20924 67250 01186E−04
2	3.43023 29449 18514E−04
x_5	1.59675 85555 60262E+01
0	2.06658 43246 58470E−06
1	−2.51151 97278 44657E−06
2	1.34221 54771 75663E−06
x_6	2.51284 54457 88476E+01
0	4.90350 90267 98168E−10
1	−7.54621 72123 61932E−10
2	3.97882 96930 18253E−10
x_7	3.86869 29895 70813E+01

	Laguerre Weight(Cont.)
0	2.11247 02320 86343E−15
1	−3.88705 39056 98067E−15
2	2.05338 74356 35854E−15
	n = 8, m = 2
x_1	1.70279 63230 51010E−01
0	3.69188 58934 16375E−01
x_2	9.03701 77679 93799E−01
0	4.18786 78081 43430E−01
x_3	2.25108 66298 66131E+00
0	1.75794 98663 71718E−01
x_4	4.26670 01702 87659E+00
0	3.33434 92261 21565E−02
x_5	7.04590 54023 93466E+00
0	2.79453 62352 25673E−03
x_6	1.07585 16010 18100E+01
0	9.07650 87733 58213E−05
x_7	1.57406 78641 27800E+01
0	8.48574 67162 72532E−07
x_8	2.28631 31736 88926E+01
0	1.04800 11748 71510E−09
	n = 8, m = 4
x_1	2.46368 55257 90876E−01
0	5.42345 42133 49683E−01
1	3.62527 03593 36114E−02
2	8.32360 01101 70972E−03
x_2	1.59937 96281 16687E+00
0	3.98887 36095 45477E−01
1	−4.01303 97974 40404E−02
2	3.89171 17017 54320E−02
x_3	4.20548 76441 22826E+00
0	5.68375 08705 13553E−02
1	−2.12442 75804 98174E−02
2	1.31289 43832 49867E−02
x_4	8.17401 00320 35467E+00
0	1.91605 90619 92721E−03
1	−1.33172 95979 90876E−03
2	7.43741 04046 47436E−04
x_5	1.36998 33321 30164E+01
0	1.36347 46985 61898E−05
1	−1.38971 16606 61216E−05
2	7.46271 92960 07821E−06
x_6	2.11337 10459 41020E+01
0	1.51949 44243 77953E−08
1	−2.00811 84886 77467E−08
2	1.05894 02244 09306E−08
x_7	3.11779 13307 88391E+01
0	1.42583 61254 01618E−12
1	−2.27582 81225 15954E−12
2	1.19116 73824 75763E−12
x_8	4.56858 86298 19770E+01
0	2.28413 07902 94169E−18
1	−4.24736 88444 08217E−18
2	2.23288 69406 77115E−18
	n = 9, m = 2
x_1	1.52322 22773 18082E−01
0	3.36126 42179 79625E−01
x_2	8.07220 02274 22558E−01
0	4.11213 98042 39844E−01
x_3	2.00513 51556 19347E+00
0	1.99287 52537 08856E−01
x_4	3.78347 39733 31233E+00

	Laguerre Weight(Cont.)
0	4.74605 62765 65160E−02
x_5	6.20495 67778 76613E+00
0	5.59962 66107 94583E−03
x_6	9.37298 52516 87576E+00
0	3.05249 76709 32106E−04
x_7	1.34662 36911 09209E+01
0	6.59212 30260 75352E−06
x_8	1.88335 97788 99170E+01
0	4.11076 93303 49548E−08
x_9	2.63740 71890 92738E+01
0	3.29087 40303 50708E−11
	n = 9, m = 4
x_1	2.19834 92205 94328E−01
0	5.00438 31691 15375E−01
1	3.10522 24967 84981E−02
2	6.06794 90922 73108E−03
x_2	1.42513 57280 33522E+00
0	4.18333 38764 73694E−01
1	−2.99382 61516 29166E−02
2	3.26095 10998 70847E−02
x_3	3.73682 92929 69678E+00
0	7.71962 22695 32319E−02
1	−2.28734 54779 15059E−02
2	1.45128 79381 04233E−02
x_4	7.22990 19337 37983E+00
0	3.97747 00866 72074E−03
1	−2.27296 14989 98705E−03
2	1.28420 52007 22529E−03
x_5	1.20317 97641 92781E+01
0	5.44360 90278 66002E−05
1	−4.68793 27128 34346E−05
2	2.53223 59989 47699E−05
x_6	1.83546 21304 33343E+01
0	1.66486 92663 64780E−07
1	−1.90039 32171 24683E−07
2	1.00459 35049 29719E−07
x_7	2.65679 85569 10755E+01
0	8.18891 44548 43159E−11
1	−1.14725 15725 70101E−10
2	5.99830 85805 53080E−11
x_8	3.74012 62748 79278E+01
0	3.37587 31609 45429E−15
1	−5.53527 10848 67681E−15
2	2.88210 34522 32548E−15
x_9	5.27687 66947 10135E+01
0	2.21995 47471 32278E−21
1	−4.16174 02663 08668E−21
2	2.17970 21595 57879E−21
	n = 10, m = 2
x_1	1.37793 47054 04924E−01
0	3.08441 11576 50201E−01
x_2	7.29454 54950 31705E−01
0	4.01119 92915 52736E−01
x_3	1.80834 29017 40316E+00
0	2.18068 28761 18094E−01

	Laguerre Weight(Cont.)
x_4	3.40143 36978 54900E+00
0	6.20874 56098 67775E−02
x_5	5.55249 61400 63804E+00
0	9.50151 69751 81101E−03
x_6	8.33015 27467 64497E+00
0	7.53008 38858 75388E−04
x_7	1.18437 85837 90007E+01
0	2.82592 33495 99566E−05
x_8	1.62792 57831 37810E+01
0	4.24931 39849 62686E−07
x_9	2.19965 85811 98076E+01
0	1.83956 48239 79631E−09
x_{10}	2.99206 97012 27389E+01
0	9.91182 72196 09009E−13
	n = 10, m = 4
x_1	1.98459 89648 55396E−01
0	4.64257 36116 75606E−01
1	2.68033 15083 76837E−02
2	4.55844 46723 26685E−03
x_2	1.28527 24641 60369E+00
0	4.30738 73593 50617E−01
1	−2.20555 65718 23314E−02
2	2.74115 86320 78406E−02
x_3	3.36337 82573 58604E+00
0	9.78158 71992 19426E−02
1	−2.33432 43663 01337E−02
2	1.52214 64966 08219E−02
x_4	6.48664 60030 15374E+00
0	7.02909 88817 21789E−03
1	−3.34092 69217 19411E−03
2	1.91057 19339 80845E−03
x_5	1.07436 07524 68832E+01
0	1.57932 41325 55459E−04
1	−1.15787 63973 17979E−04
2	6.29487 92317 63510E−05
x_6	1.62743 03555 43137E+01
0	9.98154 34632 45664E−07
1	−9.89035 65142 61929E−07
2	5.24625 75379 11778E−07
x_7	2.33035 21691 88152E+01
0	1.45551 42478 41691E−09
1	−1.79850 82324 19956E−09
2	9.41407 07318 59928E−10
x_8	3.22160 61440 73498E+01
0	3.45554 77417 61117E−13
1	−5.05387 33325 22763E−13
2	2.62678 89221 42585E−13
x_9	4.37648 98673 76546E+01
0	6.78480 88889 84564E−18
1	−1.13574 03290 35132E−17
2	5.89059 68626 58232E−18
x_{10}	5.99201 03669 10753E+01
0	1.98000 38059 78796E−24
1	−3.73606 76158 58268E−24
2	1.95091 40128 75742E−24

References

[1] M. Abramowitz and I. A. Stegun, "Handbook of Mathematical Functions with Formulas, Graphs, and Mathematical Tables", National Bureau of Standards, *Applied Mathematics Series* **55**, 1964.

[2] H. Amann, *"Lectures on Some Fixed Point Theorems"*, IMPA, Rio de Janeiro, 1974.

[3] M. Avriel, *"Nonlinear Programming: Analysis and Methods"*, Prentice Hall, 1979.

[4] D. L. Barrow, On multiple node Gaussian quadrature formulae, *Math. Comput.* **32** (1978), 431–439.

[5] S. Bernstein, Sur les polynomes orthogonaux relatifs à un segment fini, *J. Math. Pure Appl.* **9** (1930), 127–177.

[6] B. D. Bojanov, Extremal problems in the set of polynomials with fixed multiplicities of zeros, *Comput. Rend. Acad. Bulgare Sci.* **31** (1978), 377–400.

[7] B. D. Bojanov, Oscillating polynomials of least L_1-norm, in *"Numerical Integration"* (G. Hammerlin, ed.), pp. 25–33, ISNM **57**, Birkhäuser, Basel, 1982.

[8] B. D. Bojanov, On a quadrature formula of Micchelli and Rivlin, *J. Comput. Appl. Math.* **70** (1996), 349–356.

[9] B. D. Bojanov, Gaussian quadrature formulae for Tchebycheff systems, *East J. Approx.* **3** (1997), 71–88.

[10] B. D. Bojanov, D. Braess, and N. Dyn, Generalized Gaussian quadrature formulas, *J. Approx. Theory* **48** (1986), 335–353.

[11] K. Borsuk, Drei Sätze über die n-dimensionale euklidische Sphäre, *Fund. Math.* **20** (1933), 177–191.

[12] P. Borwein and T. Erdélyi, *"Polynomials and Polynomials Inequalities"*, Springer-Verlag, New York, 1995.

[13] P. L. Chebyshev, Sur les valeurs limites des intégrales, *J. Math.*(2) **19** (1874), 157–160.

[14] E. W. Cheney, *"Introduction to Approximation Theory"*, McGraw-Hill Book Co., New York, 1966.

[15] F. Deutsch, G. Nürenberger, and I. Singer, Weak Chebyshev subspaces and alternation, *Pacific J. Math.* **89** (1980), 9–31.

[16] Z. Ditzian and V. Totik, "Moduli of Smoothness", *Springer Series in Computational Mathematics*, Vol. 9, Springer-Verlag, New York and Berlin, 1987.

[17] C. F. Dunkl and Y. Xu, *"Orthogonal Polynomials of Several Variables"*, Cambridge University Press, Cambridge, 2001.

[18] P. Erdős, On the uniform distribution of the roots of certain polynomials, *Ann. Math.* **43** (1942), 59–64.

[19] P. Erdős and G. Frued, On orthogonal polynomials with regularly distributed zeros, *Proc. London Math. Soc.* **29** (1973), 521–537.

[20] P. Erdős and P. Turán, On interpolation I. Quadrature- and mean-convergence in the Lagrange-interpolation, *Ann. Math.* **38** (1937), 142–155.

[21] P. Erdős and P. Turán, On interpolation II. On the distribution of the fundamental points of Lagrange and Hermite interpolation, *Ann. Math.* **39** (1938), 703–724.

[22] P. Erdős and P. Turán, On interpolation III. Interpolatory theory of polynomials, *Ann. Math.* **41** (1940), 510–553.

[23] P. Erdős and P. Vértesi, On the almost everywhere divergence of Lagrange interpolatory polynomials for arbitrary system of nodes, *Acta Math. Hungar.* **36** (1980), 71-89; **38** (1981), 263.

[24] G. Freud, *"Orthogonal Polynomials"*, Akad. Kiadó, Budapest, Pergamon, Oxford, 1971.

[25] G. Freud, On Hermite-Fejér interpolation processes, *Studia Sci. Math. Hungar.* **7** (1972), 307–316.

[26] W. Gautschi, How and how not to check Gaussian quadrature formulae, *BIT* **23** (1983), 209–216.

[27] W. Gautschi and G. V. Milovanović, S-orthogonality and construction of Gauss-Turán-type quadrature formulae, *J. Comput. Appl. Math.* **86** (1997), 205–218.

[28] A. Ghizzetti and A. Ossicini, Su un nuovo tipo di sviluppo di una funzione in serie di polinomi, *Atti Acad. Naz. Lincei Rend. Cl. Sci. Fis. Mat. Natur.* **43** (1967), 21–29.

[29] A. Ghizzetti and A. Ossicini, *"Quadrature Formulae"*, Akademie Verlag, Berlin, 1970.

[30] A. Ghizzetti and A. Ossicini, Polinomi s-ortogonali e sviluppi in serie ad essi collegati, *Mem. Accad. Sci. Torino Cl. Fis. Mat. Natur.* **18** (1974), 1–16.

[31] A. Ghizzetti and A. Ossicini, Generalizzazione dei polinomi s-ortogonali e dei corrispondenti sviluppi in serie, *Atti Acad. Sci. Torino. Cl. Sci. Fis. Mat. Natur.* **109** (1975), 371–379.

[32] A. Ghizzetti and A. Ossicini, Sull'esistenza e unicitat delle formule di quadrature Gaussiane, *Rend. Mat.* **5** (1975), 1–15.

[33] L. Gori Nicolo-Amati, On the behaviour of the zeros of some s-orthogonal polynomials, in *"Orthogonal Polynomials and Their Applications"*, Second International Sumposium, Segovia, 1986, *Monogr. Acad. Cienc. Exactas, Fis., Quim., Nat.*, Zaragoza, 1988, pp. 71–85.

[34] L. Gori and M. L. Lo Cascio, On an invariance property of the zeros of some s-orthogonal polynomials, in *"Orthogonal Polynomials and Their Applications"* (C. Brezinski, L. Gori and A. Ronveaux, Eds.), *IMACS Ann. Comput. Appl. Math.*, Vol. **9**, J. C. Baltzer AG, Scientific Publ. Co., Basel, 1991, pp. 277–280.

[35] L. Gori, The Posse-Markov-Stieltjes inequality for Turán type quadrature sums, in *"Progress in Approximation Theory"* (P. Nevai and A. Pinkus, eds.), pp. 453-466, Academic Press, New York, 1991.

[36] L. Gori and C. A. Micchelli, On weight functions which admit explicit Gauss-Turán quadrature formulas, *Math. Comput.* **65** (1996), 1567–1581.

[37] I. S. Gradshteyn and I. M. Ryzhik, *"Table of Integrals, Series, and Products"*, Academic Press, New York, 1980.

[38] S. Guerra, Polinomi generati da successioni peso e teoremi di rappresentazione di una funzione in serie di tali polinomi, *Rend. Ist. Mat. Univ. Trieste* **8** (1976), 172–194.

[39] S. Guerra, Su un determinante collegato ad un sistema di plinomi ortogonali, *Rend. Ist. Mat. Univ. Trieste* **10** (1978), 66–79.

[40] C. R. Hobby and J. R. Rice, A moment problem in L_1-approximation, *Proc. Amer. Math. Soc.* **16** (1965), 665–670.

[41] S. Karlin and A. Pinkus, Gaussian quadrature formulae with multiple nodes, In *"Studies in Spline Functions and Approximation Theory"* (S. Karlin et al., eds.), pp. 113–141, Academic Press, New York, 1976.

[42] S. Karlin and A. Pinkus, An extremal property of multiple Gaussian nodes, In *"Studies in Spline Functions and Approximation Theory"* (S. Karlin et al., eds.), pp. 143–162, Academic Press, New York, 1976.

[43] S. Karlin and W. J. Studden, *"Tchebycheff Systems: with Applications in Analysis and Statistics"*, Wiley, New York, 1966.

[44] B. Khalilova, On some estimates for polynomials, *Izv. Akad. Nauk. Azerb. SSR* **2** (1974), 46–55 (in Russian).

[45] O. Kis, Remark on mechanical quadrature, *Acta Math. Acad. Sci. Hungar.* **8** (1957), 473–476 (in Russian).

[46] K. Kitahara, Best L^1-approximations, in *"Progress in Approximation Theory"* (P. Nevai and A. Pinkus, eds.), pp. 505–512, Academic Press, New York, 1991.

[47] M. G. Krein, The ideas of P. L. Chebyshev and A. A. Markov in the theory of limiting values of integrals and their further developments, *Uspehi Mat. Nauk* **6** (1951), 3–120; *Amer. Math. Soc. Transl., Ser. 2*, **12** (1951), 1–122.

[48] X. Li, E. B. Saff, and Z. Sha, Behavior of best L_p polynomial approximants on the unit interval and on the unit circle, *J. Approx. Theory* **63** (1990), 170–190.

[49] G. G. Lorentz, K. Jetter, and S. D. Riemenschneider, *"Birkhoff Interpolation"*, Addison Wesley Pub. Co., Reading, Massachusetts, London, 1983.

[50] D. S. Lubinsky, A survey of general orthogonal polynomials for weights on finite and infinite intervals, *Acta. Appl. Math.* **10** (1987), 237–296.

[51] D. S. Lubinsky, An extension of the Erdős-Turán inequality for the sum of successive fundamental polynomials, *Ann. Numer. Math.* **2** (1995), 305–309.

[52] D. S. Lubinsky and E. B. Saff, Strong asymptotics for L_p extremal polynomials ($1 < p \leq \infty$) associated with weights on $[-1,1]$, *Lecture Notes in Masthematics*, Vol. **1287**, pp. 83–104, Springer-Verlag, Berlin, 1987.

[53] A. A. Markov, Démonstration de certaines inégalités de M. Tchébycheff, *Math. Ann.* **24** (1885), 172–180.

[54] A. A. Markov, *"Differenzenrechnung"*, Leipzig, 1896.

[55] G. Mastroianni and P. Vértesi, Mean convergence of Lagrange interpolation on arbitrary system of nodes, *Acta Sci. Math. (Szeged)* **57** (1993), 429–441.

[56] A. Máté, P. Nevai, and V. Totik, Necessary conditions for weighted mean convergence of Fourier series in orthogonal polynomials, *J. Approx. Theory* **46** (1986), 314–322.

[57] A. Máté, P. Nevai, and V. Totik, Strong and weak convergence of orthogonal polynomials, *Amer. J. Math.* **109** (1987), 239–282.

[58] C. A. Micchelli and T. J. Rivlin, Turán formulae and highest precision quadrature rules for Chebyshev coefficients, *IBM J. Res. Develop.* **16** (1972), 372–379.

[59] C. A. Micchelli and A. Sharma, On a Problem of Turán: multiple nodes Gaussian quadrature, *Rend. Math. Serie VII* **3** (1983), 529–552.

[60] G. V. Milovanović, Construction of s-orthogonality and Turán quadrature formulae, in *"Numerical Methods and Approximation Theory III"* (G. V. Milovanović, ed., Niš, 1987), pp. 311–328, Univ. Niš, Niš, 1988.

[61] G. V. Milovanović, Quadratures with multiple nodes, power orthogonality, and moment-preserving spline approximation, *J. Comput. Appl. Math.* **127** (2001), 267–286.

[62] G. V. Milovanović and M. M. Spalević, Construction of Chakalov-Popoviciu's type quadrature formulae, *Rend. Circ. Mat. Palermo (2)* II (Suppl. 52) (1998), 625–636.

[63] G. V. Milovanović and M. M. Spalević, Quadrature formulae connected to σ-orthogonal polynomials, *J. Comput. Appl. Math.*, to appear.

[64] P. Nevai, Mean convergence of Lagrange interpolation, *J. Approx. Theory* **18** (1976), 363–377.

[65] P. Nevai, "Orthogonal Polynomials", *Memoirs of the Amer. Math. Soc.*, Vol. **213**, Amer. Math. Soc., Providence, RI, 1979.

[66] P. Nevai, Distribution of zeros of orthogonal polynomials, *Trans. Amer. Math. Soc.* **249** (1979), 341–361.

[67] P. Nevai, Géza Freud, Orthogonal polynomials and Christoffel functions: A case study, *J. Approx. Theory* **48** (1986), 3–167.

[68] G. Nürnberger and M. Sommer, Weak Chebyshev subspaces and continuous selections for the metric projection, *Trans, Amer. Math. Soc.* **238** (1978), 129–138.

[69] A. Ossicini, Costruzione di formule di quadratura di tipo Gaussiano, *Ann. Mat. Pura Appl.* **72** (1966), 231–237.

[70] A. Ossicini, Le funzioni di infuenza nel problema di Gauss sulle formule di quadratura, *Matematiche (Catania)* **23** (1968), 7–30.

[71] A. Ossicini and F. Rosati, Funzioni caratteristiche nelle formule di quadratura gaussiane con nodi multipli, *Boll. Un. Mat. Ital.* **11** (1975), 224–2371 (in Italian).

[72] A. Ossicini and F. Rosati, Comparison theorems for the zeros of s-orthogonal polynomials, *Calcolo* **16** (1979), 371–381 (in Italian).

[73] A. Ossicini and F. Rosati, Numeri di Christoffel e polinomi s-ortogonali, in *"E. B. Christoffel"* (P. L. Butzer and F. Fehér, Eds.), Birkhäuser, Basel, 1981, pp. 148–157.

[74] A. Ossicini and F. Rosati, S-orthogonal polynomials Jacobi polynomials, *Rend. Mat. Appl.* **12** (1992), 309–403 (in Italian).

[75] A. Pinkus, A simple proof of the Hobby-Rice theorem, *Proc. Amer. Math. Soc.* **60** (1976), 80–82.

[76] G. Pólya, Sur une algorithm toujours convergent pour obtenir les polynomes de meilleure approximation de Tchebysheff pour une fonction continue quelconque, *Comptes Rendus* **157** (1913), 840–843.

[77] T. Popoviciu, Aspura unei generalizari a formulei de integrare numerica a lui Gauss, *Acad. Republ. Popul. Romine Studii Cerc. Mat.* **6** (1955), 29–57.

[78] M. Riesz, Eine trigonometrische Interpolationformel und einige Ungleichungen für Polynome, *Jahr. D. Math. Verein.* **23** (1914), 354–368.

[79] P. O. Runck, Bemerkungen zu den approximationssätzen von Jackson und Jackson-Timan, in *"ISNM 10, Abstract Spaces and Approximation, Proc. of Conf. Oberwolfach, 1968"* (P. L. Butzer and Sz. Nagy, eds.), pp. 303–308, Birkhäuser, Basel, 1969.

[80] E. B. Saff and V. Totik, Weighted polynomial approximation of analytic functions, *J. London Math. Soc. (2)* **37** (1988), 455–463.

[81] E. B. Saff and V. Totik, *"Logarithmic potentials with external fields"*, Springer-Verlag, Berlin, Heidelberg, 1997.

[82] Y. G. Shi, A class of functions having the same best approximation for all L_p norms $(1 \leq p \leq \infty)$, *Approx. Theory Appl.* **5** (1989), 69–73.

[83] Y. G. Shi, On critical order of Hermite-Fejér type interpolation, in *"Progress in Approximation Theory"* (P. Nevai and A. Pinkus, eds.), pp. 761-766, Academic Press, New York, 1991.

[84] Y. G. Shi, The exact order of $\sum_{k=1}^{n} |x-x_k|^s |\ell_k(x)|^t$, *Approx. Theory Appl.* **8** (1992), 1–10.

[85] Y. G. Shi, Bounds and inequalities for general orthogonal polynomials on finite intervals, *J. Approx. Theory* **73** (1993), 303–333.

[86] Y. G. Shi, Bounds and inequalities for arbitrary orthogonal polynomials on finite intervals, *J. Approx. Theory* **81** (1995), 217–230.

[87] Y. G. Shi, A solution of Problem 26 of P. Turán, *Science in China (Series A)* **38** (1995), 1313–1319.

[88] Y. G. Shi, On Problem 26 of P. Turán, *Acta Sci. Math. (Szeged)* **61** (1995), 285–292.

[89] Y. G. Shi, Mean convergence of interpolatory processes on an arbitrary system of nodes, *Acta Math. Hungar.* **70** (1996), 27–38.

[90] Y. G. Shi, Orthogonal polynomials on infinite intervals and Problem 54 of P. Turán, *Approx. Theory Appl.* **12** (1996), 53–60.

[91] Y. G. Shi, An analogue of Problem 26 of P. Turán, *Bull. Austral. Math. Soc.* **53** (1996), 1–12.

[92] Y. G. Shi, Necessary conditions for L_p convergence of Lagrange interpolation on an arbitrary system of nodes, *J. Approx. Theory* **85** (1996), 16–26.

[93] Y. G. Shi, The generalized Markov-Stieltjes inequality for Birkhoff quadrature formulas, *J. Approx. Theory* **86** (1996), 229–239.

[94] Y. G. Shi, Necessary conditions for mean convergence of Lagrange interpolation on an arbitrary system of nodes, *Acta Math. Hungar.* **72** (1996), 251–260.

[95] Y. G. Shi, Generalized Gaussian Kronrod-Turán quadrature formulas, *Acta Sci. Math. (Szeged)* **62** (1996), 175–185.

[96] Y. G. Shi, Mean convergence of truncated Hermite interpolation on an arbitrary system of nodes, *Acta Math. Hungar.* **76** (1997), 45–58.

[97] Y. G. Shi, Optimal Lebesgue function type sums, *Acta Math. Hungar.* **77** (1997), 287–300.

[98] Y. G. Shi, On the exact order of Lebesgue function type sums of Hermite interpolation, *Acta Math. Hungar.* **78** (1998), 93–98.

[99] Y. G. Shi, On L_p extremal polynomilas, *Constr. Approx.* **14** (1998), 187–194.

[100] Y. G. Shi, General Gaussian quadrature formulas on Chebyshev nodes, *Adv. Math. (China)* **27** (1998), 227–239.

[101] Y. G. Shi, On Turán quadrature formulas for the Chebyshev weight, *J. Approx. Theory* **96** (1999), 101–110.

[102] Y. G. Shi, Generalized Gaussian quadrature formulas with Chebyshev nodes, *J. Comput. Math.* **17** (1999), 171–178.

[103] Y. G. Shi, On Gaussian quadrature formulas for the Chebyshev weight, *J. Approx. Theory* **98** (1999), 183–195.

[104] Y. G. Shi, Turán quadrature formulas and Christoffel type functions for the Chebyshev polynomials of the second kind, *Acta Math. Hungar.* **85** (1999), 253–263.

[105] Y. G. Shi, On some problems of P. Turán concerning L_m extremal polynomials and quadrature formulas, *J. Approx. Theory* **100** (1999), 203–220.

[106] Y. G. Shi, Generalized Gaussian quadrature formulas for Tchebycheff systems, *Far East J. Appl. Math.* **3** (1999), 153–170.

[107] Y. G. Shi, A kind of extremal problem of integration on an arbitrary measure, *Acta Sci. Math. (Szeged)* **65** (1999), 567–575.

[108] Y. G. Shi, On Christoffel type functions for L_m extremal polynomials. I, *J. Approx. Theory* **103** (2000), 281–291.

[109] Y. G. Shi, On Hermite interpolation, *J. Approx. Theory* **105** (2000), 49–86.

[110] Y. G. Shi, Convergence of Gaussian quadrature formulas, *J. Approx. Theory* **105** (2000), 279–291.

[111] Y. G. Shi, Convergence of (0,1,2,3) interpolation on an arbitrary system of nodes, *J. Comput. Math.* **19** (2001), 151–156.

[112] Y. G. Shi, On Christoffel type functions for L_m extremal polynomials. II, *J. Approx. Theory* **111** (2001), 59–80.

[113] Y. G. Shi, Erdős-Turán type inequality for the sum of successive fundamental polynomials of Hermite interpolation, *J. Approx. Theory* **111** (2001), 81–90.

[114] Y. G. Shi, Truncated Hermite interpolation on an arbitrary system of nodes, *Acta Math. Sinica* **21** (2001), 9–20.

[115] Y. G. Shi, On Normal pointsystems of Hermite-Fejér interpolation of arbitrary order, *J. Approx. Theory* **112** (2001), 44–60.

[116] Y. G. Shi, New characterizations of ratio asymptotics for orthogonal polynomials, *J. Approx. Theory* **115** (2002), 1–8.

[117] Y. G. Shi, Bounds and inequalities for L_m extremal polynomials, *J. Approx. Theory* **115** (2002), 167–185.

[118] Y. G. Shi, Mean convergence of Lagrange type interpolation on an arbitrary system of nodes, *Acta Math. Appl. Sinica, English Series* **18** (2002), 277–288.

[119] Y. G. Shi, A survey on Hermite interpolation of higher order, in *"Paul Erdős and His Mathematics. I"* (G. Halás, et al., eds.), pp. 653–686, Bolyai Society Mathematical Studies, **11**, Springer-Verlag, Berlin and New York, 2002.

[120] Y. G. Shi, On ratio asymptotics for general polynomials, *J. Approx. Theory* **120** (2003), 355–370.

[121] Y. G. Shi, L_m extremal polynomials associated with generalized Jacobi weights, *Acta Math. Appl. Sinica, English Series* **19** (2003), 205–218.

[122] Y. G. Shi, Convergence of Hermite-Fejér type interpolation of higher order on an arbitrary system of nodes, *J. Approx. Theory* **123** (2003), 173–187.

[123] Y. G. Shi, *"Theory of Birkhoff Interpolation"*, Nova Science Publishers, Inc., New York, 2003.

[124] Y. G. Shi, Convergence of generalized Gaussian quadrature formulas for an arbitrary measure, *manuscript*.

[125] Y. G. Shi Christoffel type functions for m-orthogonal polynomials, *J. Approx. Theory* **137** (2005), 57–88.

[126] Y. G. Shi, Asymptotics of polynomials having asymptoyic recurrence relation, Submitted to *Acta Math. Appl. Sinica, English Series*.

[127] Y. G. Shi and G. Xu, Construction of σ-orthogonal polynomials and Gaussian quadrature formulas, *Adv. Comput. Math.*, to appear.

[128] I. Singer, *"Best Approximation in Normed Linear Spaces by Elements of Linear Subspaces"*, Springer-Verlag, Berlin, 1970.

[129] H. Stahl and V. Totic, Nth root asymptotic behavior of orthogonal polynomials, In *"Orthogonal Polynomials: Theory and Practice"* (P. Nevai, ed.), pp. 395–417, Kluwer Academic Pub., Dordrecht/Boston/London, 1990.

[130] D. D. Stancu and A. H. Stroud, Quadrature formulas with simple Gaussian nodes and multiple fixed nodes, *Math. Comp.* **17** (1963), 384–394.

[131] T. J. Stieltjes, Quelques recherches sur la théorie des quadratures dites méchaniques, *Ann. Sci. École Norm. Sup.*(3) **1** (1884), 409–426.

[132] A. H. Stroud and D. D. Stancu, Quadrature formulas with multiple Gaussian nodes, *SIAM J. Numer. Anal.* **2** (1965), 129–143.

[133] J. Szabados, On the order of magnitude of fundamental polynomials of Hermite interpolation, *Acta Math. Hungar.* **61** (1993), 357–368.

[134] J. Szabados and A. K. Varma, On (0,1,2) interpolation in uniform metric, *Proc. Amer. Math. Soc.* **109** (1990), 975–979.

[135] J. Szabados and P. Vértesi, *"Interpolation of Functions"*, World Scientific, Singapore, 1990.

[136] J. Szabados and P. Vértesi, A survey on mean convergence of interpolatory processes, *J. Comput. Appl. Math.* **43** (1992), 3–18.

[137] G. Szegő, "Orthogonal Polynomials", *Amer. Math. Soc. Colloq. Publ.*, Vol. **23**, Amer. Math. Soc., Providence, RI, 1939.

[138] L. Tchakaloff, Über eine allgemeine Quadraturformel, *C. R. Acad. Bulgar. Sci.* **1** (1948), 9–12.

[139] L. Tchakaloff, Generalized quadrature formulae of Gaussian type, *Bulgar. Akad. Nauk. Izv. Mat. Institut* **2** (1954), 67–84 (in Bulgarian).

[140] P. Turán, On the theory of mechanical quadrature, *Acta Sci. Math. (Szeged)* **12** (1950), 30–37.

[141] P. Turán, On orthogonal polynomials, *Analysis Math.* **1** (1975), 297–311.

[142] P. Turán, On some open problem of approximation theory, *J. Approx. Theory* **29** (1980), 23–86.

[143] J. L. Ullman, On the regular behaviour of orthogonal polynomials, *Proc. London Math. Soc.* **24** (1972), 119–148.

[144] W. Van Assche, Some results on the asymptotic distribution of the zeros of orthogonal polynomials, *J. Comput. Appl. Math.* **12&13** (1985), 615–623.

[145] W. Van Assche and J. L. Teugels, Second order asymptotic behaviour of the zeros of orthogonal polynomials, *Rev. Roumaine. Math. Pures Appl,* **32** (1987), 15–26.

[146] P. Vértesi, On some problems of P. Turán, *Acta Math. Hungar.* **29** (1977), 337–353.

[147] P. Vértesi, Hermite-Fejér type interpolations. II, *Acta Math. Hungar.* **33** (1979), 333–343.

[148] P. Vértesi, On sums of Lebesgue function type, *Acta Math. Hungar.* **40** (1982), 217–227.

[149] P. Vértesi, Hermite-Fejér type interpolations of higher order. I, *Acta Math. Hungar.* **54** (1989), 135–152.

[150] P. Vértesi, Recent results on Hermite-Fejér interpolations of higher order (uniform metric), *Israel Math. Conf. Proc.* **4** (1991), 267–271.

[151] P. Vértesi, Lebesgue function type sums of Hermite interpolations, *Acta Math. Hungar.* **64** (1994), 341-349.

[152] P. Vértesi and Y. Xu, Truncated Hermite interpolation polynomials, *Studia Sci. Hungar.* **28** (1993), 205–213.

[153] Shijun Yang, "On the Quadratures of Gauss-Type", *PhD Dissertation*, Zhejiang University, Hangzhou, Zhejiang, China, 2002.

[154] Shijun Yang, Quadrature formulas for Fourier-Chebyshev coefficients, *J. Zhejiang Univ. SCIENCE* **3** (2002), 326–331.

[155] Shijun Yang and Xinghua Wang, New quadrature formulas based on the zeros of the chebyshev polynomials of the second kind, *Computers Math. Appl.* **45** (2003), 1611–1618.

[156] Shijun Yang and Xinghua Wang, Fourier-Chebyshev coefficients and Gauss-Turán quadrature with chebyshev welght, *J. Comput. Math.* **21** (2003), 189–194.

[157] Shijun Yang and Xinghua Wang, Remarks on some Gauss-Turán quadratures, *Math. Numer. Sinica* **25** (2003), 199–208 (in Chinese).

[158] Shijun Yang and Xinghua Wang, Some indentities concerning expansions of divided diffences with multiplicity (in Chinese), *manuscript*.

[159] Shijun Yang and Xinghua Wang, New quadrature formulas based on Chebyshev nodes, *manuscript*.

[160] G. Xu and Y. G. Shi, Progressive computation and numerical tables of generalized Gaussian quadrature formulas, *Chinese J. Numer. Math. Appl.* **27** (2006), 9–23 (in Chinese).

[161] C. Y. Zhou, Convergence of Gaussian quadrature formulas on infinite intervals, *J. Approx. Theory* **123** (2003), 280–294.

[162] A. Zygmund, *"Trigonometric Series"*, Vol. 1, Cambridge Univ. Press, Cambridge, 1977.

Index